MW00452336

Profiles *in the* History *of the*
U.S. SOIL SURVEY

Edited by

Douglas **Helms**,

Anne B. W. **Effland**,

Patricia J. **Durana**

Iowa State Press
A Blackwell Publishing Company

DOUGLAS HELMS received his Ph.D. in American history from Florida State University. He is the historian of the Natural Resources Conservation Service, U.S. Department of Agriculture.

ANNE B. W. EFFLAND received her Ph.D. from Iowa State University. She is a social science analyst with the Economic Research Service, U.S. Department of Agriculture.

PATRICIA J. DURANA received her B.S. degree from the University of Maryland. She is a natural resource manager with the Animal Husbandry and Clean Water Programs Division, Natural Resources Conservation Service, U.S. Department of Agriculture.

© 2002 Iowa State Press
A Blackwell Publishing Company
All rights reserved

Iowa State Press
2121 State Avenue, Ames, Iowa 50014

Orders: 1-800-862-6657
Office: 1-515-292-0140
Fax: 1-515-292-3348
Web site: www.iowastatepress.com

Authorization to photocopy items for internal or personal use, or the internal or personal use of specific clients, is granted by Iowa State Press, provided that the base fee of $.10 per copy is paid directly to the Copyright Clearance Center, 222 Rosewood Drive, Danvers, MA 01923. For those organizations that have been granted a photocopy license by CCC, a separate system of payments has been arranged. The fee code for users of the Transactional Reporting Service is 0-8138-2759-0/2002 $.10.

Printed on acid-free paper in the United States of America

First edition, 2002

Library of Congress Cataloging-in-Publication Data

Profiles in the history of the U.S. Soil Survey/edited by Douglas Helms, Anne B.W. Effland, and Patricia J. Durana.—1st ed.
    p.   cm.—
  Includes bibliographical references (p.   ).
  ISBN 0-8138-2759-0 (alk. paper)
  1. Soil surveys—United States—History. 2. Soil scientists—United States—History. 3. United States. Division of Soils—History. I. Helms, Douglas. II. Effland, Anne B. W. III. Durana, Patricia J.
  S599.A1 P76 2002
  631.4'973—dc21   2001005865

The last digit is the print number: 9 8 7 6 5 4 3 2 1

Contents

CONTRIBUTORS

E. Arthur Bettis (B.A., M.S., Iowa State University; Ph.D., University of Iowa) is assistant professor of geoscience at the University of Iowa. He served as a research soil scientist at Iowa State University and as a research geologist for the Iowa Department of Natural Resources–Geological Survey Bureau, working on a wide range of problems in alluvial, loess, and till stratigraphy, soils, landscape evolution, and nonpoint-source contamination of groundwater. He has published extensively on a variety of topics including soil geomorphology, landscape evolution, Quaternary stratigraphy, and geoarchaeology.

Peter W. Birkeland (B.S., University of Washington; Ph.D., Stanford University, 1962) was a professor in soils at the University of California-Berkeley (1962–1967) and a professor in geology at the University of Colorado (1967–1997), where he taught courses in soil geomorphology and Quaternary geology. He is now professor emeritus from the University of Colorado. He has published articles and book chapters on Quaternary geology, geomorphology, and soils. He won the Kirk Bryan Award (Quaternary Geology & Geomorphology) of the Geological Society of America in 1988 for the second edition of *Soils and Geomorphology*. The third edition was published in 1999.

Patricia J. Durana (B.S., Biology, University of Maryland, 1989) is a natural resource manager with the Animal Husbandry and Clean Water Programs Division, Natural Resources Conservation Service, U.S. Department of Agriculture. Ms. Durana served as a senior analyst and project director for the U.S. Congress, Office of Technology Assessment (1984–1995), conducting policy analyses on natural resource and agriculture issues. She has authored or coauthored numerous publications related to U.S. and foreign agriculture, conservation, and development issues.

Anne B. W. Effland (A.B., History, Hood College, 1979; M.A., History, West Virginia University, 1983; Ph.D., Agricultural History and Rural Studies, Iowa State University, 1991) is a social science analyst with the Economic Research Service, U.S. Department of Agriculture. She joined the service in 1990, where her work and publications as a historian and policy analyst have focused primarily on agricultural and rural development policy, farm labor, and rural women and minorities. She has also coauthored several articles on the history of soil science and the USDA's Soil Survey Division and has had extensive experience as an editor. She serves as the assistant secretary of the Agricultural History Society.

Klaus W. Flach (B.S., General Agriculture, Technische Hochschule, Munich, Germany, 1950; USDS scholar, Washington State College (now University) 1951–1952; M.S., Soils and Mineralogy, Cornell University 1955; Ph.D., Cornell University 1955, 1958) held numerous positions with the U.S. Department of Agriculture, Soil Conservation Service. He was a soil scientist specializing in mineralogy at the Soil Survey Laboratory in Beltsville, Maryland (1958–1962). Dr. Flach went on to serve as the head of the Soil Survey Laboratory in Riverside, California (1962–1972); director of the Soil Survey Investigations Division, Washington, D.C. (1972–1976); assistant administrator for the Soil Survey (1976–1980); associate deputy chief, Natural Resources Assessment (1980–1984); and special assistant to the chief for Science and Technology (1984–1988). Dr. Flach was also a senior research scholar with Colorado State University from 1988 to 1990, and lecturer with Johns Hopkins University between 1991 and 1993.

Douglas Helms (A.B., History, University of North Carolina–Chapel Hill, 1967; Ph.D., American History, Florida State University, 1977) is the historian of the Natural Resources Conservation Service (formerly the Soil Conservation Service). He was an archivist specializing in agricultural records at the National Archives in Washington, D.C., from 1973 to 1981. Dr. Helms became the historian of the Soil Conservation Service in 1981. He has written widely in the history of agriculture and soil and water conservation, including *Readings in the History of the Soil Conservation Service*, 1992. He was elected president of the Agricultural History Society in 1999.

Vance T. Holliday   (B.A., University of Texas; M.A., Texas Tech University; Ph.D., University of Colorado) is professor of geography at the University of Wisconsin–Madison. He teaches courses in general physical geography and advanced classes on soil geomorphology and on the Quaternary period. He has published journal articles, book chapters, an edited volume, and two research books on various aspects of soils, Quaternary landscape evolution, Paleo-Indian geoarchaeology, and soils applications in archaeological research. He is a fellow of the Geological Society of America (GSA), past president of the American Quaternary Association, and recipient of the George "Rip" Rapp Award (Archaeological Geology) and the Kirk Bryan Award (Quaternary Geology & Geomorphology) of the GSA.

C. Steven Holzhey   (B.S. and M.S., University of Idaho; Ph.D., University of California–Riverside) had a 39-year career in the soil survey, during which he mapped soils and worked as a research soil scientist. He worked in the soil survey laboratory system of the Soil Conservation Service for nearly 20 years, and was the first head of the National Soil Survey Laboratory and later the first head of the National Soil Survey Center, Lincoln, Nebraska. He was adjunct professor at the University of Nebraska, and authored or coauthored more than 40 publications.

Berman D. Hudson   (B.S., M.S., West Virginia University; Ph.D., North Carolina State University) is a senior soil scientist in the Soil Survey Division, Natural Resources Conservation Service, U.S. Department of Agriculture. He has served in nearly all of the field and supervisory level positions in the National Cooperative Soil Survey, including field soil scientist, soil survey project leader, state soil scientist, and national leader. He also has worked in private industry as a soil survey consultant to large paper and timber companies. Dr. Hudson has published numerous articles on the theory and practice of soil survey.

Maxine J. Levin   (B.S., Soils and Plant Nutrition, University of California–Berkeley, 1975) is a national program manager with the Soil Survey Division, Natural Resources Conservation Service, U.S. Department of Agriculture in Washington, D.C. She has worked as a field soil

scientist, soil survey project leader, soil correlator, and soil survey interpretations specialist with the National Cooperative Soil Survey in California, Maryland, New Jersey, Arizona, Tennessee, and Utah.

M. Dewayne Mays (B.S., Agronomy, A.M.& N. College/University
of Arkansas–Pine Bluff, 1968; M.S., Soil Management, Kansas State
University, 1971; Ph.D., Agronomy, University of Nebraska, 1982) is
the head of the Soil Survey Laboratory, Natural Resources Conservation Service, U.S. Department of Agriculture. He was a field soil scientist in Nacogdoches and Athens, Texas (1972–1977) and soil
scientist/supervisory soil scientist at the National Soil Survey Laboratory (1977–1989); and became soil scientist, National Soil Interpretation Staff (1989–1994). He developed the Fuzzy Logic methodology
that is currently used by the National Cooperative Soil Survey for
interpretations. He has also conducted research in soil carbon
methodology, soil phosphorus, and soil dispersion, and has published
numerous scientific articles and papers. He has participated in several
short-term foreign assignments.

Leslie D. McFadden (B.A., Stanford University; M.S., Ph.D., University of Arizona) is a professor of geology at the University of New
Mexico and currently the chair of the Department of Earth and Planetary Sciences. He teaches upper division and graduate courses in soil
geomorphology, soil stratigraphy, soil classification, and soil genesis.
He has published articles in journals and book chapters on aspects of
soil-forming processes in drylands with applications in paleoclimatic
and environmental studies. Dr. McFadden is a past chair of the Quaternary Geology and Geomorphology Division of the Geological Society of America (GSA) and a GSA Fellow.

James Muhn (B.A. History, Western State College–Gunnison, 1975;
M.A., History, Montana State University–Bozeman, 1986) is currently
a senior research associate for Morgan, Angel, and Associates—a
Washington, D.C., public policy consulting group. Mr. Muhn was land
law historian for the Bureau of Land Management for more than 20
years. He has written widely on the history of public lands, the Bureau
of Land Management, and the General Land Office, and often was an
expert witness in public land cases.

Joe D. Nichols (B.S., M.S., Soils, Oklahoma State University) was reared on a farm in eastern Oklahoma. He joined the Soil Conservation Service in Oklahoma, where he mapped soils and eventually became the state soil correlator. He was also state soil scientist in Colorado. He spent over 21 years at the National Technical Service Center at Fort Worth, Texas, where he was principal soil correlator for the southern states for 12 years. He is author or coauthor of 18 publications.

Steven E. Phillips (B.A., George Washington University; Ph.D., Georgetown University) is assistant professor of history at Towson University, Towson, Maryland. He served as an historian with the Soil Conservation Service (SCS), where he wrote a 170-page book on the SCS response to the 1993 floods in the Midwest, the historical background of flood control in the Upper Mississippi region, and such contentious environmental issues as levee repair and wetlands. Dr. Phillips' study of post-World War II Taiwan politics will be published by Stanford University Press.

Dennis Roth (B.A., Anthropology, University of Chicago; Ph.D., Anthropology, University of Oregon) has worked as an historian and social scientist for the U.S. Forest Service, U.S. Department of Agriculture since 1978. He is the author of books and articles on Forest Service history, conservation history, and environmental awareness, including *The Wilderness Movement and the National Forests* (Intaglio Press, 1995) and *Rhythm Vision* (Intaglio Press, 1990).

Horace Smith (B.S., Agronomy and Soils, Virginia State University, 1964; M.S., Soil Genesis and Classification, Ohio State University, 1972) is former director of the Soil Survey Division, Natural Resources Conservation Service, U.S. Department of Agriculture, Washington, D.C. Mr. Smith provided leadership for the National Cooperative Soil Survey and the National Soil Survey Center and the Soil Quality Institute. He has served in practically all of the field and supervisory level positions in the soil survey, including field soil scientist, soil survey project leader, state soil scientist, assistant principal soil correlator, and division director. He has written numerous articles in soil science journals, and his soil survey of the District of Columbia established new methods and standards for urban mapping.

John P. Tandarich (B.S., Anthropology, Loyola University of Chicago, 1971; M.A., Anthropology, Archaeology, University of Iowa, 1975; M.S., Soil Science, University of Missouri-Columbia, 1982; Ph.D., Natural Resources and Environmental Sciences, University of Illinois at Urbana-Champaign, 2001) is soil scientist and archaeologist with Hey and Associates, Inc. of Libertyville and Chicago, Illinois; research soil scientist for Wetlands Research, Inc., Chicago, Illinois; and consulting soil scientist for the Wetlands Initiative, Chicago, Illinois. He has worked on problems in hydric soil formation and classification, geopedology, geoarchaeology, remote sensing, and historical-archival research. He has published on hydric soil formation and distribution, wetland restoration, presettlement landscape assessment, and the histories of archaeology and soil science.

INTRODUCTION
Douglas Helms

Soil surveying and soil science in the United States are both quite young. They developed only a little more than a century ago, beginning with several early state and regional surveys of "agricultural soils." Investigators made slow progress in understanding soils from laboratory analysis and from field studies, observations, and research. They came to recognize the complexity of the soil and began to understand that the creation of parent material by geological processes was only one of several interrelated processes giving rise to individual, unique soil bodies. To understand the complexity and move beyond trying to relate soils to one or two genetic factors, soil science had to bring the tools of chemistry, physics, biology, climatology, geography and other sciences to the study of the soil itself. At the same time, agricultural chemists discovered some of the keys to plant growth. While they demonstrated that chemical soil amendments increased plant growth, field studies showed that soils differed in their response to fertilizers. Much still needed to be learned about soil properties and their importance to agriculture.

In the last decade of the nineteenth century, the U.S. Department of Agriculture (USDA) and the state agricultural experiment stations, which began receiving federal support in 1887, commenced building a vast agricultural research system. The study of the soil was only one of many initiatives for scientific research that had the utilitarian purpose of developing American agriculture. The U.S. Soil Survey, initiated in 1899 at the early stages in scientific investigation of the origins and properties of soil, eventually succeeded in the objective of providing maps and descriptive materials to aid in making wise land-use decisions. As the scientists of the soil survey developed their understanding of soils for mapping, they also made important contributions to the development of soil science and soil classification. Field experience and observations could confirm or raise questions about theories developed in the laboratory. Soil scientists and soil surveyors could see relationships in the field that contributed to soil classification and to greater understanding of soil properties. Surveying, laboratory, and plot research worked in an iterative process, each influencing the other to the benefit of both soil science and soil survey.

Few historians have turned their attention to this crucial earth science, though more may be expected to in time, as the intriguing connections between soils, the environment, and human history become evident. Meanwhile, soil scientists themselves have produced valuable historical monographs and articles on various aspects of soil survey history in soil science professional journals. Especially valuable are the works of Roy W. Simonson and David R. Gardner (for citations see Appendix B, "History of the U.S. Soil Survey: A Bibliography").

This volume examines the history of soil surveying, the area of soil science that focuses on mapping, analysis, and description of soils as found on the earth's surface. Most chapters deal with the cooperative program of soil surveying begun in the USDA in 1899, but others describe the soil survey and soil science programs of other federal agencies and the applications of soil survey interpretations to other fields and applications.

The first chapter in this volume, by Douglas Helms, Anne B. W. Effland, and Steven E. Phillips, examines the influences of key individuals and the state experiment stations in the establishment of federal support and coordination of soil surveys. Helms also contributes a second chapter profiling these and other individuals and institutions that have been instrumental in the development of the soil survey.

From early in its history, the soil survey operated a laboratory to supply information needed in soil surveys. The contributions made by the laboratory to soil science and the soil survey are described by Klaus W. Flach and C. Steven Holzhey, who worked in the laboratory for many years. Another career-long soil scientist, Joe D. Nichols, contributes an autobiographical account of life in the field, from county soil survey party to regional supervisory positions in correlation and soil survey operations. The career journey Nichols recounts represents experiences shared by many soil scientists who spent their careers with the soil survey.

A chapter by Maxine J. Levin reviews the slow opening of careers in soil science and soil survey to women and the subsequent rapid growth of interest among women in careers in soil science. M. Dewayne Mays, Horace Smith, and Douglas Helms examine the role of African-American soil scientists in the National Cooperative Soil Survey and the critical importance of the 1890 land-grant universities in preparing students for careers in soil survey and in supporting faculty research in the field of soil science.

While the USDA's Bureau of Soils and later the Soil Conservation Service oversaw the soil surveys made on private lands, the federal land management agencies operated their own soil survey programs. Dennis Roth and James Muhn describe the soil survey programs of the U.S. Forest Service and the U.S. Bureau of Land Management.

Three chapters examine aspects of the utility of the soil survey to other fields of research and to land-use planning. Vance T. Holliday, Leslie D. McFadden, E. Arthur Bettis, and Peter W. Birkeland explore the value of the work of the U.S. Soil Survey to their discipline of soil geomorphology, which seeks to understand aspects of soil landscape development. Patricia Durana and Douglas Helms review the varied interpretations of soil survey data and the myriad uses to which they are applied in land-use decision making. Finally, the former director of Soil Survey Division in USDA's Natural Resources Conservation Service, Horace Smith, and senior soil scientist Berman D. Hudson discuss the future of the U.S. Soil Survey and the new directions both the survey and its uses will take.

The editors wish to thank particularly Roy W. Simonson, former director of Soil Classification and Correlation, who provided very detailed comments on the chapter "Early Leaders of the Soil Survey," and Peter F. Smith and Berman Hudson of the Natural Resources Conservation Service (NRCS), who commented on drafts of all the chapters. The editors would also like to thank Paul Reich of the NRCS, who assisted with the photographs that appear in the volume, and the following individuals and institutions who provided research assistance and additional photographs: Edward McCarter, Nicholas Natanson, and Joe Schwartz of the National Archives, College Park, Maryland; Susan Snyder, the Bancroft Library, University of California, Berkeley; Aaron D. Purcell, Special Collections, University Libraries, University of Tennessee, Knoxville; in the Wilson Library of the University of North Carolina, Chapel Hill, Alice Cotten of the North Carolina Collection, Jerry Cotten of the Photographic Archives, and John White and

Richard Shrader of the Southern Historical Collection; the staff of the Special Collections and University Archives, D. H. Hill Library, North Carolina State University; the staff of the Special Collections Department, University of Virginia Library; and the staff of the Ferdinand Hamburger, Jr., Archives, Johns Hopkins University.

1

FOUNDING THE USDA's DIVISION OF AGRICULTURAL SOILS: CHARLES DABNEY, MILTON WHITNEY, AND THE STATE EXPERIMENT STATIONS

Douglas Helms, Anne B. W. Effland, and Steven E. Phillips

INTRODUCTION

Creation of the Division of Agricultural Soils in the United States Department of Agriculture (USDA) in 1894 marked the beginning of a national commitment to soil science and soil survey activities. By 1899, soil surveyors had commenced field investigations for the first USDA soil surveys; one century later, soil surveys have been published for most counties in the United States. The surveys—inventories consisting of narrative descriptions, tables, and maps—include a wealth of interpretive information about soil characteristics and properties and about their limitations and potentials. These interpretations of how soils respond to different uses assist land users in making appropriate decisions. And while Congress and the federal government have supported the soil survey effort to help guide agricultural development, surveys have wide application in a variety of environmental and land management uses. The development of the national soil survey program grew out of the intellectual curiosity and personal ambition of two men in particular, Charles W. Dabney, Jr. and Milton Whitney. But it was the research at state agricultural experiment stations that created the favorable climate for Dabney's and Whitney's successes. This state-level research grew from a desire by state agricultural leaders to understand their soils and the relationship of the characteristics of those soils to plant growth. Ultimately, they believed, knowledge would contribute to a healthy agricultural economy. Thus, even before the passage of the

Hatch Experiment Station Act of 1887 that authorized federal funding to establish such stations, some states had begun to set them up on their own. With the financial support brought by passage of the Hatch Act, state agricultural experiment stations emerged nationwide.

Soon experiment station staffs joined with the land-grant universities to form an association for discussing issues of mutual interest. Those discussions became a forum to press for improvements. While each experiment station studied soils in its respective state, the leaders and scientists of those stations felt the need for a coordinated national effort. Collectively, these efforts led directly to the creation of a soils research unit within the USDA.

Although many leading scientists of the time contributed to the establishment of a soils research unit in the USDA, the appointment of Charles Dabney as assistant secretary of agriculture in 1894 marked the beginning of a federal commitment to soil science and soil survey activities as part of the USDA's scientific work. In fact, Dabney would have considerable influence not only on the soils work of the USDA but also on the scientific work of the entire department. Milton Whitney, whom Dabney appointed chief of the new Division of Agricultural Soils, in turn, influenced the scientific and interpretive direction of the new division, and as a result, the direction of the newly initiated national soil survey efforts.

DABNEY AND WHITNEY IN NORTH CAROLINA

Charles William Dabney, Jr.'s devotion to education and agriculture are evident in his lifetime of work, first as director of the North Carolina Agricultural Experiment Station, then as president of the University of Tennessee, assistant secretary of the USDA, and as president of the University of Cincinnati. Born at Hampden–Sydney, Virginia, on 19 June 1855, Dabney graduated from the college there in 1873. He studied chemistry at the University of Virginia under Dr. John William Mallet and took classes in physics, geology, Latin, French, and German. Mallet, from Great Britain, had earned a Ph.D. in Chemistry at the University of Göttingen in Germany and became, in Dabney's words, "my scientific father" (Dabney Memoir n.d.). Emory and Henry College appointed Dabney as professor in 1877, based on Mallet's recommendation.

Dabney taught there for a year before resuming his academic training in Europe in 1878, where he studied chemistry in Germany at the universities of Berlin, Heidelberg, and Göttingen. Aspiring chemists of the nineteenth century, including those of the subdiscipline of agricultural chemistry, valued the graduate-level training available at German universities. Dabney earned his doctorate in chemistry at the University of Göttingen in 1880. While in Germany, he accepted a professorship in Chemistry at the University of North Carolina, based once again on Professor Mallet's recommendation.

As he prepared to return to the United States, Dabney could count himself among the elite of science educators (Powell 1979–1996). Yet when he arrived in Chapel Hill, he discovered that his new job did not provide for a laboratory. Coincidentally, however, the position of state chemist of the Department of Agriculture and head of the North Carolina Agricultural Experiment Station was vacant, following the departure of its first director. The station was established several years earlier by the North Carolina State Department of Agriculture Act of 1877, with a charge to analyze fertilizers, suppress fraudulent sales of ineffective fertilizers, and investigate plant growth and nutrition and fertilizers and crops suited to particular soils. The experiment station boasted a small laboratory on the campus in Smith Hall, and perhaps more importantly, access to fertilizer tax funds to purchase additional equipment. Dabney seized the opportunity to run a laboratory and pursue practical research. He turned down his teaching post at the university and instead became the second director of the second state agricultural experiment station established in the United States (Dabney Memoir n.d.).

Dabney found much to admire in the work of the state chemist and first director of the agricultural experiment station, Dr. Albert R. Ledoux (Schaub 1955). Ledoux and his assistants had university training, and in some cases graduate training in the United States or Europe. In fact, Ledoux and Dabney shared a common background, having studied chemistry, including agricultural chemistry, at Göttingen (Powell 1979–1996; Schaub 1955). Ledoux did not have an experimental farm, however, though field tests were carried out with cooperating farmers.

When Ledoux left to establish a private laboratory in New York, the state experiment station had established a good reputation. But the prospects did not seem bright for improving the quarters on the University of North Carolina campus (Dabney Memoir n.d.). The young Dr. Dabney, however, soon displayed the ability and personal persuasiveness that was needed to expand and organize operations. He convinced the board of agriculture to provide better facilities with money collected from the fertilizer analyses. In late 1881, the experiment station moved from the university to the old National Hotel in Raleigh, which had been outfitted with modern laboratories and equipment (Battle 1966; Dabney Memoir n.d.).

Dabney's next goal was an experimental farm. In December 1885, the board of agriculture purchased 10 acres of land west of the center of Raleigh on the Hillsborough Road near the fair grounds. Not an unalloyed blessing, Dabney remembered that "it was as poor a piece of land as could be found in North Carolina" (Dabney Memoir n.d.). Nevertheless, North Carolina now truly had an experiment station in fact as well as in name. Two years later, using funds acquired through the Hatch Act, Dabney improved the station by adding a small barn, soil laboratory, and cottage for the manager.

Little in the way of written guidance on the operation of an experiment station was available to directors at that time. Dabney shaped the North Carolina program based on his experiences and advice from other experts in the United States and Europe. In Germany he visited agricultural experiment stations and learned about their work. According to his memoirs, he received much inspiration from observing the work at the Weender Agricultural Station near Göttingen. Dabney also knew the major figures in experiment station work and sought their assistance. He was in contact with Sir John Bennett Lawes of the pioneering Rothamstead experiment station in England. He wrote to Dabney advising him on establishing permanent plots (North Carolina Agricultural Experiment Station 1887). He visited Dr. Samuel W. Johnson, director of the experiment station at New Haven, Connecticut, and a pioneer in the American experiment station movement (Dabney Memoir n.d.). Dabney likely came to select the station's first experimental farm superintendent, Milton Whitney, as a result of this visit to Johnson. Johnson's lab, where Whitney was working at the time, provided many of the nascent experiment stations with their first staff (Dabney Memoir n.d.; Rossiter 1975).

Whitney arrived in Raleigh to organize the experimental farm on 1 April 1886. While he stayed at the farm for only a short time, the influential role he would play in the future direction of soils research began to take shape in those years (North Carolina Agricultural Experiment Station 1887). Agricultural science in the second half of the nineteenth century profoundly influenced Whitney's experimental designs, which encompassed a complex quest for answers to the host of scientific riddles behind crop productivity that could not yet be seen or measured, including the characteristics of soil. Traditional views of soil fertility, as Whitney put it in the experiment station farm's first annual report, held "that a chemical analysis of a particular plant would show what that plant required for perfect development, and that a like analysis of a soil would show what was lacking in the latter to produce a maximum crop of any particular kind" (North Carolina Agricultural Experiment Station 1887, 84).

At first enamored of this view, American scientists soon learned that the fertilizers developed to supply the chemical shortcomings of various soils were only one key to productivity. In his memoirs, Dabney recalled that the search for the keys to soil fertility was his motivation for starting the experiment station farm and hiring a manager with experience. "Knowing that the chemical analysis of soils was not the only thing in determining their productive capacity, we desired to make experiments on the physical properties of soils, their capacity to hold water and to nurture the bacteria which convert the vegetable matter into muck" (Dabney Memoir n.d.). Dabney placed a great deal of faith in the value of understanding soil fertility and management for assisting the state's farmers; thus, he became highly supportive of Whitney's efforts.

Whitney reported a "modest start" in his first season as superintendent of the experimental farm. He conducted experiments on cotton grown on manured and unmanured land and on root development of cotton, peas, and some forage plants. He also planned experiments for pasture and hay grasses and some small grains. Whitney began pioneering evapotranspiration studies on cotton and pot culture research related to moisture and plant growth (McCracken 1989).

But Whitney's other line of investigation at the farm—studies of air and soil temperatures, moisture in the soil, and a record of sunshine—would prove to be the foundation of his work relating soil physical characteristics to fertility. Whitney's soil studies initially focused on recording the temperatures of soils at different depths. He believed the readings would be correlated to soil type. After the relationship of a particular soil to temperature had been determined, it was correlated with observations on rainfall, soil moisture, and air moisture in order to produce information of use to the farmer. The first few months of study proved promising enough that the station planned to continue the studies for several years. Whitney made it clear that he needed a weather station at the experimental farm to collect meteorological information, so at the request of the North Carolina Board of Agriculture, the U.S. Signal Corps stationed W. O. Bailey at the experimental farm to staff a weather station.

Dabney found the young Whitney "to be energetic and capable, awake to the scientific problems of correct farming. . . ." Writing years later, near the end of his life, Dabney claimed some credit for North Carolina's giving Whitney his start on the study of physical properties of soils. Whitney did make some studies at the experiment station in Connecticut, but according to Dabney, "North Carolina deserves the credit of having given him the facilities for making the first really important investigations on the subject made in this country"(Dabney Memoir n.d.; Dabney 1927a, 1927b).

Dabney's committed leadership of the experiment station and energy were evident in its expanded operations during his tenure. Dabney supervised the laboratories in Raleigh, the experimental farm, the state's agricultural survey, the state museum, various agricultural exhibitions, the state weather service, and farmers' institutes and other educational enterprises. Dabney would have been a candidate, perhaps the leading candidate, for the presidency of the state's new agricultural college, North Carolina College of Agriculture and Mechanic Arts. He had been centrally involved in pressing for its establishment. But the governor opposed the new school and was not inclined to promote any further advancement of Dabney in the wake of that battle (Dabney to Eggleston 1942, 1944; Dabney 1936, 185–188). His record of accomplishment did bring him an offer of president of the University of Tennessee in 1887, which he accepted at age 32. He resigned from the experiment station effective 1 September 1887 (North Carolina Agricultural Experiment Station 1888).

Whitney's career stalled temporarily. The state legislature did not provide funds for 1888, expecting newly authorized federal Hatch Act support to take the place of state funding in the upcoming year. At the end of 1887, all experimental fieldwork at the farm was to be discontinued. Whitney resigned in July and left the North Carolina Agricultural Experiment Station later that year to become vice-director of the newly established South Carolina Experiment Station located on the campus of the University of South Carolina at Columbia (Schaub

1955; Williams n.d.). The soils work conducted at the experiment station during that first year reflected Whitney's interests; station staff made a thorough analysis of the soils on each of the three experiment station farms at Spartanburg, Columbia, and Darlington.

In South Carolina, Whitney continued his work on the mechanical analysis of soil particles and his studies of soil temperature and meteorology. His thinking about how to develop a method of relating analysis of soils to crop production continued to evolve. Whitney collected samples of soils that were being used for growing rice and sea island cotton. He recorded observations on the mode of cultivation, manuring, and general relation to crop production of these soils, but he did not have time to make a thorough examination of their physical properties. Whitney "devoted considerable time to a study of methods for their (soils) examination with a view to extending the investigation in time to a study of our typical soils and staple crops." When Whitney found that the agricultural experiment station was to be moved, he wrote that there was "no present prospect of being able to extend the investigation to our different soils and crops as proposed . . ." (South Carolina Experiment Station 1890, 50). He would have to find that opportunity elsewhere.

The productive and scientifically impressive, albeit small, soils staff at the South Carolina station soon dispersed. Robert Loughridge, analyst of soils and seeds, accepted an offer from Eugene W. Hilgard to join the faculty at the University of California. Hilgard was ready to consider Whitney, had Loughridge declined, but Whitney, even at this early date in his career, had given Hilgard the impression that he had aspirations for employment in the USDA (Jenny 1961). Whitney returned to his native state of Maryland, where he became a soil physicist at the Maryland Agricultural Experiment Station. There he continued to pursue his studies of the relationship of soils to agriculture.

Major Henry E. Alvord, president of the Maryland Agricultural College and director of the Maryland Agricultural Experiment Station, had made the soil survey a priority. In describing his plans for the survey, Alvord wrote, "Soils will be studied with regard to the varieties to be found in different sections of the State. . . . These will be examined with reference to their natural condition and the effects of different modes of treatment, tillage, drainage, etc. . . ." (Maryland Agricultural Experiment Station 1888, 13). A few years later, he seized upon the geological survey of Maryland being made by the U.S. Geological Survey as an opportunity to make "a systematic study of the soils of the State, with a view to their classification, description, defining the boundaries of the typical formations and explaining local variations" (Maryland Agricultural Experiment Station 1891, 85). The Maryland Agricultural College was already working with the U.S. Geological Survey on a geological map of the state, creating data that could be used for the soil investigations. Professors in the departments of chemistry, geology, and physics offered to help solve some of the scientific questions.

In Whitney, Alvord found a man well qualified to lead this soils investigation work. Johns Hopkins University, a cooperator in Maryland's soil survey efforts, provided Whitney with offices and laboratories at the Hopkins Mansion in Baltimore, which he occupied in June 1891. Alvord took pride in proclaiming Maryland as the first agricultural experiment station to establish a division with a staff working exclusively on soil investigations (Maryland Agricultural Experiment Station 1892).

THE ASSOCIATION OF AMERICAN AGRICULTURAL COLLEGES AND EXPERIMENT STATIONS AND THE U.S. WEATHER BUREAU

Soon after passage of the Hatch Act in 1887 and the subsequent emergence of experiment stations nationwide, the Association of American Agricultural Colleges and Experiment Stations was formed. Whitney and Dabney, who served as director of the Tennessee Agricultural Experiment Station as well as president of the University of Tennessee, regularly attended the annual conventions of the association. Dabney quickly became a leader in the group, eventually serving a term as president and hosting an annual convention at Knoxville. The association used its influence to promote the interests of scientific research at state institutions and in the USDA. The association's prodding was a positive influence in the blossoming of scientific research in the department in the 1890s. To ensure that state governments and experiment station staff devoted each station's funds to their proper use, the association successfully sought "inspection" of experiment station expenditures by the secretary of agriculture (Association 1894, 64; Association 1895, 17–18, 63).

The small soils contingent within the association was quite active. Whitney and Eugene Hilgard consistently attended and reported on various aspects of soil investigations, and both often made presentations on soil studies at the annual convention. At the 1891 meeting, they appeared on the program in a session titled "State soil surveys, how far are they practicable and how should they be conducted?"

Bailey Willis of the U.S. Geological Survey "addressed the conference on the work of the survey classifying soils" at the 1889 meeting (Association 1890, 105–106). After the presentation, the conference appointed a committee to investigate soil classification, and Whitney was among those selected (Association 1890, 106). In that same year, the convention encouraged the USDA to publish the results of domestic and international soil investigations (Association 1890, 111).

The association directly advocated soil investigations within the USDA. As part of that effort, Henry Alvord, who also held influential posts in the association including president, championed shifting the U.S. Weather Bureau to the Department of Agriculture. Alvord and his colleagues argued that the alignment would be mutually advantageous: The state experiment stations could collect weather data for the Weather Bureau, and a USDA-based Weather Bureau would provide an opportunity to expand weather data analysis, making it of greater assistance to farmers. As part of their association with the Weather Bureau, the land grant universities could train climatologists and undertake the research needed to produce such analysis (Association 1892, 40–61).

The Weather Bureau moved to the USDA in 1891, and soon after an association committee, of which Alvord was a member, pressed the bureau to add the study of "physics, conditions, and changes of agricultural soils" to its work (Association 1892, 60). In November 1891, Milton Whitney wrote to Mark Harrington, chief of the Weather Bureau, to request $150 per month through July 1892 to prepare a report on soil moisture and the movement of water in the soil. He stressed the relevancy of this work, writing, "Crop production is not directly limited by the amount of rainfall, but by the moisture in the soil" (Whitney to Harrington 1891). Whitney did not act alone; Alvord sent a copy of Whitney's letter to Harrington to the secretary of agriculture, along with letters from the directors of the Pennsylvania and Wisconsin experiment stations that heartily endorsed Whitney's work (Alvord to Harrington 1891).

Whitney urged that he be authorized to begin as quickly as possible. On 30 November 1891, Harrington appointed Whitney as a special agent of the Weather Bureau charged with investigating "water in soil and its movements" (Harrington to Secretary of Agriculture 1891). In 1892, the Bureau published Whitney's bulletin on the relationship of physical properties of soils to moisture and crops, as well as a bulletin written by Eugene Hilgard on the relationship of soils to climate (Association 1893, 106).

Whitney's relationship with the Weather Bureau expanded after the Maryland legislature established a state weather service in 1892, and Alvord appointed Whitney on 26 April as its secretary and treasurer (Alvord to Harrington 1892). Whitney requested Weather Bureau funds for eight to ten stations to measure soil moisture and temperature, as well as the temperatures just above the soils. He also requested an extension of Weather Bureau support for his soil moisture and water movement studies. He stressed to Harrington, "This work is attracting very widespread interest and I am receiving many encouraging letters from all over the country." Whitney reminded Harrington that support was vital "so that we may keep in the lead and point out the direction the work should take" (Whitney to Harrington 1892a).

Whitney's correspondence with Harrington revealed much about his theories of soils and agriculture. His theories eventually shaped his ideas of soil classification and soil survey. Based on his career experience in four states (Connecticut, North Carolina, South Carolina, and Maryland), he called for greater understanding of the relationship of soils to crop productivity. He complained to Harrington, "There has been a long period of inaction after the failure to show the relation of plant growth to the chemical composition of soils, but the time now seems ripe for this line of physical research." He emphasized the importance of physical properties of soil, reporting to Harrington on differences in wheat productivity in regions of states where the difference in climate was scant and the chemical composition of the soils similar. "I have shown that these soils differ very greatly in their relationship to moisture and heat, according to their texture as shown by mechanical analysis. . . ." He already had a theory to explain increased productivity through adding fertilizers. "I have worked out the texture of these soil types and the relationship of these soils to water, or the movement of the rainfall after it enters the ground, and I am working now on the effect of fertilizers in changing the texture of soil and their relationship to moisture and heat" (Whitney to Harrington 1892a).

Harrington, however, was prepared to end funding at the end of July 1892 as was originally agreed. Whitney reiterated the need for continued work on soil physical properties, particularly the role played by texture in soil water retention, and warned that without Weather Bureau support the Maryland state effort to establish soil stations would collapse. In August, in fact, he expanded his request to include funding for examining samples sent from around the country. As he explained to Harrington, "While the Department may not be ready for this at present I find that there is a very great interest in this work throughout the country, and from nearly all of the States I get either voluntary offers of samples or I get very ready responses to requests for samples . . ." (Whitney to Harrington 1892b). Although Harrington remained unconvinced, the presidential election brought a change in leadership at the Department of Agriculture that would work in Whitney's favor. Whitney would soon be reunited with his former sponsor Charles Dabney in Dabney's new role as assistant secretary of agriculture.

DABNEY AS ASSISTANT SECRETARY OF AGRICULTURE

The Department of Agriculture established the new position of assistant secretary, with the primary responsibility of supervising the department's scientific work, in 1889. Henry Alvord and the leaders of the Association of Agricultural Colleges and Experiment Stations were successful in influencing selection of the candidate for the new position. Their choice, Eugene Hilgard, received cabinet approval but withdrew prior to Senate confirmation, much to the exasperation of the association leadership. The episode brought to a close Hilgard's campaign for a national agricultural survey, which he had first proposed be placed in the U.S. Geological Survey (Amundson and Yaalon 1995, 11). Edwin Willets was named to the position instead, which he held until the election of Grover Cleveland as president in 1892.

In October 1893, President Cleveland offered Charles Dabney the position. Dabney accepted, and he took a leave of absence from the University of Tennessee on 1 January 1894. Dabney learned from President Cleveland that a number of officials of agricultural colleges and experiment stations, with whom he had met in February 1893 to discuss the scientific work of the department as well as the relations of the colleges and stations with the federal government, had promoted his candidacy (Association 1894, 21).

Eugene Woldemar Hilgard, 1874. (Courtesy of the Bancroft Library, University of California, Berkeley—UARC PIC 13:282)

Dabney served under secretary of agriculture J. Sterling Morton, but he received his portfolio and charge directly from the president. Dabney was to be directly in charge of "the entire direction of all the scientific divisions, and likewise of the Office of Experiment Stations, of the Office of Irrigation Inquiry, of the Office of Fiber Investigation, and of the Museum" (USDA 1895, 24). When Cleveland met with Dabney, he explained his desires for the work of the USDA and said, "that of all the departments of the government, the Department of Agriculture should be the freest from political interference and 'spoils;' and he expected me to see that all its bureaus should be kept free from such influences, and to see that all its officers were competent, scientific men, protected from such influences and put in the Civil Service" (Dabney Memoir n.d.). The president instructed Dabney to come to the White House at a predetermined time each day if he had any business. Dabney availed himself of the opportunity and often was instructed to write the section on agriculture for the president's report to Congress (Dabney Memoir n.d.).

Charles W. Dabney (on right) as assistant secretary of agriculture, 1894. (Courtesy of Special Collections, University Libraries, University of Tennessee, Knoxville, Tenn.)

Throughout his career, Dabney's efforts validated his belief that knowledge of soils was prerequisite to agricultural improvement. After accepting his USDA post, Dabney quickly moved to turn his ideas into concrete organizations. He appointed Whitney chief of a new Division of Agricultural Soils and indicated his support for Whitney's research direction, writing in his appointment letter, "I beg to congratulate you upon the prospects for interesting and valuable work that lie before you and to express my hearty interest in all your plans" (Dabney to Whitney 1894).

Dabney's intent to play a central role in the division, located within the Weather Bureau and financed by $2,000 from the Weather Bureau budget, was reflected in the authority he secured from Secretary Morton over the division's spending. Weather Bureau chief Mark Harrington had "requested to be relieved from all responsibility attending the work and expenditures relating to said division" (Morton to Evans 1894), and Morton ordered all financial matters concerning the new

Milton Whitney, first chief of the Division of Agricultural Soils, n.d. (Courtesy of the
National Archives at College Park, Maryland; Print 16-G; Record Group 16, Records of
the Office of the Secretary of Agriculture)

division go directly through the assistant secretary (Morton to Harrington 1894). Dabney expanded that authority to include all correspondence to the division (Dabney to Harrington 1894).

Even as the Division of Agricultural Soils was formally created,
Dabney was counseling the secretary of agriculture to make it independent from the Weather Bureau. Dabney emphasized his preference to
representatives of the state experiment stations in his 1894 address,
"Scientific Work of the Department." He framed his discussion of the
division in the context of other new organizations, including the Division of Forestry, the Division of Vegetable Pathology, and the Division
of Agrostology. In explaining his reasoning behind support for an independent Division of Agricultural Soils, Dabney reminded his audience,
"I speak this morning to many agricultural chemists, so that I need not
take time to explain the disappointments that we have all felt with regard to the results of the chemical analysis of soils." He stressed that
the chemical work was of limited value to farmers. Endorsing Whitney's plan to investigate soil physical properties beyond chemical

analysis in soil-crop productivity investigations, Dabney announced, "Our Department had decided, therefore to attack this old problem from two different sides; first, from the physical side, by studying its relation to heat, moisture, etc.; and second, from the biological side, by studying its nitrifying organisms, etc." The former goal necessitated the creation of the Division of Agricultural Soils in order to "develop methods of these investigations and to encourage an extensive study of soils of the country by State stations and colleges . . ." (Association 1895, 63–69). The division would do work that the states could not, such as collect samples from around the country and compare them. The Division of Agricultural Chemistry would continue to collect soils for its own analysis, based on its congressional mandate.

CONCLUSION

In retrospect, had Hilgard accepted the position of assistant secretary of agriculture in 1889, the research direction pursued by the Division of Soils might have been far different. Hilgard was a champion of chemical analysis as a means of understanding the fertilizer needs of each soil. Conversely, Whitney, in his most exaggerated assertions, held that soil texture and moisture determined productivity and that annual crops could be grown indefinitely on most soils. As it turned out, additional scientific discoveries in the twentieth century and improvements in the chemical analysis of soils largely confirmed Hilgard's beliefs and revealed the fallacies of some of Whitney's extreme hypotheses. Nevertheless, Whitney's efforts to establish a national soil survey and to expand the range of soils investigations in the USDA enhanced the value of the division's contributions to soil science.

During the decade and a half after Charles Dabney returned to the United States after receiving his German Ph.D. in chemistry, he had been consistent in his view that knowledge of soils was key to improving agricultural productivity. His work to create a Division of Agricultural Soils in the Department of Agriculture marked the culmination of his efforts to make the study of soil a key component of scientific agriculture. In 1899, five years after the creation of the Division of Agricultural Soils, the division embarked on the field investigations for the first publication of its series of soil surveys, *Field Operations*.

And while the study of agricultural soils did not begin with the creation of the state experiment stations, they had a profound influence on the emergence of a national soils research effort, both as training and staging grounds for leaders in the development of soil analysis. From the development of the state agricultural experiment stations, there emerged an awareness of the need for a unifying federal presence and support, which ultimately resulted in the establishment of the Division of Agricultural Soils and the soil survey.

The soil survey that is produced today is far different from the early surveys produced in 1900. Much is owed to scientific advances that have helped illuminate the properties and functions of soils. But the soil scientists have also learned much from experience and from observing the reactions of soils to differing uses and treatments. The unifying theme between the early steps toward knowledge of soils in the nineteenth century and the current soil survey is the promise that understanding soils will assist land users.

REFERENCES

Alvord, Henry E. to Mark W. Harrington. 4 November 1891. Letter 1098; Letters Received, 1891–1894; Record Group 27, Records of the Weather Bureau. National Archives at College Park, Maryland.

Alvord, Henry E. to Mark W. Harrington. 26 April 1892. Letter 2389; Letters Received, 1891–1894; Record Group 27, Records of the Weather Bureau. National Archives at College Park, Maryland.

Amundson, Ronald, and Dan H. Yaalon. 1995. E. W. Hilgard and John Wesley Powell: Efforts for a joint agricultural and geological survey. *Soil Science Society of America Journal* 59: 4–13.

Association of American Agricultural Colleges and Experiment Stations (Association). 1890. *Proceedings of the Third Annual Convention, 1889*. Office of Experiment Station Miscellaneous Bulletin No. 2, U.S. Department of Agriculture, Washington, D.C.

Association of American Agricultural Colleges and Experiment Stations (Association). 1892. *Proceedings of the Fifth Annual Convention, 1891*. Experiment Station Bulletin No. 7, U.S. Department of Agriculture, Washington, D.C.

Association of American Agricultural Colleges and Experiment Stations (Association). 1893. *Proceedings of the Sixth Annual Convention, 1892*. Office of Experiment Stations Bulletin No. 16, U.S. Department of Agriculture, Washington, D.C.

Association of American Agricultural Colleges and Experiment Stations (Association). 1894. *Proceedings of the Seventh Annual Convention, 1893*. Office of Experiment Stations Bulletin No. 20, U.S. Department of Agriculture, Washington, D.C.

Association of American Agricultural Colleges and Experiment Stations (Association). 1895. *Proceedings of the Eighth Annual Convention, 1894*. Office of Experiment Stations Bulletin No. 24, U.S. Department of Agriculture, Washington, D.C.

Battle, H. B. 1966. *Forty Years After.* History Series No. 3. Agricultural Experiment Station, North Carolina State University, Raleigh, N.C.

Dabney Memoir. n.d. A memoir by Charles W. Dabney, with an introduction by J. D. Eggleston. Joseph Dupuy Eggleston Papers, Special Collection, University of Virginia, Charlottesville, Va. Microfilm. This is a more complete version of Charles W. Dabney, Jr.'s memoir. He wrote several drafts that were never published. Portions of drafts are among the Charles W. Dabney Papers in the Southern Historical Collection, University of North Carolina, Chapel Hill, N.C.

Dabney, Charles W. 1927a. The founding and the first ten years of the North Carolina Agricultural Experiment Station. Special Collections, North Carolina State University, Raleigh, N.C. Typescript.

Dabney, Charles W. 1927b. Notes for historical sketch of North Carolina Agricultural Experiment Station from founding until 1888. Special Collections, North Carolina State University, Raleigh, N.C. Typescript.

Dabney, Charles W. 1936. *Universal Education in the South.* Vol. 1. Chapel Hill, N.C.: University of North Carolina Press.

Dabney, Charles W. to Joseph D. Eggleston. 30 December 1942. Eggleston Family Papers, Virginia Historical Society, Richmond, Va.

Dabney, Charles W. to Joseph D. Eggleston. 18 February 1944. Eggleston Family Papers, Virginia Historical Society, Richmond, Va.

Dabney, Charles W. to Mark W. Harrington. 23 March 1894. Letter 44, Vol. 16; Letters Sent by Assistant Secretary of Agriculture, 1889–1929; Record Group 16, Records of the Office of the Secretary of Agriculture. National Archives at College Park, Maryland.

Dabney, Charles W. to Milton Whitney. 6 May 1894. Letters Received 1894—1901 by the Division of Soils; Record Group 54, Records of the Bureau of Plant Industry, Soils, and Agricultural Engineering. National Archives at College Park, Maryland.

Harrington, Mark W. to the Secretary of Agriculture. 30 November 1891. Letter 1107; Letters Received, 1891–1894; Record Group 27, Records of the Weather Bureau. National Archives at College Park, Maryland.

Jenny, Hans. 1961. *E. W. Hilgard and the Birth of Modern Soil Science.* Pisa, Italy: Collana Della Rivista "Agrochimica."

McCracken, Ralph J. 1989. The history of soil science at North Carolina State University. Special Collections, North Carolina State University. Raleigh, N.C. Typescript.

Maryland Agricultural Experiment Station. 1888. *History, Organization and Work.* Bulletin No. 1, College Park, Md.

Maryland Agricultural Experiment Station. 1891. *Third Annual Report of the Maryland Agricultural Experiment Station for the Year 1890.* College Park, Md.

Maryland Agricultural Experiment Station. 1892. *Fourth Annual Report of the Maryland Agricultural Experiment Station for the Year 1891.* College Park, Md.

Morton, J. Sterling to F. I. Evans. 26 March 1894. File 5587; Letters Received 1870–84; 1888–1895; Record Group 27, Records of the Weather Bureau. National Archives at College Park, Maryland.

Morton, J. Sterling to Mark W. Harrington. 12 March 1894. File 52; Letters Received 1870–84; 1888–1895; Record Group 27, Records of the Weather Bureau. National Archives at College Park, Maryland.

North Carolina Agricultural Experiment Station. 1887. *Annual Report, 1886.* Raleigh, N.C.

North Carolina Agricultural Experiment Station. 1888. *Tenth Annual Report of the North Carolina Agricultural Experiment Station for 1887.* Raleigh, N.C.

Powell, William S., ed. 1979–1996. *Dictionary of North Carolina Biography.* Chapel Hill, N.C.: University of North Carolina Press.

Rossiter, Margaret W. 1975. *The Emergence of Agricultural Science: Justus Liebig and the Americans, 1840–1880.* New Haven, Conn.: Yale University Press.

Schaub, I. O. 1955. *North Carolina Agricultural Experiment Station: The First 60 Years, 1877–1937.* Bulletin 390, North Carolina Agricultural Experiment Station, Raleigh, N.C.

South Carolina Experiment Station. 1890. *Second Annual Report of the South Carolina Experiment Station for the Year Ending December 31, 1889.* Columbia, S.C.

U.S. Department of Agriculture (USDA). 1895. *Yearbook of the United States Department of Agriculture, 1894.* Washington, D.C.

Whitney, Milton to Mark W. Harrington. 4 November 1891. Letter 1098; Letters Received, 1891–1894; Record Group 27, Records of the Weather Bureau. National Archives at College Park, Maryland.

Whitney, Milton to Mark W. Harrington. 15 April 1892a. Letter 2339; Letters Received, 1891–1894; Record Group 27, Records of the Weather Bureau. National Archives at College Park, Maryland.

Whitney, Milton to Mark W. Harrington. 18 June 1892b. Letter 2754; Letters Received, 1891–1894; Record Group 27, Records of the Weather Bureau. National Archives at College Park, Maryland.

Williams, C. B. n.d. History and achievements of research in agronomy in North Carolina during fifty years (1878–1927). Special Collections, North Carolina State University, Raleigh, N.C. Typescript.

2

EARLY LEADERS OF THE SOIL SURVEY
Douglas Helms

INTRODUCTION

For centuries people have acted upon empirical observations made about the values, or limitations, of particular soils. In fact, many soil and land-use decisions are still made on the basis of empirical experience that does not incorporate a full understanding of the processes at work in the soil. Beginning in the nineteenth century and accelerating rapidly in the twentieth, science has enhanced our understanding of the soil. Scientists, farmers, engineers, and many others not trained in the science of soils have contributed to soil science and interpretations. But it is the province of the soil scientist and of the U.S. Soil Survey particularly to provide a map so that users of the soil might apply the knowledge of soil science to a particular place and a particular soil.

Much of the tradition of the soil survey has been utilitarian, with the intention of helping people use the soil more wisely, efficiently, and safely. But that is not the sole motivation for the many individuals who choose to make soil science a lifelong pursuit. Most are also captivated by the wonders and complexity of the soil and its inhabitants—from the microbe to the mole. This chapter describes some of the people whose careers and lives have been intertwined in the history of the soil survey, with added emphasis on those who have been neglected in the historical literature.

Because the soil survey, the discipline of soil science, and the development of land-grant university curricula in soil science grew in an interdependent fashion, a number of the individuals profiled represent both university-based soil science as well as the soil survey. Growth of soil science departments relied in part on graduates receiving employment from soil survey programs. The interpretive value of soil surveys, especially in the agronomic realm, was enhanced by research in the land-grant universities and experiment stations. Many of the soil scientists in the soil survey of the last 75 years were trained as soil scientists or earned agricultural or natural resources degrees with a heavy emphasis on soils. But the first generation of soil survey employees was drawn from students trained in a mixture of geology, agronomy, and agricultural chemistry. In the early years of the discipline, particular professors and universities helped guide students toward an interest in soils and the soil survey, from which emerged informal networks that persisted for generations.

MILTON WHITNEY

CAREER

Milton Whitney, a Marylander who was to become the first chief of the Division of Agricultural Soils, was born 2 August 1860. A resident of "Riverview" Waterbury, Anne Arundel County, he attended private schools and spent a year at St. Johns College in Annapolis. For the most part, however, he was trained at home, "my health not having permitted close confinement" as he explained to Johns Hopkins University on requesting admittance in 1879. He explained, "My object in wishing to come to the University is to perfect myself in the study of analytical chemistry desiring to take that up as a profession, not being physically able to go through an entire college or university course, I would confine myself as I have done for some time to the study of chemistry & those branches deemed necessary with it" (Whitney application 1879).

Professor of chemistry Ira Remsen did not altogether like the idea of admitting Whitney, but thought him "an intelligent lad" who would probably study if admitted (Whitney application 1879). Despite his lack of formal education, Johns Hopkins University allowed him to undertake a three-year special course of study with Professor Remsen. Following his course of study, which did not lead to formal graduation, Whitney joined the staff of the Connecticut Agricultural Experiment Station at New Haven as an assistant chemist in 1883 (Cattell and Cattell 1926; 1926). The director of the station, Samuel Johnson, had established one of the best laboratories in the nation. Johnson was at the forefront of the movement to bring science, particularly chemistry, to agriculture and was also developing the American variant of the agricultural experiment station—an institution he had seen while a student in Europe (Rossiter 1975).

Milton Whitney, first chief of the Division of Agricultural Soils, n.d. (Courtesy of the National Archives at College Park, Maryland; Negative 16-G-M-2378; Record Group 16, Records of the Office of the Secretary of Agriculture)

Charles W. Dabney, the second head of the second agricultural experiment station in North Carolina, subsequently recruited Whitney to establish the station's experimental farm. In North Carolina and in later positions in South Carolina and Maryland, Whitney studied the relationship of soil and soil management to agriculture. In Maryland, Whitney turned his interests to mapping large areas of agricultural soils. When Dabney was appointed assistant secretary of agriculture in 1894, with a charge to buildup a professional, nonpartisan scientific corps in the U.S. Department of Agriculture (USDA), he selected Whitney to head the Division of Agricultural Soils. (In its early years, the U.S. Soil Survey was the primary responsibility and work activity, successively, of the Division of Agricultural Soils [1894–1897]; the Division of Soils [1897–1901]; and the Bureau of Soils [1901–1927]. The soil survey, often organizationally known as the Soil Survey Division, has since been in the Bureau of Chemistry and Soils [1927–1938]; Bureau of Plant Industry, Soils, and Agricultural Engineering [1938–1952]; Soil Conservation Service [1952–1994]; and the Natural Resources Conservation Service [since 1994].) Whitney continued his studies of soil texture and moisture conditions of soils and investigated the suitability of soils for high value crops such as tobacco. In 1899, under Whitney's leadership, field parties of division personnel began mapping and describing soils, and issued their written reports as soil surveys.

Since agriculture was the predominant use of soils, agricultural scientists of Whitney's generation sought to understand and explain scientifically what observation had shown, that agricultural crops grew better and produced more on some soils than on others. The rather imprecise methods of chemical analysis available, however, hampered their quest to understand the relationship of soil chemistry to plant growth. Even when equal amounts of fertilizer were applied on different soils, plants grew differently. Thus, they were aware that fertilizer was not the sole answer to productivity, and that soil characteristics were variable and played a role in productivity.

Whitney took this observation and in searching for causation, arrived at the conclusion that temperature and moisture were the most important attributes that could be related to productivity and that texture of the soil was the best index of these conditions (Whitney 1893; Whitney and Cameron 1903; Tanner and Simonson 1993). Soil texture—the ratios of clay-, silt-, and sand-sized particles in a soil—did affect soil moisture conditions and the availability of water for plants, but how did one account for the observed effects of fertilizers? As Whitney wandered further out on the theoretical limb, he asserted that fertilizers affected the circulation and retention of the moisture in the soil. Furthermore, he insisted that most soils contained sufficient and inexhaustible amounts of nutrients for plant growth and that differences in productivity were the result of differences in the texture of soils.

Whitney developed an elaboration of the theory that most soils contained enough nutrients for plant growth in Bureau of Soils Bulletin 22, *The Chemistry of Soil As Related to Crop Production*, a hypothesis that Roy Simonson summarized as follows:

> (i) the concentration of plant nutrients in the soil solution is the same in productive and unproductive soils, (ii) that concentration is maintained by natural processes, (iii) all soils suitable for crop production contain enough nutrients for satisfactory plant growth, and (iv) the supply will last indefinitely. (Tanner and Simonson 1993, 290)

Whitney's ideas, now published in Bureau of Soils bulletins, took on the air of official policy of the Bureau of Soils and the Department of Agriculture. Respected agricultural chemists of the day such as Eugene W. Hilgard of the University of California and Cyril G. Hopkins of the University of Illinois publicly contested these ideas and eventually sought to have Whitney removed (Hopkins 1910; Jenny 1961). But this episode was rather more denouement than first act, as Whitney and Hilgard had sparred over the issue a decade earlier. Whitney retained his position as chief of the bureau, but the verdict of history and chemistry eventually favored Hilgard and Hopkins.

Whitney has rightly been criticized for clinging too long and too ardently to his theories of the relationship of soil texture to productivity. The imbroglio over these ideas strained relations with some cooperators, irreparably broke ties with others, and ultimately retarded development of the survey. The episode should not, however, blind one to Whitney's positive leadership characteristics in establishing the soil survey so that it might flourish later. He was energetic in building and expanding the soil survey operation.

Knowing that expansion of the soil survey depended on congressional support, he unabashedly reminded cooperators to seek the support of their congressional delegation at appropriations time. He established national standards and methods for the survey. Agricultural experiment stations would be cooperators, but the published soil survey would be issued by the Bureau of Soils, and the survey methods would conform to Whitney's, not the individual station's, standards. The series of published surveys would have a uniform format and be recognized as part of a national inventory.

The early soil surveyors tried to identify soils that were suited to particular crops, especially for high value crops like tobacco. They focused their attention on particular soil bodies or soil types. The potential for recommendations and interpretations led early soil surveyors to concentrate on small areas, such as counties or valleys, as opposed to making broader state level or regional studies. Curtis Marbut, whom Whitney hired to supervise the soil survey, contended that this specificity—the scale of map—distinguished these soil surveys from any previous mapping efforts on broader scales (Weber 1928; 92).

LEGACY

Scientists of the Early Division

Whitney contacted soil science colleagues such as Hilgard, Thomas Chrowder Chamberlain of the University of Chicago, and Nathaniel Southgate Shaler of Harvard University for recommendations about potential employees. William Bullock Clark recommended several people to Whitney. As the division's soil chemist in the early period, Whitney recruited Franklin Kenneth Cameron, who earned a Ph.D. in chemistry at Johns Hopkins University in 1894 and joined the division in 1898 after several assistant positions at Cornell University and Catholic University. Cameron coauthored the bulletin *Chemistry of Soil as Related to Crop Production* that created so much consternation over its claims that most soils were inherently and unalterably fertile. Whether because of that development or some other, Cameron resigned from government service in 1915 to become a chemical engineer; he believed he "could further the problems of big public interest on which I was engaged better from the outside" (Cameron to Ball 1916).

Whitney also recruited Lyman J. Briggs as the bureau's soil physicist. Briggs was a Ph.D. graduate of Johns Hopkins University in physics, with degrees also from Michigan Agricultural College and the University of Michigan. He and another of the first group of soil surveyors, Thomas H. Means, devised electrolytic methods for measuring soil

Employees of the Bureau of Soils, November 1922. (Courtesy of the National Archives at College Park, Maryland; Negative 16G-2619; Record Group 16, Records of the Office of the Secretary of Agriculture)

moisture, soil temperature, and soluble salts in soils (Bonsteel to Harding 1942). Later, Briggs moved to the Bureau of Plant Industry, also in the USDA, where he was classified as a biophysicist and worked on evapotranspiration and water requirements of plants to aid agriculture in dryland areas. During World War I, Briggs moved to the Bureau of Standards in the Department of Commerce to aid in the war effort. The temporary assignment became permanent and Briggs remained with the Bureau of Standards where he eventually became director (Briggs n.d.).

Whitney hired Frank Gardner, Clarence W. Dorsey, and Thomas Herbert Means to accomplish the soil survey fieldwork. Means had earned a B.S. (1898) and M.S. (1901) degree in geology from George Washington University, where leading geologist George P. Merrill taught. Gardner was an agronomist from the University of Illinois who studied the relationship of vegetation to soils and had made some of the earliest soil surveys. Dorsey, a Harvard University graduate with a background in geology, followed the practice of many early soil surveyors in relating soils to underlying geology (Bonsteel to Harding 1942).

Whitney, representing the USDA, and William Bullock Clark as state geologist of Maryland agreed in 1899 to survey the soils of Maryland at the rate of one county per year. Under the agreement, Dorsey began work on Cecil County with the assistance of Jay Allan Bonsteel, a geology graduate student at Johns Hopkins (Whitney 1900, 15). Bonsteel, who became a long-time Division of Soils and Soil Conservation Service employee, next surveyed St. Mary's County. His Division of Soils publication *Soil Survey of St. Mary's County* also served as his Ph.D. dissertation, with the added subtitle *Showing the Relationship of Geology to the Soils* (Bonsteel 1905a). Clark and Whitney had a momentary disagreement over publications when Whitney asserted the division's editorial and publishing control. Eventually, Dorsey's and Bonsteel's soil surveys appeared as USDA publications and as chapters in the Maryland Geological Survey publications on St. Mary's and Cecil Counties (Bonsteel 1901, 1905a)

Franklin Hiram King Whitney had an eye for accomplished scientists and hired them when he could. He was familiar with Franklin Hiram King, the agricultural physicist at the University of Wisconsin's agricultural experiment station, through King's correspondence with Whitney, particularly on the methods and equipment for measuring soluble salts in soil (King to Whitney 1899, 1900). Whitney hired King to head a Division of Soil Management, but found to his dismay that King's findings contradicted his own theories of soil fertility. Whitney allowed publication of three studies, collectively titled *Investigations in Soil Management*, but felt compelled to add a disclaimer stating that the basic, quantifiable data was being made available to the scientific community but that the bureau was in no way endorsing King's theories and interpretations of the data (King 1905). By the time of publication, King had resigned. He later privately published another three bulletins that Whitney had refused to publish (Tanner and Simonson 1993).

King remained a very active, innovative soil scientist and earned a reputation as one of the pioneering soil physicists in the United States. One authority dubbed King the father of soil physics in the United States (Gardner 1977). He studied his favorite topic, soil management, in China where he wrote *Farmers of Forty Centuries: Or, Permanent Agriculture in China, Korea and Japan*. Decades later, the book so impressed Robert Rodale, disciple of organic farming, that his Rodale Press reprinted the 1911 publication (King 1911, 1973).

Whitney had a similar conflict later with George Nelson Coffey, a Bureau of Soils employee who was in charge of the soil classification work from 1905 to 1908. Whitney published Coffey's ideas about soil classification and genesis while stating they were Coffey's personal opinions, not official policy. In both the Coffey and King cases, Whitney's personal theories were "official" bureau policy and theories of subordinates were accorded the secondary status of "personal" opinion (Coffey 1912). Despite the outcome, Whitney's publication of King's research on soil management illustrated Whitney's commitment to fulfilling his promise of making soil surveys useful by providing interpretations to the basic data. Even after the conflict with King, he continued to emphasize interpretations, and in 1907 he created a Division of Utilization of Soil Resources to discover the "great fundamental facts of crop adaptation, soil management, and soil fertilization . . ." (Whitney 1908, 19).

But the scientific discoveries that were needed to make interpretations continued to undermine Whitney's faulty theories. The episode of King's resignation was only one of several partings based on scientific or theoretical discord. Whitney hired some bright, innovative people, but found it difficult to accept subordinates' contrasting views, views that conflicted with his own public and published ideas. In any case, a number of other division members were in accord with Whitney and wrote in a similar vein. Jay Bonsteel and W. J. McGee, for example, also wrote about the inexhaustibility of soil and likely did so out of conviction (Bonsteel 1905b, 103; McGee 1909b, 131).

Soil Conservation in the Early Bureau of Soils

William John McGee The early Bureau of Soils heightened the awareness of soil erosion as a problem facing American agriculture. But the bureau was also active in the wider progressive conservation movement through William John McGee, one of the major scientific figures in the federal government in the nineteenth and early twentieth centuries. McGee variously listed his occupations as geologist, ethnologist, anthropologist, and hydrologist (McGee n.d.), and indeed he had justifiable claims to all of those titles. When the largely self-taught McGee joined the Bureau of Soils in 1907, he was already a man of importance in the infant conservation movement. Whitney placed McGee in charge of a unit on "Soil Erosion Investigations."

The son of an Irish immigrant farmer, McGee was born on 17 April 1853, near Farley in Dubuque County, Iowa. He left school at 14, but benefited from tutoring in Latin, German, mathematics, and astronomy by an older brother who had attended college. He learned black-

smithing and built and sold agricultural implements when not exploring the countryside with his brothers. In 1878 he published papers on
glacial drift and prehistoric burial mounds. From 1877 to 1881, he carried out his own topographic and geological survey of 12,000 square
miles in northeastern Iowa, published as "The Pleistocene History of
Northeastern Iowa."

John Wesley Powell hired McGee as a permanent employee of the
U.S. Geological Survey in 1883, and published McGee's survey as the
first generalized geologic map of the United States (Nelson 1999).
When Powell became director of the Bureau of American Ethnology in
1893, McGee followed him and eventually published some 30 reports
on native peoples from 1894 to 1903. McGee was appointed to the
Bureau of Soils in 1907 following two years (1903–1904) in charge of
the anthropological exhibit at the St. Louis Exposition and two years
as director the St. Louis Public Museum (Shor 1999). Whitney recommended McGee for appointment for the "purpose of enabling the
bureau to take up the important study of soil erosion or wash, and
sedimentation which has not hitherto been fully investigated for inabil-
ity to obtain a man with the necessary training and attainments"
(McGee n.d.). Whitney also informed the secretary that President
Theodore Roosevelt had only recently appointed McGee to the Inland
Waterways Commission, where he would be working with the U.S.
Forest Service, the Engineering Department of the Army, and the Hy-
drographic Service of the Department of the Interior. This position
would afford McGee an "opportunity to push these investigations with
the assistance and advice from these other branches of the Government
service, whose work is really dependent upon and made necessary to a
large extent, by the erosion of the soil" (Whitney to Secretary of Agriculture 1907).

McGee's understanding of the interrelated nature of resources and
resource issues was advanced for his time. As a member of the Waterways Commission, he and a few compatriots pushed for a natural
resources conference. Finding that the Lakes-to-the-Gulf Deep Waterways Association planned to call together a score or more governors
for a conference restricted to waterway improvement needs and water
resources development, McGee and his colleagues won President
Theodore Roosevelt's pledge to call the Conference of Governors on
Conservation of Natural Resources. McGee, employed by the Bureau
of Soils, and Gifford Pinchot, chief of the Forest Service, shaped the

conference, which was held at the White House in May 1908, but Pinchot recalled that it was McGee "who pulled the laboring oar" (Pinchot 1947, 346).

The governors were allowed to speak briefly, but the substance of the published proceedings rested on the presentations by the resource experts, whom McGee had selected. The governors' conference, along with the publication of its speeches, called attention to the need for conservation and was a seminal event in the history of the conservation movement (McGee 1909a). Pinchot's assessment of McGee's status in the conservation movement was unqualified, "W J McGee was the scientific brains of the Conservation movement all through its early critical stages" (Pinchot 1947, 359). Historian Samuel Hays, who examined what he termed the "progressive conservation movement" spanning 1890–1920, concurred, calling McGee the "chief theorist of the conservation movement" (Hays 1959, 102).

McGee acquired his interest in and concern about erosion during his studies for the U.S. Geological Survey. While studying erosion as a geological process, he became a prescient observer of human-induced, accelerated erosion. In studying Mississippi's coastal plain, he found soils "adapted to distinct crops and special modes of tillage; and they are differently affected by old-field erosion, which has already wrought lamentable destruction in different portions of the coastal plain, and is progressing with ever-increasing rapidity" (McGee 1892, 106). McGee also produced a Bureau of Soils bulletin, *Soil Erosion*, which was the bureau's most complete treatment of the issue to that point (McGee 1911).

During the later part of his career, McGee studied groundwater, what he called subsoil water. He authored bulletins, which were published after his death, that correctly identified the need to view soils and water resources as a unit. But McGee is not remembered for his Bureau of Soils groundwater bulletins, mainly because they set forth theories of capillary action, hydrology, and water cycle and consumption that further scientific investigation has found wanting (McGee 1913a, 1913b). However, McGee remains as a central figure among federal employees in the progressive conservation movement. He was still employed by the bureau when he died on 4 September 1912, in his quarters at the Cosmos Club in Washington, D.C.

Erosion Identification by Early Soil Surveyors McGee's prestige brought attention to the bureau's role in soil conservation, but he by no means originated it. A cadre of young soil scientists with some concern about the effects of soil erosion developed in the early Bureau of Soils. Published soil surveys during the early years increasingly referred to soil erosion and the need for soil conservation, along with some suggested lines of action (Gardner 1998, 70). The early soil surveyors had taken notice of soil erosion from the beginning of their work, both as a factor in soil classification and for soil management recommendations. They were developing what we now call the soil type—soil bodies that share significant soil properties.

Soil surveyors began seeing separations based on erosion. Clarence W. Dorsey surveyed the area around Lancaster, Pennsylvania, in 1900, and described the Hagerstown clay, "These soils may be said to be the Hagerstown loam from which the top covering of loam has been removed, exposing the clay subsoil, . . ." (Dorsey 1901, 71–72). Jay A. Bonsteel, who surveyed St. Mary's County, Maryland, the same year, noted that cultivating slopes of Leonardtown loam resulted in "scalds or washes" that needed permanent sod (Bonsteel 1901, 129). Bonsteel, while jointly serving as a surveyor with the Division of Soils and Professor of Soil Investigations at Cornell University early in the century, examined the so-called worn-out soils around Ithaca. Like Whitney and others of the period, Bonsteel was among the ranks of those questioning Justus Liebig's theory that repeated cropping diminished the available plant food in the topsoil. Bonsteel believed many farmers around Ithaca cultivated a subsoil far different from the topsoil cultivated by their ancestors. The stone fences where the topsoil lodged provided the evidence. Reacting perhaps too strongly to Liebig's thesis, he averred that erosion was "one of the agencies totally destroying the validity of the hypothesis of soil deterioration by removal of crops." Further, he cited the effects of wind erosion in the northeast as a "greatly underestimated" factor in the alteration of the soil (Bonsteel 1905b, 103).

In 1910, the surveyors began to identify "eroded" phases of established soil types (Gardner 1998, 58). As the soil survey matured, it adopted a nomenclature that grouped soil types into a soil series. The series combined a place name followed by a texture designation, as in Jordan sandy loam. In time, the surveys added slope and degree to the soil type designation (Simonson 1987, 4). The 1911 surveys of Farfield County, South Carolina, identified a large area of "rough gullied land" (Gardner 1998, 58).

Making a soil survey, 1914. (Courtesy of the Natural Resources Conservation Service, U.S. Department of Agriculture)

A. T. Sweet samples soil in a Piedmont gully, Carroll County, Georgia, March 1921. (Courtesy of the Natural Resources Conservation Service, U.S. Department of Agriculture)

Hugh Hammond Bennett Hugh Hammond Bennett, who had joined the survey in 1903, began to relate recommendations to particular soil types. For instance, concerning the Orangeburg sandy loam of Lauderdale County, Mississippi, he wrote, "If the gentler slopes are not terraced and the steep situations kept in timber, deep gorgelike gullies or 'caves' gradually encroach upon cultivated fields, eventually bringing about a topographic situation too broken for other than patch cultivation" (Bennett et al. 1912). Bennett's *The Soils and Agriculture of the Southern States* highlighted erosion and advised that some soil types were unsuitable for cultivation or in need of conservation measures if used for agriculture (Bennett 1921). Later, as head of the USDA's Soil Conservation Service, Bennett and colleagues used susceptibility to erosion as a key element in the land capability classification (Helms 1997).

Hugh Hammond Bennett, who had joined the bureau fresh out of the University of North Carolina (UNC) at Chapel Hill in 1903, became the most recognizable link of the soil conservation movement to the early Bureau of Soils. Rather than being a lone voice, however, Bennett was in fact among compatriots. Though not mentioning McGee specifically, Bennett made clear the importance of the atmosphere within the bureau that recognized the problem of erosion and the need for soil conservation, created in part by McGee. Bennett, a half century after the event, recalled that it was the paper on "Soil Wastage" given at the 1908 governors' conference on conservation by Thomas Nelson Chamberlain, head of the Department of Geology at the University of Chicago, that "fixed my determination to pursue that subject to some possible point of counteraction" (Bennett 1959, 13). Bennett did not say specifically that he attended the conference; more likely he read the version of the paper published in the conference proceedings (McGee 1909a). One of Bennett's college classmates, R. O. E. (Royall Oscar Eugene) Davis, a chemist in the Bureau of Soils, wrote bulletins titled *Soil Erosion in the South* and *Economic Waste From Soil Erosion* (Davis 1914, 1915).

While McGee's, Bennett's, and Davis's bulletins gave erosion by water preeminence as a conservation concern, Edward E. Free produced a classic treatment of wind erosion in *The Movement of Soil Material by Wind* (Free 1911). In 1894, Franklin Hiram King, before joining the Bureau of Soils, wrote one of the earliest bulletins about wind erosion and its amelioration for the Wisconsin Agricultural Experiment Station (King 1894). His later pioneering work in soil man-

Hugh Hammond Bennett, n.d. (Courtesy of the National Archives at College Park, Maryland; Negative 114-G-90002; Record Group 114, Records of Natural Resources Conservation Service)

agement addressed soil conservation and maintenance in a broad sense, not just as the need for halting erosion. Davis also wrote a bulletin on a different type of soil degradation, *The Effect of Soluble Salts on the Physical Properties of Soils* (Davis 1911). Lyman Briggs, the bureau's soil physicist, and early surveyors in the West, Thomas Means and Frank Gardner, built on the work of Hilgard in developing methods to identify soluble salts. Thus, they assisted in another type of soil conservation, the avoidance of salt accumulation on the surface and in the upper profile through the use of irrigation water management, including drainage systems or wise land-use decisions.

Bennett gradually moved his campaign for soil conservation beyond the confines of the Soil Survey Division to educate the public and politicians through writing and speaking engagements. He identified areas where the combination of geography and agricultural systems caused serious erosion. As a first step in attacking the problem, he wanted research on erosion conditions and conservation measures. Based largely on his campaign, Congress authorized a series of soil erosion experiment stations. Bennett selected the locations for experiment stations, where interdisciplinary teams of researchers established plots to measure erosion conditions under different types of crops, soils, rotations, and various agricultural management practices and structures. A few state experiment station staff had carried out similar experiments, but the federal impetus led eventually to building up nationwide data on erodibility of soils. The origins of the erodibility data that currently support conservation planning tools, such as the Universal Soil Loss Equation (and the revised equation), reach back to these pioneering studies (Lyles 1985; Meyer and Moldenhauer 1985).

Bennett and colleagues in the early Bureau of Soils pursued soil conservation with considerable success, despite the fact that soil erosion was not recognized as a substantial concern by the soil survey leadership of that time. With the creation of the Soil Conservation Service in the U.S. Department of Agriculture in 1935, with Bennett as its first chief, the interpretation of soils for soil and water conservation was firmly established and accepted.

THE UNIVERSITY OF NORTH CAROLINA CONNECTION

Hugh Hammond Bennett, a UNC graduate, is well known today as the father of soil conservation, but he was only one of several UNC graduates who joined the soil survey in its first decade. The story of these graduates is one that converged the state government's desire to improve agriculture with UNC professor Collier Cobb's ambition to build up academic programs and provide employment for students.

In 1899, as Whitney was beginning soil survey field operations in the Division of Soils, B. W. Kilgore, state chemist in the North Carolina Department of Agriculture, wrote to Whitney about North Carolina's plans for evaluating soils. Kilgore informed Whitney that the Board of Agriculture had directed him "to take up the study of soils and to make experiments in different portions of the State, with reference to the best fertilizers for the crops and soils of the different sections, as well as to the growth of different crops" (Kilgore to Whitney 1899). The state Department of Agriculture would use the survey to place "test farms" in various areas of the state. Kilgore and Whitney agreed upon a plan for fieldwork during 1900, with Whitney paying salaries and the state paying the fieldwork expenses (Kilgore to Whitney 1900a). But Kilgore soon realized that this new type of survey would move more slowly than he and the department wished. He wrote to Whitney, "At the present rate, with only one man in the field, it will take forty or fifty years to survey and map our State and we can not very well proceed any faster than you will let us, as we are dependent on your Division for trained men. They are not to be had elsewhere" (Kilgore to Whitney 1900c).

Kilgore sought to expedite the survey. He hired men who had had experience in the U.S. Geological Survey to make base maps. Soon, with the help of Joseph A. Holmes, state geologist, he arranged for the state to help fund U.S. Geological Survey topographical surveys, which could be used as base soil survey maps, in the areas where soil surveys were planned (Kilgore to Whitney 1900e, 1901; Holmes to Whitney 1900a, 1900b, 1900c, 1900d). And he hired a young geology graduate from UNC, George Nelson Coffey, who was to work with William G. Smith of Whitney's staff to gain experience and help expedite the survey. Whitney soon hired Coffey to continue working in North Carolina as part of the federal soil survey (Kilgore to Whitney 1900b and 1900d).

Coffey's geology professor at UNC, Collier Cobb, almost immediately seized upon the expanding Soil Survey Division as a source of employment for his graduates. The soil survey satisfied not only this need but also his personal interests. A North Carolina native and UNC graduate himself, Cobb returned to teach at the university in 1892,

Collier Cobb, ca. 1902. (Courtesy of the North Carolina Collection, University of North Carolina Library at Chapel Hill, N.C.)

George Nelson Coffey, 1900. (Courtesy of the North Carolina Collection, University of North Carolina Library at Chapel Hill, N.C.)

having earned a master's degree at Harvard University under Nathaniel Southgate Shaler. As author of *Origin and Nature of Soils*, Shaler was one of the handful of geologists and agricultural chemists, including Thomas Chrowder Chamberlain and Eugene Woldemar Hilgard, who wrote about and mapped soils (Tandarich 1998).

As Whitney received additional appropriations to expand soil survey work, Cobb supplied him with a continuing stream of graduates in geology and chemistry. By May 1902, when the soil survey was still quite small, Cobb could boast to his father that during his decade of teaching he had placed seven graduates in the USDA soil survey work, seven in the U.S. Geological Survey, and another seven in economic positions in the federal government (Cobb to Father 1902). Yet he clearly felt the soil survey work to be more satisfying. "We have men in all branches of the Government Scientific Work, but I believe that those in the Soil Survey get rather more pleasure out of life than the others" (Cobb to Whitney 1903).

To give students qualifications for the work, Cobb initiated a "Special Course in Soil Investigation" in 1905. Soon thereafter students could earn a "Bachelor of Science in Soil Investigation" in the School of Applied Science after taking the courses in origin and nature of soils, agricultural soils, soils of the United States, soil mapping, and the soil seminar—a course of reading in soil literature and writing reports. Cobb named one of the courses "Origin and Nature of Soils," after the title of his mentor's treatise (Shaler 1891; University of North Carolina 1905–1913). At Cobb's request, Whitney granted Coffey and another Cobb protégé and career-long soil survey employee, Williamson Edward Hearn, permission to visit Chapel Hill and instruct UNC students in soil surveying methods (Cobb to Whitney 1910). Coffey evidently made some presentations on soils to the classes from 1905 to 1910, and beginning in 1906, the course catalogue annually listed Hearn as the teacher, along with Cobb, of the soils courses (Cattell and Brimhall 1923; University of North Carolina 1905–1913).

A number of Cobb's students, including Thomas D. Rice, Williamson Edward Hearn, and Robert Campbell Jurney had long careers in the soil survey, as did chemistry students Royall Oscar Eugene Davis, William Henry Fry, and Joshua John Skinner. Edward Parrish Carr and Adolphus Williamson Mangum left the survey but continued in agricultural-related careers. James LaFayette Burgess turned to plant breeding for the North Carolina Agricultural Experiment Station and eventually became state botanist (Grant 1924).

The earliest of the UNC recruits, George Nelson Coffey, deserves further discussion because of his innovations in the soil science discipline, especially in the areas of soil genesis and soil classification. But appreciation among soil scientists for his contributions was slow to come, primarily because he left the soil survey in midcareer. Soil scientist and historian of the soil survey Roy Simonson informed Charles Kellogg, head of the soil survey from 1935 to 1971, of Coffey's significance. Together they gradually made soil scientists aware of Coffey's place in the history of soil science.

Coffey worked his way up in the soil survey, and had an opportunity to see soils in a wide variety of locations in the United States. While working in the soil survey, Coffey earned an M.S. and Ph.D. degree in 1908 and 1911 from George Washington University under George P. Merrill, whose *Rocks, Rock Weathering and Soils* was a primary text of the day. By the time Coffey wrote his M.S. thesis in 1908, he had encountered the ideas of the Russian soil scientist Vasily Vasilyevich Dokuchaev through the writings of Nikolai Sibirtzev. Coffey's thesis included citations to the French version of the report from the Seventh International Geologic Congress in St. Petersburg (1897), "Etude des sols de la Russie", as well as English translations by Peter Fireman in the *Experiment Station Record* (Coffey 1908; Sibirtzev 1899, 1901).

Coffey took the Russian theories, which outlined multiple factors of soil formation to create unique, natural soil bodies, and with his innovative mind applied them to the soils he had seen and studied throughout the United States. Having reviewed the domestic and international soil classification systems, Coffey concluded that they focused too much on one or more genetic processes. Coffey added the novel, and now accepted, concept that soil classification should be built upon the soil characteristics as observed and measured, not on genetic processes. As much as he had learned from the Russian ideas, he was no unquestioning acolyte, as he faulted the Russians for basing their classification ideas too much on one genetic factor, climate. He also referred to Hilgard's overreliance on vegetation as an indicator of soil characteristics (Coffey 1911, 1912).

Coffey could not dislodge Whitney and the Bureau of Soils from their geologic moorings. In 1911, George Washington University accepted Coffey's dissertation, and it was published as a bureau bulletin. Once again Whitney found himself prefacing a bureau publication, adding this disclaimer to the reprinting of Coffey's dissertation: "In publishing it, however, the Bureau of Soils does so for the purpose of offering it to the scientific world . . . without endorsing the scheme of classification proposed. . . ." According to Whitney, it was not possible "at the present time and with our limited knowledge, to construct a general map based upon the characteristics of the soil itself" (Coffey 1912, letter of transmittal).

By the time of publication in 1912, Coffey had resigned to join the Ohio Agricultural Experiment Station. Since joining the survey in 1900, Coffey had worked his way up to inspector and then specialist in soil classification. After earning a Ph.D., he found himself a candidate to supervise the soil surveys, with final selection to be made by Whitney. But Curtis Marbut soon eclipsed him. It is one of the great ironies of the history of soil science in the United States that Marbut arrived in the Washington office in 1910 and was appointed head of the soil survey in 1911 during the period in which Coffey's ideas had been rejected, but that Marbut later became familiar independently with the Russian ideas and is acclaimed for introducing them into the United States.

Another North Carolina student of Collier Cobb's had a happier fate in the Department of Agriculture, but not without a struggle. Hugh Hammond Bennett studied both chemistry and geology, and Cobb thought so highly of his work in chemistry that he recommended him for a laboratory position. But in 1903 Bennett had an opportunity to work on the soil survey of Davidson County, Tennessee, where he worked until he was available to travel to Washington to take the position (Bennett 1959). Bennett found the fieldwork to his liking and eventually became inspector of surveys for the southern states. As noted above, his early published surveys included recognition of erosion. An active writer, he produced *The Soils and Agriculture of the Southern States* in 1921, in which he identified soil types that should be left in trees or pastures and types needing terraces or other conservation treatments (Bennett 1921). His campaign for soil conservation through speaking and writing led eventually to the creation of the Soil Conservation Service in 1935, with Bennett as the first chief and the acknowledged "father of soil conservation."

Collier Cobb's own son, William Battle Cobb, made soil science his career. He joined the soil survey in 1913, served in World War I, and then embarked on a teaching career, beginning at Louisiana State University. When the younger Cobb joined the faculty of North Carolina State College of Agriculture and Engineering, North Carolina's land-grant university, in 1924, the school already had a full complement of soils courses, including soil survey. In the academic year 1924–1925, Cobb taught a course titled "Pedology," which one might consider a descendant, albeit with greatly advanced understanding of soil genesis, of his father's course on the "origin and nature of soils" (Grant 1924; North Carolina State College of Agriculture and Engineering 1924, 114).

This father and son case illustrates a critical element of the development of the new science. Collier Cobb eventually gave up his soil curriculum at UNC. In some states, like North Carolina, it had been necessary to establish a land-grant university, separate from the established state university, to teach the agricultural and mechanical subjects. This separation, however, had the unintended and unfortunate effect of hindering the acceptance of soil science as a basic science to be taught in all universities. Since the main interpretations and the economic contributions of soil science and soil survey were in agriculture, the new science became associated with the land-grant universities where agronomy was taught. Cobb's son, a soil scientist, and others at North Carolina State built one of the outstanding soil science departments in the country. Unfortunately, many private universities and many state universities outside the land-grant system do not teach soil science as a basic science, despite the fact that it is critical to earth studies (Hudson 2000).

THE EARLHAM COLLEGE CONNECTION

A number of graduates of Earlham College, a Society of Friends (Quaker) college in Richmond, Indiana, also had prominent careers in the early soil survey. Allen D. Hole, the college's sole geology teacher, graduated from Earlham, taught there, and attended the University of Chicago, where he eventually earned a Ph.D. in geology and paleontology. Hole took part in a summer field geology course at Chicago and instituted a similar course at Earlham. Hole also enlisted the aid of his students when he worked for the state of Indiana on soil surveys. Some of Hole's students joined the early USDA soil survey (Tandarich et al. 1988). Mark Baldwin worked at most levels in the survey and rose to a major supervisory responsibility. Another graduate, Earl Fowler, spent part of his career on soil surveys in the coastal plain, where he is remembered for identifying the soils now known to contain plinthite (Fowler 1928; Gamble 1999).

Hole helped students find summer work and experience on soil surveys. Allen Hole's son, Francis D. Hole, worked one summer on the survey of Franklin County, Indiana, under the supervision of Oliver Rogers, another Earlham graduate and employee of the Soil Survey Division (Hole 1999). The younger Hole earned a Ph.D. in geology at the University of Wisconsin, where he remained as a professor and trained soil scientists. He also remained active in the soil survey and emphasized field study. His sense of duty included bringing an appreciation of the wonders of the soil to the public. During graduate school, when his father had fallen ill, he returned briefly to Earlham to teach soils and geology. During that time, he inspired another future soil scientist, Ralph McCracken. Following a career in the Soil Science Department at North Carolina State University, McCracken was a deputy chief in the Soil Conservation Service, where his area of administration included the soil survey operation.

James Thorp, another Earlham graduate who joined the soil survey, was asked by the government of China to map and write a soil survey for the country. The English version of Thorp's *Geography of the Soils of China* was published in 1936 (Thorp 1936). After his retirement from the Soil Survey Division, Thorp returned to Earlham College and taught for a decade. One of Thorp's M.A. students, Erling Gamble, collaborated with him on studies of the origin of clays in midwestern soils. A controversy swirled around whether clay rich B horizons had formed in place or had been formed by translocation of clay particles. Thorp and Gamble favored the translocation mechanism. Gamble's career in the Soil Survey Division of the Soil Conservation Service combined work on the southeastern soil geomorphology with later specialization in explaining soil-landscape relationships to help guide soil survey fieldwork (Gamble 1999).

The legacy of the Earlham College group is reaffirmed whenever the field soil scientist receives *Soil Survey Horizons* in the mail. With the demise of the *American Soil Survey Association, Bulletin,* Francis Hole recognized that field soil scientists needed another means to communicate. The new *Soil Science Society of America Proceedings* transmitted exciting developments in the field of soil science, but it did not provide a forum for the field soil scientist. Stanley W. Buol, a graduate student at the University of Wisconsin, helped Hole assemble the first issue of *Soil Survey Horizons* in 1960 in the Soils Building at the University of Wisconsin. Buol recalled the rationale behind their effort to guide communications among field scientists, "Ideas, doubts, experiences, and reevaluations should not be confined to individuals, they belong to all" (Buol 1997). Buol, like the other Hole students, managed to combine a career of teaching and research with involvement in promoting the soil survey.

CURTIS FLETCHER MARBUT

Curtis Marbut, who became head of the soil survey in 1911, had his first formal association with the soil survey as a field agent appointed to map Missouri soils in 1909. At the time, Marbut was a professor of geology at the University of Missouri. In 1910, he took leave from the university to join the Washington office of the Bureau of Soils. Only a year later, Milton Whitney placed Marbut in charge of the soil survey section of the Bureau of Soils. Whitney remained as chief of the Bureau of Soils until 1927, however, and so maintained his strong influence on the survey and soil classification. Marbut remained in charge of the Soil Survey Division until 1934, but never headed the entire Bureau of Soils or the reorganized Bureau of Chemistry and Soils (Whitney to Secretary of Agriculture 1909, 1912).

Curtis Marbut was born on 19 July 1863, in Lawrence County in the Missouri Ozarks. Marbut graduated from the University of Missouri in 1889. From 1890 to 1893 he worked in the state's geological survey, until he entered Harvard University, where he received his master's degree in 1894. From 1895 to 1910, he taught geology at the University of Missouri, as instructor, assistant professor, and eventually professor for the last 11 years (Kellogg 1935).

Pedologists and historians of soil science remember Marbut for introducing American soil scientists to the Russian concepts of soil genesis. The Russian theory emphasized not only geologic origins (parent material), but also climate, plants and animals, landscape position, and time as factors which, in combination, created unique soils. But when Marbut first arrived at the bureau, he held a strong belief that knowledge of geology (parent material) was the key to understanding soils. The western inspector of the soil survey, Macy Lapham, recounted in his autobiography *Crisscross Trails* that when he learned Marbut was critical of soil surveyors' limited knowledge of geology, he immediately commenced studying the standard geology texts of the day to ensure his job. The irony of Marbut's conversion to a belief in multiple factors

Curtis F. Marbut, 1936. (Courtesy of National Archives at College Park, Maryland; Negative 114-G-90702; Record Group 114)

of soil formation was not lost on Lapham, who never told Marbut of his attack of anxiety about his job security (Lapham 1949, 94–95).

Marbut's conversion began when he read about the Russian efforts at soil mapping and classification in K. D. Glinka's *Die typen der Bodenbuidung*, published in 1914 (Glinka 1914). This approach emphasized studying the soil itself, rather than the geology underlying the soil (Simonson 1987; Weber 1928). The new theories of soil genesis informed Marbut's supervision of the soil survey and his soil classification efforts. Largely through Marbut's influence, American soil scientists began to appreciate soils as individual natural bodies, the study of which should begin with the soil itself, not external factors of formation. He took great pride in the acceptance of the published soil survey as a new scientific document and in the acceptance of pedology, or the study of the soils, as a scientific discipline. At the same time, he de-emphasized interpretation in the interest of refining soil classification and establishing soil science as an independent discipline.

According to Charles Kellogg, Marbut's successor as head of the soil survey, Marbut appreciated the contributions of the Russians, particularly their emphasis on climatic factors in defining broad regions and their establishment of the soil profile as a distinct entity. But Marbut stuck to the "inductive" method of scientific study and emphasized soils as the starting point. In eulogizing him, Kellogg encapsulated the Marbut credo of soil surveying: "Describe the soils as you find them. Get the facts first and philosophize about the genesis of the soil later. Soils must be studied and classified as soils, not as geological products, climatic products or from the point of view of anything outside the soils itself" (Kellogg 1935).

This approach meshed well with the American concentration on detailed, small area studies for practical use, rather than surveys of soil, climatic, and plant relationship over broad areas. Marbut's emphasis on the soil and what characteristics and properties should be observed, mapped, and described, and his work on classification, led the way in developing a system where the "facts and data had meaning and significance" (Kellogg 1935). Although he stepped aside as head of the soil survey, Marbut never retired from the study of the soil, which he continued up to his death. President Roosevelt, at the department's request, granted Marbut three annual exemptions from the mandatory retirement age.

In 1935, Marbut attended the Third International Congress of Soil Science in Oxford, England, and then embarked on railroad travel to Peking, China, where he was to meet James Thorp, the Bureau of Soils employee working on a soil survey of China. But Marbut grew ill on the train and had to disembark at Harbin, Manchuria, where he was taken to a hospital. Summoned from Peking, Thorp arrived before Marbut died of double pneumonia on 24 August (25 August, Washington time) (Marbut n.d., personnel file; Tandarich et al. 1988).

CHARLES E. KELLOGG

CAREER

Charles E. Kellogg, as a young professor of soil science at North Dakota Agricultural College (now North Dakota State University), had an established reputation for innovative use of soils information when he joined the soil survey in 1934. Kellogg was appointed as a soil scientist with a special assignment in soil survey in the USDA in February, and on 1 July 1934, he became acting chief of the Division of Soil Survey when Curtis Marbut stepped aside. One year later, on 1 July 1935, Kellogg succeeded Curtis Marbut as chief of the Division of Soil Survey.

A native of Ionia County, Michigan, Kellogg earned an undergraduate and Ph.D. degree in soil science at Michigan State University. Kellogg's dissertation, "Preliminary Study of the Profiles of the Principal Soil Types of Wisconsin" (Kellogg 1930), was firmly in the new field of pedology. L. R. Schoemann, a professor and one of the architects of a landmark in land classification, the "Michigan Land Economic Survey," inspired Kellogg's interest in the soil. In appreciation, Kellogg later dedicated *The Soils that Support Us* to Schoemann as the man "Who First Showed Me the Soil" (Kellogg 1943).

While a student at Michigan State, Kellogg received a fellowship from the state highway department, which had experienced buckling problems with expensive new roads. Kellogg was asked to apply soil survey field and laboratory techniques to the location and design of important highways to eliminate the potential for buckling problems. After graduation, in addition to his first teaching job at North Dakota Agricultural College, Kellogg was one of the leaders of a team making a soil survey of McKenzie County. The survey was to be used as a basis for classification and tax assessment of rural lands. By the time Kellogg moved to the USDA's Soil Survey Division he had already participated in three of the seminal events in soil survey interpretations: the interpretation of soil data for highway construction and tax assessment and the integration of soil survey into a broader interdisciplinary land-evaluation system.

Kellogg appreciated Marbut's contributions to soil survey. As a student, he, like many others, had been fascinated by explanations of soil genesis in understanding soil morphology. But Marbut, according to Kellogg, was something of a "continental soil scientist," who devoted most of his time to broad classification (Kellogg 1963). Kellogg's stewardship of the soil survey marked a fundamental departure in emphasis. Kellogg focused as a student and young professor on the uses, or interpretations, of the soil survey data to assist people. He continued that emphasis in directing the activities of the soil survey. Upon his retirement, when asked about his proudest accomplishments, he said, "In general, they were efforts that helped people." He cited the interpretations of soil survey information for highways and the rural land classification. In his capacity as head of the soil survey, he cited the opportunity to "stimulate and help many young men make a career in basic soil science and its application to the soil survey for improving the selection and management of soils for the uses needed by people in both town and country" (Kellogg n.d., press release).

His orientation to interpretation in no way diminished his support of scientific research and investigations. In fact, the need to improve interpretations required refinement and improvement in surveying methodology and soil classification. Kellogg initiated the effort that eventually resulted in a new soil classification system, eventually published as *Soil Taxonomy* (Soil Survey Staff 1975). The soil survey staff released a draft, the 7th Approximation (Soil Survey Staff 1960), in 1960 for testing in the field. The first published edition of *Soil Taxonomy* appeared in 1975 as *Soil Taxonomy: A Basic System of Soil Classification for Making and Interpreting Soil Surveys*. The emphasis placed upon quantification and laboratory analysis in the *Soil Taxonomy* were not solely for the purpose of soil classification, but also for the ultimate objective of improved interpretations.

The early years of Kellogg's leadership focused on tools to improve and advance the soil survey. Improving the mapping and identification of individual soils, with their imprecise boundaries as they graded into one another, was the first order of business. Kellogg explained his objective: "To perfect this concept of the soil individual our greatest need in 1935 was for more precision in soil definition. This resulted in the first edition of the *Soil Survey Manual*" (Kellogg 1963). The manual was published in 1937, about three years after Kellogg assumed leadership, and was the first major revision of field instructions in 23 years (Kellogg 1937).

Marbut had published a soil classification in the *Atlas of American Agriculture* in 1935 shortly before his death (Marbut 1935). Mark Baldwin, Kellogg, and James Thorp published a major revision of the classification in the 1938 Yearbook of Agriculture, *Soils and Men* (Baldwin et al. 1938). In addition to specific alterations, the revised classification revealed another of Kellogg's characteristics, which was to conduct collaborative work and to use the division's personnel as a team. This approach was unlike Marbut's classification, which was single-authored. Following Kellogg's example in working with Mark Baldwin and James Thorp, the soil classification became a survey division product.

Kellogg's experience with both the 1938 classification system and the 1949 soil classification revision published in *Soil Science* revealed anomalies of individual soil bodies that did not fit neatly into the upper-level classification categories. To resolve these inconsistencies, Kellogg selected Guy D. Smith, an Iowa native with a Ph.D. from the University of Illinois, to lead an effort in the critical job of developing a new soil classification system, eventually called Soil Taxonomy. Smith taught at the University of Illinois and headed the land acquisition activities for region three of the Resettlement Administration from 1934 to 1936 before becoming a soil correlator in the Soil Survey Division. His work on the *Soil Taxonomy* necessitated collaboration with soil scientists internationally, and its successful completion brought him an international reputation (Jaques Cattell Press, Vol. 5 1972; Smith 1983).

The new classification system built upward from qualitative and quantitative definitions and descriptions of the soil characteristics of individual soil series. The greater level of precision brought to the description of characteristics of the soil series, including laboratory analysis, helped in building interpretations for safe and efficient uses of the soil. The *Soil Taxonomy* has been regarded as one of the more important scientific contributions of the soil survey. But it is well to remember that it is merely part of a continuum, and in some regards the capstone, of the utilitarian reorientation that Kellogg brought to the survey when he assumed leadership in 1935. Kellogg explained the rationale for a new classification as follows:

But the best classification that can be devised is still only a tool to help us understand soils better and to apply soil science to achieve better systems of soil use. No system of classification should ever become so sacred or so classical that the system becomes an end in itself. It should always remain a tool for use, to be sharpened or to be replaced as the attainment of our objectives in applying soil science demands. (Kellogg 1963)

Kellogg consistently proclaimed as his ultimate accomplishment the use of soil science and the soil survey to assist people in using soils. He explained that his and the survey's contributions to science were a means to that end. Upon his retirement, Kellogg reminisced that a colleague would occasionally suggest that perhaps his career would have been more satisfying had he continued to teach in the university. Kellogg replied that he had a great many opportunities to teach in his USDA job, through counseling, lectures, and writing (Kellogg n.d., press release).

Indeed he did; Kellogg wrote prodigiously and well, and influenced not just the young people of the soil survey but Congress, government officials, and fellow scientists. His messages reached the generalist and the specialist, and his range of subjects was not limited to soil survey. Early writing in the university years focused on typical topics for a young soil scientist, such as the morphology and genesis of particular soils and an appreciation of soils as part of the biological complex (Kellogg 1934a, 1934b). But soon he looked beyond profiles to the landscapes of interactions between soils and society. *The Soils That Support Us*, first published in 1941 and reprinted numerous times, acquired a general audience. To use Kellogg's colloquialism, he examined "how soil and people got on together" (Kellogg 1943, vii). His interest in historical interactions of soils and people was of a piece, not a separate activity, from Kellogg's utilitarian efforts to refine soil survey interpretations for the soil user.

Undoubtedly, he influenced more young soil scientists in the soil survey than he would have as a university professor, and his audience of inservice soil scientists greatly expanded after the merger of the soil survey into the Soil Conservation Service in 1952. In addition to his published articles, Kellogg wrote papers on subjects of immediate interest to the agency's soil scientists, such as interpretations and soil classification, which were mimeographed and distributed to field soil scientists.

In the role of mentor and counselor, he tried to help soil scientists continue their education through reading. A prodigious reader himself, Kellogg held strong opinions on good literature and on the need for soil scientists to be well-read. Kellogg believed that to understand soil science "one must also have some knowledge of literature, history, anthropology, geology, botany, chemistry, and other arts and sciences" (Kellogg 1943, 342). It was a theme to which Kellogg frequently returned when he discussed the damage done by fragmentation of the body of knowledge for the convenience of college courses and professional societies.

When reminiscing, field soil scientists of the Kellogg era will often inquire with a combination of appreciation and amusement, "Do you remember the reading list?" meaning Kellogg's *Reading for Soil Scientists, Together with a Library* (Kellogg 1971). The recommendations for the well-read soil scientist included not just classics in soil science but also a liberal sprinkling of history and literature. Kellogg compiled the list between 1930 and 1940, at first mimeographing it for students at North Dakota Agricultural College and finally publishing it in the *Journal of the American Society of Agronomy* in 1940. By the time he had issued processed revisions in 1956, 1964, and 1971, it had grown to include far more citations to fiction, drama, poetry, and history than to soil science (Simonson 1999a).

Understandably, Kellogg's interest in reading and writing extended to the history of his own discipline. His lead article "We Seek; We Learn" in the 1957 Yearbook of Agriculture titled *Soil* presented the public with an overview of the basic history of soil science (Kellogg 1957). In other periodicals he wrote on Russian contributions to soil science, discoursed on conflicting soil science theories, and instructed the larger scientific world on the status of soil science (Kellogg 1946, 1948a, 1948b). That the 1951 edition of the *Soil Survey Manual* included an introductory chapter on the discipline's history demonstrates that he thought the field soil scientist should be familiar with the subject (Soil Survey Staff 1951).

Although Kellogg left university teaching, he did not forsake his interest in education, particularly in agricultural education, on which he held strong beliefs. With David C. Knapp he coauthored *The College of Agriculture: Science in the Public Service*, which imparted some of the same ideas he impressed upon his students in soil science at North Dakota Agricultural College. As a college professor, he insisted upon the basic sciences as the foundation for a career in soil science (Cline

1998). The same applied to other areas of agricultural sciences. In his book with Knapp he wrote, "The foundation of undergraduate education for the agricultural specialist is founded on a knowledge of the principles of the natural and social sciences and mathematics out of which new and better systems of agriculture grow" (Kellogg and Knapp 1966, 94).

To his mind, too many courses had been organized around temporary agricultural issues and systems, which would be of little use to a professional several years later when circumstances had changed. A firm grounding in science, instead, laid the foundation for learning, innovating, and adapting to new conditions. His attitude extended to field soil scientists, some who may not have concentrated sufficiently in science in undergraduate school. The soil survey staff, along with the faculties at Cornell University and Iowa State University hosted the Soil Science Institute for agency soil scientists. During a six-week period, agency soil scientists were given intensive training in soil classification, plant physiology, soil microbiology, soil chemistry, and soil physics (Zwerman et al. 1968).

Kellogg advised governments and the international scientific community, especially in developing countries. After World War II, many agriculturists unthinkingly recommended the highest level of western-developed technology for food-poor areas of the world. Kellogg recognized the necessity of working within existing systems for improvement, rather than trying to export entirely new agricultural systems to countries that did not have agricultural infrastructure, institutions, and trained specialists to support western-style agriculture. Kellogg's sophisticated understanding of the interplay of history, soils, and agriculture is best reflected in his warning against automatically condemning shifting cultivation. Speaking of the Congo, he wrote, "Outright condemnation of shifting cultivation is irresponsible. Much more research is urgently needed to work out effective systems. Then, as the tropical countries have more industry, chemical fertilizers can be used to substitute in part, at least, for forest fallow, *provided* that organic matter and shade are maintained as needed for soil structure" (Kellogg and Davol 1949, 68).

Shifting cultivation was neither inherently good nor bad. Under certain conditions of culture, population, economy, and lack of industrial development, it might be the most suitable system. Under a more developed economic and industrial system and higher level of technical expertise, agriculture that is more dependent on commer-

Charles E. Kellogg (center) interviews farmers at Oyeko, Gold Coast (now Ghana) about cocoa production, August 1954. Left of Kellogg is the chief farmer of the Gold Coast, Kwame Poku. (Courtesy of the Natural Resources Conservation Service, U.S. Department of Agriculture)

cial fertilizers could be beneficial. Following a reconnaissance survey of the Belgian Congo in the 1940s, Kellogg recognized the advantages of the corridor system of shifting cultivation. He warned against indiscriminately using commercial fertilizers, especially lime, on the Latosols until research had investigated the potential for interference with the availability of minor nutrients (Kellogg and Davol 1949).

It is perhaps ironic that Kellogg, one of the best scientists his profession had to offer, saw the challenge of meeting world food needs as one of human relationships rather than of scientific riddles. His travels in tropical areas convinced him that the scientific challenges to developing efficient production systems were dwarfed by the socioeconomic, cultural, and political challenges to sustaining those systems. He wrote of the challenge in feeding the world: "The social, economic, and political problems are many and difficult. The technical problems of soils, plants, and animals, great as they are, are small by comparison" (Kellogg and Orvedal, 1969).

LEGACY

Kellogg's influence endured through the people he selected to work on the soil survey. In his early years in charge of the survey, quite a number of recruits were former students whom he had inspired to follow a career in soil science. They continued to shape the survey for another decade past Kellogg's retirement. Roy W. Simonson took classes from Kellogg in the early 1930s as an undergraduate student at the North Dakota Agricultural College. After earning a Ph.D. in soil science at the University of Wisconsin in 1938, Simonson taught at Iowa State College from 1938 to 1942. He then began a career in the Soil Survey Division, where he specialized in soil classification and correlation. Among his contributions to pedology, he is best remembered for the article "Outline of a Generalized Theory of Soil Genesis." This general theory held that processes within soils could be combined into four general categories: gains, losses, translocations, and transformations. The scientist might infer the original state of minerals, compare it to the current state, and then estimate the combinations and rates of processes (Jaques Cattell Press, Vol. 5 1972; Arnold 1983; Simonson 1959).

Simonson and his generation participated in the creation of the discipline now termed soil science. For 18 years Simonson served as editor-in-chief for *Geoderma*, an international journal of soil science. He not only made contributions as a scientist but also became historian of the soil survey. Soil scientists are indebted to him for his writings on the historical aspects of soil classification and the soil survey that have appeared in American and international journals (Simonson 1985, 1987, 1989a, 1989b, 1991). David Rice Gardner, who graduated from the University of California at Berkeley, was not a former Kellogg student, but Kellogg recognized his ability and potential to advance in the profession. With Kellogg's blessing, Gardner took an opportunity to earn a Ph.D. in public administration from Harvard University. Gardner selected the history of the soil survey for his doctoral thesis, which he completed in 1957. His and Simonson's writing are some of the best historical works on the history of the soil survey in the United States (Gardner 1998).

Marlin G. Cline, another North Dakota Agricultural College undergraduate, earned a Ph.D. in soils at Cornell University in 1942. Cline remained at Cornell as a faculty member but retained a cooperator's role with the soil survey, first in actively making soil surveys and then increasingly in national leadership roles. One of his major contributions to the soil survey was to host the Soil Science Institute, Kellogg's innovative school for expanding scientific training of soil survey staff. Field soil scientists could not only enhance their field experiences with intensive training in areas they might not have studied in college but also could learn new scientific information from the rapidly evolving field of soil science (Jaques Cattell Press, Vol. 1 1971; Zwerman et al. 1968).

Cline's other notable contributions to the national soil survey effort included establishing the procedures for use by the World Soil Geography Unit in collecting and interpreting information and leading the effort in the late 1960s to revise the 1951 edition of the *Soil Survey Manual* (Arnold 1999). One of Cline's students, Klaus Flach became head of the soil survey, and he was succeeded by former Cornell University faculty member Richard Arnold. Cline and Simonson, just two of the students from Kellogg's brief tenure at North Dakota State, were awarded honorary degrees from foreign universities, Cline from Ireland and Simonson from Norway.

A. Clifford Orvedal also studied under Kellogg at North Dakota and in 1938 earned an M.S. degree in soil science from Michigan State University, while on leave from the soil survey, which he had joined in 1935. Orvedal headed the World Soil Geography Unit from 1946 to 1966 and later worked with the president's panel on world food supply. He wrote and published information about potentially arable soils in the world (Orvedal n.d., biographical file; Kellogg and Orvedal 1969). Andrew "Andy" Aandahl earned an undergraduate and master's degree from North Dakota Agricultural College and a Ph.D. in agricultural economics from Iowa State College (Jaques Cattell Press 1965). Equipped with training in soil science and economics, Aandahl made pioneering contributions in economic interpretations of soil surveys by producing studies on the productivity of soils and land evaluation, particularly of Iowa's principal soil groups (Baumann et al. 1955; Ottoson et al. 1954). After retiring from the Soil Conservation Service, Aandahl wrote *Soils of the Great Plains* (Aandahl 1982).

Roy Simonson recalls that he and Cline, Orvedal, and Aandahl were all majoring in other curricula when they first took a soil science course with Kellogg and subsequently switched to major in soil science. Years later Kellogg commented that all too often land-grant university graduates had told him that they found soil science boring, their teacher uninspiring, and that they took as few courses in the subject as possible. Kellogg believed that soil science should never be a boring topic, devoid of the human element, and wrote, "The general soil scientist has more human-interest material to draw on than any other teacher in a college of agriculture" (Kellogg 1961, 422). Evidently, he followed his own prescription and was an inspiring teacher. His numerous articles and the book *The Soils that Support Us*, which examined the interaction of humans and soils, demonstrate that he personally taught natural science in the context of human involvement.

William M. Johnson also studied briefly with Kellogg and earned an undergraduate degree at North Dakota Agricultural College and a master's degree at the University of Wisconsin. Johnson returned to his North Dakota alma mater in 1938 as a soil science professor. But Johnson had been introduced to the soil survey much earlier. His father, Martin "M. B." Johnson, was a farmer in McKenzie County, North Dakota. Martin also served as an extension agent in ranch economics, working cooperatively between the state agricultural experiment station and the USDA's Bureau of Agricultural Economics. The elder Johnson encouraged McKenzie County commissioners to take advantage of a state law that allowed them to appropriate money for surveying and classifying rural land for the purpose of tax assessment. Charles Kellogg, then a young professor of soil science at North Dakota Agricultural College, supervised the classification of McKenzie County's rural lands, and Johnson advised Kellogg on the grazing values of particular range sites.

The survey of McKenzie County was a seminal event in the use of soil surveys for land classification for taxation purposes, as well as in the broader area of soil survey interpretations. His father's enthusiasm for Kellogg's work influenced William Johnson to study soils when he entered North Dakota State. Following World War II, William Johnson joined the Soil Survey Division as a soil correlator in several locations. He eventually succeeded Kellogg as head of the Soil Survey Division (Kellogg 1933; Johnson 1982).

Kellogg first encountered another soil survey scientist, J. Kenneth Ableiter, when both assisted Andrew Robeson Whitson with the soil survey of Bayfield County, Wisconsin, from 1928 to 1929. Kellogg championed his hiring at North Dakota Agricultural College and later hired him again. This time he was hired into the Soil Survey Division, where he worked on, among other things, crop yield data by soil types and land classification for rural tax assessment (Jaques Cattell Press 1965).

When Kellogg's Soil Survey Division in the Bureau of Plant, Industry, Soils, and Agricultural Engineering was merged into the Soil Conservation Service (SCS) as part of a departmental reorganization in 1952, Kellogg utilized the talents of the SCS soil scientists. Roy Hockensmith, who had been the SCS's head of conservation surveys, acted as an assistant to Kellogg in supervising field operations. Albert A. Klingabiel revised the land capability classification, and later continued as leader in developing the soil survey interpretations. After the Soil Survey Division was transferred into the Soil Conservation Service, Kellogg was also reunited with a few former students at various levels of SCS on soil conservation survey operations (Simonson 1999b).

Kellogg clearly controlled the combined operations of both his former staff and his new SCS personnel who were making conservation surveys. Marlin Cline believed that one of Kellogg's critical decisions during the transition was to insist that all of the field staff be classified as soil scientists (Cline 1998). Even if the reality fell short of the ideal in some particular cases, the overall recognition inspired a sense of professionalism and status of soil scientists within a large line and staff organization.

CONCLUSION

The soil has beckoned a stream of young scientists to study its wonders. The waves of young people entering the soil survey have differed through the decades. Some training or experience in geology seemed a prerequisite for the early surveyors, who combined an interest in agricultural soils with the tradition of surveying that grew out of geology. The promise that the surveys, the maps, and descriptive materials be useful, particularly to agriculturists, placed a premium on understanding the relationship of soil properties to plant growth. Soil survey, and soil science, developed a systemic attachment to the land-grant universities, which trained students in soil science and the related sciences necessary to interpretations, particularly agronomy.

At its height in the 1970s, the National Cooperative Soil Survey in the SCS employed as many as 1,500 soil scientists. In the years since the early scientists joined the soil survey, the objectives and substance of the surveys have changed markedly. And that is a trend that is likely to continue. Nonagricultural interpretations of soil surveys are hardly new; they too have a long history. But the audience and users of soil surveys are growing in diversity and numbers. While agriculturists have long appreciated the value of the survey, many among the new audiences profess an interest in the natural world. They have acquired a growing appreciation that inventorying soils, along with the plant and animal life of terrestrial and marine communities, can bring greater understanding of the natural system.

Subjects such as global warming and the role of the soil in carbon sequestration cause spikes in appreciation of soil science and soil surveys. But rather than being the totality of the shift in objectives, the interest in carbon sequestration is but one manifestation of the larger trend. The survey will evolve in unpredictable ways, but we can feel assured that it will be in ways that will continue to attract young scientists to study it as a critical component of the environment in which we live. Future generations will continue to be fascinated by the soil and will perceive a need to display what they have learned spatially, through the work and products of the soil survey.

ACKNOWLEDGMENTS

The author wishes to thank Dr. Roy W. Simonson of Oberlin, Ohio, for his thorough review and suggestions.

REFERENCES

Aandahl, Andrew Russell. 1982. *Soils of the Great Plains: Land Use, Crops, and Grasses*. Lincoln, Neb.: University of Nebraska Press.

Arnold, Richard W. 1983. Concepts of soils and pedology. In *Pedogenesis and Soil Taxonomy*, edited by L. P. Wilding, N. E. Smeck, and G. F. Hall, pp. 1–21. Amsterdam, The Netherlands: Elsevier.

Arnold, Richard W., telephone interview by author, 28 January 1999.

Baldwin, Mark, Charles E. Kellogg, and James Thorp. 1938. Soil classification. In *Soils and Men*, pp. 979–1001. Yearbook of Agriculture 1938. Washington, D.C.: Government Printing Office.

Baumann, Ross V., Earl O. Heady, and Andrew R. Aandahl. 1955. *Costs and Returns for Soil—Conserving Systems of Farming on Ida—Monona Soils in Iowa*. Research Bulletin Number 429. Iowa Agricultural Experiment Station, Ames, Iowa.

Bennett, Hugh Hammond. 1921. *The Soils and Agriculture of the Southern States*. New York: Macmillan Company.

Bennett, Hugh H. 1959. *The Hugh Bennett Lectures*. Raleigh, N.C.: Agricultural Foundation, Inc., North Carolina State College.

Bennett, Hugh H., Howard C. Smith, W. M. Spann, E. M. Jones, and A. L. Goodman. 1912. Soil survey of Lauderdale County, Mississippi. In *Field*

Operations of the Bureau of Soils, 1910, pp. 733–784. Washington, D.C.: Government Printing Office.

Bonsteel, Jay A. 1901. Soil survey of St. Mary County, MD. In *Field Operations of the Division of Soils,* 1900, pp. 125–145. Washington, D.C.: Government Printing Office.

Bonsteel, Jay A. 1905a. *The Soils of St. Mary's County, MD.: Showing the Relationship of the Geology to the Soils.* December. Baltimore, Md.: U.S. Department of Agriculture, Division of Soils.

Bonsteel, Jay A. 1905b. Worn out soils. In *Bureau of Farmers' Institutes and Normal Institutes.* pp. 99–114. Report for the Year 1904. Albany: Brandow Printing Company.

Bonsteel, J. A. to T. Swann Harding. 5 November 1942. Soil Investigations; Historian Notes; Files of J. Gordon Steel; Record Group 114, Records of the Natural Resources Conservation Service. National Archives at College Park, Maryland.

Briggs, Lyman J. n.d. Personnel file. Selected Personnel Files; Record Group 16, Records of the Office of the Secretary of Agriculture. National Archives at College Park, Maryland.

Brown, W. Norman, compiler. 1926. *Johns Hopkins Half-Century Directory.* Baltimore: Johns Hopkins University.

Buol, S. W. 1997. Beginnings of *Soil Survey Horizons. Soil Survey Horizons* 38(2):39–40.

Cameron, Franklin Kenneth to Thomas Ball. 9 March 1916. Student Files. Ferdinand Hamburger, Jr. Archives, Johns Hopkins University, Baltimore, Md.

Cattell, J. McKeen, and Dean R. Brimhall, eds. 1921. *American Men of Science.* 3d ed. Garrison, N.Y.: Science Press.

Cattell, J. McKeen, and Jaques Cattell, eds. 1927. *American Men of Science.* 4th ed. New York: Science Press.

Cline, Marlin G., telephone interview by author, 8 December 1998.

Cobb, Collier to Dear Father. 9 May 1902. Cobb Family Papers. Southern Historical Collection, University of North Carolina, Chapel Hill, N.C.

Cobb, Collier to Milton Whitney. 11 February 1903. File 3298; General Correspondence, 1901–1927; Bureau of Soils; Record Group 54, Records of the Bureau of Plant Industry, Soils, and Agricultural Engineering. National Archives at College Park, Maryland.

Cobb, Collier to Milton Whitney. 10 February 1910. File 12089; General Correspondence, 1901–1927; Bureau of Soils; Record Group 54, Records of the Bureau of Plant Industry, Soils, and Agricultural Engineering. National Archives at College Park, Maryland.

Coffey, George Nelson. 1908. The basis of soil classification. M.S. thesis, George Washington University, Washington, D.C.

Coffey, George Nelson. 1911. A study of the soils of the United States. Ph.D. diss., George Washington University, Washington, D.C.

Coffey, George Nelson. 1912. *A Study of the Soils of the United States.* U.S. Department of Agriculture Bureau of Soils Bulletin Number 85. U.S. Department of Agriculture, Washington, D.C.

Davis, Royall Oscar Eugene. 1911. *The Effect of Soluble Salts on the Physical Properties of Soils.* Washington D.C.: Government Printing Office.

Davis, R. O. E (Royall Oscar Eugene). 1914. Economic waste from soil erosion. In *Yearbook of Agriculture 1913*, pp. 207–220. Washington, D.C.: Government Printing Office.

Davis, Royall Oscar Eugene. 1915. *Soil Erosion in the South.* U.S. Department of Agriculture Bureau of Soils Bulletin Number 180. U.S. Department of Agriculture, Washington, D.C.

Dorsey, Clarence W. 1901. A soil survey around Lancaster, PA. In *Field Operations of the Division of Soils, 1900,* pp. 61–84. Washington, D.C.: Government Printing Office.

Fowler, E. D. 1928. Iron accumulation in soils of the coastal plain of the southeastern United States. *Proceedings and Papers of the First International Congress of Soil Science,* Washington, D.C., 13–22 June 1927; 4:435–441.

Free, E. E. 1911. *The Movement of Soil Material by the Wind.* U.S. Department of Agriculture Bureau of Soils Bulletin Number 68. U.S. Department of Agriculture, Washington, D.C.

Gamble, Erling, telephone interview by author, 7 January 1999.

Gardner, David Rice. 1998. *The National Cooperative Soil Survey of the United States.* Historical Notes Number 7. Natural Resources Conservation Service, U.S. Department of Agriculture, Washington, D.C.

Gardner, W. H. 1977. Historical highlights in American soil physics, 1776–1976. *Soil Science Society of America Journal* 41:221–229.

Glinka, K. D. 1914. *Die typen der Bodenbildung—ihre Klassifikation und geographische Verbreitung.* Berlin: Gebrudder Borntraeger.

Grant, Daniel Lindsey. 1924. *Alumni History of the University of North Carolina.* Chapel Hill, N.C.: General Alumni Association.

Hays, Samuel P. 1959. *Conservation and the Gospel of Efficiency: The Progressive Conservation Movement, 1890–1920.* Cambridge, Mass.: Harvard University Press.

Helms, Douglas. 1997. Land capability classification. In *History of Soil Science: International Perspectives,* edited by Dan H. Yaalon and S. Berkowicz, pp. 159–175. Advances in Geoecology 29. Reiskirchen, Germany: Catena Verlag.

Hole, Francis D., telephone interview by author, 6 January 1999.

Holmes, Joseph A. to Milton Whitney. 4 January 1900a. Letters Received; Division of Soils; Record Group 54, Records of the Bureau of Plant Industry, Soils, and Agricultural Engineering. National Archives at College Park, Maryland.

Holmes, Joseph A. to Milton Whitney. 27 April 1900b. Letters Received; Division of Soils; Record Group 54, Records of the Bureau of Plant Industry, Soils, and Agricultural Engineering. National Archives at College Park, Maryland.

Holmes, Joseph A. to Milton Whitney. 25 July 1900c. Letters Received; Division of Soils; Record Group 54, Records of the Bureau of Plant Industry, Soils, and Agricultural Engineering. National Archives at College Park, Maryland.

Holmes, Joseph A. to Milton Whitney. 24 September 1900d. Letters Received; Division of Soils; Record Group 54, Records of the Bureau of Plant Industry,

Soils, and Agricultural Engineering. National Archives at College Park, Maryland.

Hopkins, Cyril G. 1910. *Soil Fertility and Permanent Agriculture*. Boston: Ginn and Company.

Hudson, Berman, comments made to author, 2000. Berman Hudson, Soil Scientist, Natural Resources Conservation Service, called the author's attention to how the separation of soil science from the traditional state universities to the agricultural colleges of the state land-grant universities has hindered the acceptance of soil science.

Jaques Cattell Press, ed. 1965. *American Men of Science: A Biographical Directory, the Physical and Biological Sciences*. 11th ed. New York: R. R. Bowker Company.

Jaques Cattell Press, ed. 1971. *American Men and Women of Science, Formerly American Men of Science, the Physical and Biological Sciences*, Vol. 1. 12th ed. New York: Jaques Cattell Press/R. R. Bowker Company.

Jaques Cattell Press, ed. 1972. *American Men and Women of Science, Formerly American Men of Science, the Physical and Biological Sciences*, Vol. 5. 12th ed. New York: Jaques Cattell Press/R. R. Bowker Company.

Jenny, Hans. 1961. *E. W. Hilgard and the Birth of Modern Soil Science*. Pisa, Italy: Collana Della Rivista "Agrochimica."

Johnson, William M., oral history interview by author, 1982. U.S. Department of Agriculture, National Agricultural Library. Beltsville, Md.

Kellogg, Charles Edwin. n.d. Press release. Charles Kellogg File. History Office, Natural Resources Conservation Service, U.S. Department of Agriculture, Washington, D.C.

Kellogg, Charles Edwin. 1930. Preliminary study of the profiles of the principal soil types of Wisconsin. Ph.D. diss., Michigan State College of Agriculture and Applied Science, Soils Department, East Lansing, Mich.

Kellogg, Charles Edwin. 1933. A method for the classification of rural lands for assessment in western North Dakota. *Journal of Land and Public Utility Economics* 9:10–15.

Kellogg, Charles Edwin. 1934a. Morphology and genesis of the Solonetz soils of western North Dakota. *Soil Science* 38(6):483–450.

Kellogg, Charles Edwin. 1934b. The place of the soil in the biological complex. *Scientific Monthly* 34(July):46–51.

Kellogg, Charles Edwin. 1935. Curtis Fletcher Marbut. *Science* 82(2125, September 20):268–270.

Kellogg, Charles Edwin. 1937. *Soil Survey Manual*. U.S. Department of Agriculture Miscellaneous Publication Number 274. U.S. Department of Agriculture, Washington, D.C.

Kellogg, Charles Edwin. 1943. *The Soils That Support Us: An Introduction to the Study of Soils and Their Use by Men*. New York: Macmillan Company.

Kellogg, Charles Edwin. 1946. Russian contributions to soil science. *Land Policy Review* 9:9–14.

Kellogg, Charles Edwin. 1948a. Conflicting doctrines about soils. *Scientific Monthly* 66:475–487.

Kellogg, Charles Edwin. 1948b. Modern soil science. *American Scientist* 36:517–536.

Kellogg, Charles Edwin. 1957. We seek; We learn. In *Soil*, pp. 1–11. The Yearbook of Agriculture 1957. Washington, D.C.: Government Printing Office.

Kellogg, Charles Edwin. 1961. A challenge to American soil scientists on the occasion of the 25th anniversary of the Soil Science Society of America. *Soil Science Society of America Proceedings* 25(6):419–423.

Kellogg, Charles Edwin. 1963. Why a new system of soil classification? *Soil Science* 96(July):1–5.

Kellogg, Charles Edwin. 1971. *Reading for Soil Scientists, Together With a Library, Revised.* Soil Conservation Service, U.S. Department of Agriculture, Washington, D.C.

Kellogg, Charles E., and Fidelia D. Davol. 1949. *An Exploratory Study of Soil Groups in the Belgian Congo.* Serie Scientifique No. 46. Publications De L'Institut National Pour L'Etude Agronomique Du Congo, Belge.

Kellogg, Charles E., and David C. Knapp. 1966. *The College of Agriculture: Science in the Public Service.* New York: McGraw-Hill Book Company.

Kellogg, Charles E., and Arnold Clifford Orvedal. 1969. Potentially arable soils of the world and critical measures for their use. *Advances in Agronomy* 21:109–170.

Kilgore, B. W. to Milton Whitney. 17 August 1899. Letters Received; Division of Soils; Record Group 54, Records of the Bureau of Plant Industry, Soils, and Agricultural Engineering. National Archives at College Park, Maryland.

Kilgore, B. W. to Milton Whitney. 9 March 1900a. Letters Received; Division of Soils; Record Group 54, Records of the Bureau of Plant Industry, Soils, and Agricultural Engineering. National Archives at College Park, Maryland.

Kilgore, B. W. to Milton Whitney. 3 July 1900b. Letters Received; Division of Soils; Record Group 54, Records of the Bureau of Plant Industry, Soils, and Agricultural Engineering. National Archives at College Park, Maryland.

Kilgore, B. W. to Milton Whitney. 9 July 1900c. Letters Received; Division of Soils; Record Group 54, Records of the Bureau of Plant Industry, Soils, and Agricultural Engineering. National Archives at College Park, Maryland.

Kilgore, B. W. to Milton Whitney. 13 July 1900d. Letters Received; Division of Soils; Record Group 54, Records of the Bureau of Plant Industry, Soils, and Agricultural Engineering. National Archives at College Park, Maryland.

Kilgore, B. W. to Milton Whitney. 7 September 1900e. Letters Received; Division of Soils; Record Group 54, Records of the Bureau of Plant Industry, Soils, and Agricultural Engineering. National Archives at College Park, Maryland.

Kilgore, B. W. to Milton Whitney. 11 March 1901. Letters Received; Division of Soils; Record Group 54, Records of the Bureau of Plant Industry, Soils, and Agricultural Engineering. National Archives at College Park, Maryland.

King, Franklin Hiram. 1894. *Destructive Effects of Wind on Sandy Soils and Light Sandy Loams: With Methods of Protection.* Bulletin No. 42. Wisconsin Agricultural Experiment Station.

King, Franklin Hiram. 1905. *Investigations in Soil Management.* U.S. Department of Agriculture Bureau of Soils Bulletin Number 26. U.S. Department of Agriculture, Washington, D.C.

King, Franklin Hiram. 1911. *Farmers of Forty Centuries; or, Permanent Agriculture in China, Korea and Japan*. Madison, Wis.: Mrs. F. H. King.

King, Franklin Hiram. 1973. *Farmers of Forty Centuries; or, Permanent Agriculture in China, Korea and Japan*. Emmaus, Penn.: Rodale Press.

King, Franklin H. to Milton Whitney. 19 October 1899. Letters Received; Division of Agricultural Soils; Record Group 54, Records of the Bureau of Plant Industry, Soils, and Agricultural Engineering. National Archives at College Park, Maryland.

King, Franklin H. to Milton Whitney. 1 November 1900. Letters Received; Division of Agricultural Soils; Record Group 54, Records of the Bureau of Plant Industry, Soils, and Agricultural Engineering. National Archives at College Park, Maryland.

Lapham, Macy H. 1949. *Crisscross Trails: Narrative of a Soil Surveyor*. Berkeley, Calif.: Willis E. Berg.

Lyles, Leon. 1985. Predicting and controlling wind erosion. In *The History of Soil and Water Conservation*, edited by Douglas Helms and Susan Flader, pp. 103–112. Washington, D.C.: Agricultural History Society.

Marbut, Curtis F. n.d. Personnel file. Selected Personnel Files; Record Group 16, Records of the Office of the Secretary of Agriculture. National Archives at College Park, Maryland.

Marbut, Curtis F. 1935. *Soils of the United States*. Atlas of American Agriculture, Part III. U.S. Department of Agriculture, Washington, D.C.

McGee, W. J. n.d. Personnel file. Selected Personnel Files; Record Group 16, Records of the Office of the Secretary of Agriculture. National Archives at College Park, Maryland.

McGee, W. J. 1892. Report of Mr. W. J. McGee. *Thirteenth Annual Report of the U.S. Geological Survey to the Secretary of the Interior, 1891–1892*. Part 1. Washington, D.C.: Government Printing Office.

McGee, W. J. 1909a. *Proceedings of a Conference of Governors in the White House, Washington, D.C., May 13–15, 1908*. Washington, D.C.: Government Printing Office.

McGee, W. J. 1909b. Water as the basis for national prosperity. In *Official Proceedings of the Seventeenth National Irrigation Congress*, Spokane, Washington, August 9 to 14, 1909, edited by Arthur Hooker, pp. 128–134. Spokane: Shaw & Borden.

McGee, W. J. 1911. *Soil Erosion*. U.S. Department of Agriculture Bureau of Soils Bulletin Number 71. U.S. Department of Agriculture, Washington, D.C.

McGee, W. J. 1913a. *Wells and Subsoil Water*. U.S. Department of Agriculture Bureau of Soils Bulletin Number 92. U.S. Department of Agriculture, Washington, D.C.

McGee, W. J. 1913b. *Field Records Relating to Subsoil Water*. U.S. Department of Agriculture Bureau of Soils Bulletin Number 93. U.S. Department of Agriculture, Washington, D.C.

Meyer, L. Donald, and William C. Moldenhauer. 1985. Soil erosion by water: the research experience. In *The History of Soil and Water Conservation*, edited by Douglas Helms and Susan Flader, pp. 90–102. Washington, D.C.: Agricultural History Society.

Nelson, Clifford M. 1999. Toward a reliable geologic map of the United States, 1803–1893. In *Surveying the Record: North American Scientific Exploration to 1930*. Memoirs of the American Philosophical Society, Vol. 231, edited by Edward C. Carter, II, pp. 51–74. Philadelphia, Penn.: American Philosophical Society.

North Carolina State College of Agriculture and Engineering. 1924. *Catalogue*. State College Station, Raleigh, N.C.

Orvedal, A. Clifford. n.d. Biographical file. History Office, Natural Resources Conservation Service, U.S. Department of Agriculture. Washington, D.C.

Ottoson, Howard W., Andrew R. Aandahl, and L. Burbank Kristjanson. 1954. *Valuation of Farm Land for Tax Assessment*. Bulletin Number 427. Nebraska Agricultural Experiment Station, University of Nebraska, Lincoln, Neb.

Pinchot, Gifford. 1947. *Breaking New Ground*. New York: Harcourt, Brace, and Company.

Rossiter, Margaret W. 1975. *The Emergence of Agricultural Science: Justus Liebig and the Americans, 1840–1880*. New Haven, Conn.: Yale University Press.

Shaler, Nathaniel S. 1891. The origin and nature of soils. In *Twelfth Annual Report of the United States Geological Survey 1890–1891*, edited by J. W. Powell, pp. 219–344. Washington, D.C.: U.S. Government Printing Office.

Shor, Elizabeth Noble. 1999. William James McGee. In *American National Biography*, Vol. 15, pp. 47–48. New York: Oxford University Press.

Sibirtzev, Nikolai. 1899. Etude des sols de la Russie. In *Comptes Rendus 7th Congres Geologique Internationale*, St. Petersburg, Russie, pp. 73–125. St. Petersbourg: M. Stassulewitch.

Sibirtzev, Nikolai. 1901. Russian soil investigations. Translated by Peter Fireman. In *Experiment Station Record*, Vol. 12, Number 8, pp. 701–712 and 807–818. U.S. Department of Agriculture, Washington, D.C.

Simonson, Roy W. 1959. Outline of a generalized theory of soil genesis. *Soil Science Society of America Proceedings* 23:152–156.

Simonson, Roy W. 1985. Soil classification in the past—Roots and philosophies. In *Annual Report 1984*, pp. 6–18. Wageningen, The Netherlands: International Soil Reference and Information Centre.

Simonson, Roy W. 1987. *Historical Aspects of Soil Survey and Soil Classification*. Madison, Wis.: Soil Science Society of America.

Simonson, Roy W. 1989a. *Historical Highlights of Soil Survey and Soil Classification with Emphasis on the United States, 1899–1970*. Technical Paper Number 18. Wageningen, The Netherlands: International Soil Reference and Information Centre.

Simonson, Roy W. 1989b. Soil classification in the past—Roots and philosophies. In S*oil Morphology, Genesis, and Classification*, edited by Delvin Fanning and Mary Christine Fanning, pp. 139–152. New York: John Wiley and Sons.

Simonson, Roy W. 1991. The U. S. Soil Survey—Contributions to soil science and its application. *Geoderma* 48:1–16.

Simonson, Roy W. 1993. Soil color standards and terms for field use—History of their development. In *Soil Color*, edited by J. M. Bigham and E. J. Ciolkosz, pp. 1–20. Madison, Wis.: Soil Science Society of America.

Simonson, Roy W., telephone interview by author, 3 March 1999a.

Simonson, Roy W., correspondence with the author, 24 April 1999b.

Smith, Guy D. 1983. Historical development of soil taxonomy—Background. In *Pedogenesis and Soil Taxonomy*, edited by L. P. Wilding, N. E. Smeck, and G. F. Hall, pp. 23–49. Amsterdam, The Netherlands: Elsevier.

Soil Survey Staff. 1951. *Soil Survey Manual*. U.S. Department of Agriculture, Washington, D.C.

Soil Survey Staff. 1960. *Soil Classification: A Comprehensive System, 7th Approximation*. Soil Conservation Service, U.S. Department of Agriculture, Washington, D.C.

Soil Survey Staff. 1975. *Soil Taxonomy: A Basic* System *of Soil Classification for Making and Interpreting Soil Surveys*. U.S. Department of Agriculture Handbook Number 436. U.S. Department of Agriculture, Washington, D.C.

Tandarich, John P. 1998. Agricultural geology: Disciplinary history. In *Sciences of the Earth: An Encyclopedia of Events, People and Phenomena*, Vol. 1, edited by Gregory A. Good, pp. 23–29. New York: Garland Publishing Inc.

Tandarich, John P., Randall J. Schaetzl, and Robert G. Darmody. 1988. Conversations with Francis D. Hole. *Soil Survey Horizons* 29(1):9–21.

Tanner, C. B., and R. W. Simonson. 1993. Franklin Hiram King—pioneer scientist. *Soil Science Society of America Journal* 57:286–292.

Thorp, James. 1936. *Geography of the Soils of China*. Nanking, China.

University of North Carolina. 1905–1913. *The Catalogue*. Chapel Hill, N.C.: University of North Carolina.

Weber, Gustavus A. 1928. *The Bureau of Chemistry and Soils*. Service Monographs of the United States Government Number 52. Baltimore, Md.: Johns Hopkins University Press.

Whitney, Milton (Whitney application). 7 March 1879. Student Files. Ferdinand Hamburger, Jr. Archives, Johns Hopkins University, Baltimore, Maryland.

Whitney, Milton. 1893. *The Soils of Maryland*. Bulletin Number 21. Maryland Agricultural Experiment Station.

Whitney, Milton. 1900. *Field Operations of the Division of Soils, 1899*. Washington, D.C.: Government Printing Office.

Whitney, Milton. 1908. *Report of the Chief of the Bureau of Soils for 1907*. Washington, D.C.: Government Printing Office.

Whitney, Milton, and F. K. Cameron. 1903. *The Chemistry of Soil as Related to Crop Production*. U.S. Department of Agriculture Bureau of Soils Bulletin Number 22. U.S. Department of Agriculture, Washington, D.C.

Whitney, Milton to Secretary of Agriculture. 22 March 1907. W. J. McGee Personnel File; Selected Personnel Files; Record Group 16, Records of the Office of the Secretary of Agriculture. National Archives at College Park, Maryland.

Whitney, Milton to the Secretary of Agriculture. 15 April 1912. Curtis F. Marbut Personnel File; Selected Personnel Files; Record Group 16, Records of the Office of the Secretary of Agriculture. National Archives at College Park, Maryland.

Zwerman, Paul J., Gilbert Levine, and Marlin G. Cline. 1968. The soil scientist of tomorrow. *Journal of Soil and Water Conservation* 23(6):205–208.

3

HISTORY OF THE SOIL
SURVEY LABORATORIES
Klaus W. Flach and C. Steven Holzhey

INTRODUCTION

The U.S. Department of Agriculture's (USDA) enabling legislation in 1862 included provisions for a chemist, who initiated analytical work with soils in the late 1860s. Nearly 30 years later, in 1894, the USDA established a Division of Agricultural Soils within its Weather Bureau. The Agricultural Soils Division started analytical laboratory work immediately. It expanded into soil survey activities only in 1899. Thus, the soils laboratories of the department preceded the department's soil survey work by five years, and without the laboratories, the U.S. Soil Survey might never have been founded.

With the establishment of the federal soil survey in 1899, a number of states began to contribute to the soil survey. Laboratories at the state agricultural experiment stations joined in the soil survey work by making texture checks, as well as by pioneering new methods and providing ideas on how quantitative physical and chemical data could improve the quality and usefulness of soil surveys.

In the last 100 years, the federal and state laboratories have performed millions of analyses for the soil survey. But their major contribution may actually have been the ideas on soil classification and other aspects of soil science that were developed by the laboratories' scientists. The contribution of these ideas is the focus of this chapter.

SOIL LABORATORIES IN THE USDA
BETWEEN 1862 AND 1952

EARLY INFLUENCES

The first three chemists hired by the USDA performed such interesting duties as analyzing grape juice, and for the post office, the glue on postage stamps. The first two of these men did not stay long, but the third chemist, Dr. Antisell, reported to Congress in 1869 on the analysis of soils. In 1893, Congress appropriated funds for a Division of Agricultural Soils, which was established, as noted above, in the department's Weather Bureau in 1894, under the leadership of Professor Milton A. Whitney. The unit became the Division of Soils in 1897 and the Bureau of Soils in 1901.

Milton Whitney had been a professor at the University of Maryland, and published a report titled, *The Physical Properties of Soils in Relation to Moisture and Crop Distribution* in 1891, based on his work in Maryland. This work convinced him of the importance of soil texture in influencing the water supply to crops and apparently enabled him to persuade Congress to authorize the funds for a soil survey in 1898. In 1903 Whitney and his chemist, Frank K. Cameron, published Bureau of Soils Bulletin Number 22, a 71-page document in which they stressed the importance of soil moisture and also concluded that "practically all soils contain sufficient plant food for good crop yields, [and] that this supply will be indefinitely maintained and that this (sic) actual yield of plants adapted to the soil depends mainly upon the cultural methods and suitable crop rotation." They added that "a chemical analysis of the soil in itself gives no indication of the fertility of this soil or of the probable yield of a crop" (Whitney and Cameron 1903).

Whitney's conclusions were based on analyses of water extracts by field methods from some 500 soil samples collected in New Jersey, Maryland, and North Carolina; on laboratory analyses of dried samples from some 50 soils from almost all states east of Illinois; and on subjective judgments of the condition of the crop at the time of sampling.

The publication of Whitney's ideas, with their implicit denial of the value of fertilizers, in a USDA bulletin caused an uproar in the scientific community and led to acrimonious exchanges, particularly with C. G. Hopkins of the University of Illinois, E. W. Hilgard of the University of California, John Russel of Rothamsted, England (Russel 1911), and eventually to hearings by a congressional committee.

The following year, Professor F. H. King, chief of the Bureau of Soils' Division of Soil Management, submitted six manuscripts, three of which were published as Bureau of Soils Bulletin Number 26 (King 1905). Using methods similar to those used by Whitney and Cameron on similar soils, but including quantitative yield data, King showed clear correlations between the concentration of plant nutrients in the soil extract and yields. In a rather elaborate letter of transmittal to the secretary of agriculture, Whitney explained why King was denied permission to publish three of the manuscripts. He closed the letter with the announcement that "Professor King severed his connection with the Department on June 30, 1904."

Lapham informs us that Whitney edited the early soil survey reports, but we do not know to what extent he influenced the day-to-day operations of the various parts of the Bureau of Soils (Lapham 1949). In any case, attempts to have Whitney removed from his position never succeeded. He stayed on as director of the Bureau of Soils until shortly before his death in 1927.

From the beginning, the Division of Soils had a chemical laboratory under F. K. Cameron and a physical laboratory under L. J. Briggs, who later became director of the U.S. Bureau of Standards. The chemical and physical laboratories became separate divisions when the Bureau of Soils was established in 1901. Early publications of the two divisions included methods such as the centrifugal method for making mechanical soil analysis and methods for determining organic carbon, soil selenium, alkali, and the content and solubility of gypsum. In the early years, at least some soil mappers returned to headquarters in Washington during the winter and worked in the laboratories (Lapham 1949), although how long this custom persisted is not recorded.

THE TOTAL SOIL ANALYSIS PERIOD

In 1912, Congress appropriated the first funds for "chemical investigation of soil types, soil composition and soil minerals, the soil solution, solubility of soil and all chemical properties of soils in relation to soil formation, soil texture and soil productivity, including all routine chemical work in connection with the soil survey." Funds were also appropriated for physical investigations with appropriate objectives (Weber 1928).

For many years, the Soil Chemistry Division remained under the leadership of its first chief, H. C. Byer. Rather than address specific scientific problems, the division primarily performed total analyses of soils and published the results in a series of bulletins that typically described soils of geographic areas or the soils on certain parent materials. In the beginning, the laboratory analyzed the whole soil using arbitrary depth increments; later, the lab recognized horizons and added data on the composition of the clay fraction (< 5 μm) and the colloid fraction ($<$ 2 μm) in its analyses. Calculations of losses, gains, and translocations were based on changes relative to aluminum, which was assumed to be the most stable component of the soil. Oddly, the data included total nitrogen (N) but not total carbon (C) or any other measure of soil organic matter. A summary of the work of the Soil Chemistry Division appeared in 1935 as Part III of the Atlas of American Agriculture (Marbut 1935), which included mechanical and total chemical analyses of some 200 soil profiles.

Rise of Interest in Soil Mineralogy

Whitney and most soil scientists of his time believed that the colloid fractions of all soils were mineralogically identical to the coarser fractions, and differed only in amount and size. When some members of his staff began organizing the data according to soil-forming factors, they began to recognize that chernozemic soils had clays with higher ratios of silica to aluminum than lateritic soils. In 1924, Robinson and Holmes suggested that the relationship of aluminum to silica could be used for partial classification of soils. They expressed the hope that with further development of x-ray methods it might be possible to decide whether complex mixtures of soil colloids "may ultimately be composed of crystals" (Robinson and Holmes 1924).

Interest in mineralogy began when William H. Fry joined the Bureau of Soils in 1911. Soon after, W. J. McCaughey and Fry published a 100-page bulletin on the microscopic determination of soil minerals (McCaughey and Fry 1913). When in 1927 the new Bureau of Chemistry and Soils absorbed the Bureau of Soils, the new bureau included among its many branches and divisions a Fertilizer Research Division, which installed x-ray equipment and in 1928, hired Sterling Hendricks to do crystallographic work. Why the Fertilizer Research Division became the home of this work and not the Soil Chemistry or Soil Physics Divisions is not clear. Hendricks, who had crossed paths with two-time Nobel Prize winner Linus Pauling, went on to become one of the most renowned scientists in the history of the USDA.

In those early days, x-ray equipment for crystallographic work was not commercially available. A whole room full of transformers, resistors, and capacitors had to be designed and assembled in-house. Using this "homemade" equipment, Hendricks and Fry demonstrated the presence of clay minerals in soils and reported the virtual absence of crystalline primary minerals in the 23 samples of colloid they studied

(Hendricks and Fry 1930). Over the next 10 years, Hendricks and his associates published a series of papers that laid a large part of the foundation for clay mineralogy as we know it today.

With the help of the very fine instrument shop maintained by the Bureau of Chemistry and Soils, Hendricks and Alexander, a chemist in the Soil Chemistry Division, built differential thermal analysis (DTA) equipment in 1939 and demonstrated its usefulness in clay mineralogy. The original equipment used mirror galvanometers, which recorded data onto photographic paper through various mirrors. The equipment worked for some 30 years; the authors of this chapter found its precision in the late 1960s equal to instruments commercially available at that time.

There seems to have been little appreciation of this work. Not until 1938 did the work of the mineralogy group appear noticeably in the fertilizer section of the bureau chief's annual report to the secretary of agriculture. The soil chemistry and soil physics divisions continued to report on their work on colloids in soils, but they defined these colloids strictly in chemical terms. Eight years after the publication of Hendricks's work, H. C. Byers, at that time the chief chemist in the Soil Chemistry and Physics Division, wrote in a major paper published in the 1938 Yearbook of Agriculture, *Soils and Men*, titled, "General Chemistry of the Soil" that clay colloids are "for the most part" mixtures of complex silicates. He went as far as averring that colloids "sometimes have a submicroscopic crystalline structure characteristic of the clay minerals known as halloysite and kaolinite" (Byers et al. 1938a).

In a companion paper, "Formation of Soils," published in the same yearbook, Byers, Kellogg, and Anderson stated that "as yet it has been found impossible to separate [the colloid complex] into its individual components" (Byers et al. 1938b). In the whole 1938 yearbook, which is justly famous for its treatment of many aspects of soil science, neither Hendricks's work nor his name is ever mentioned. Not until 1942, the year in which Byers retired, did scientists of the soil chemistry and soil survey divisions finally reference Hendricks's work in a paper (Brown and Thorp 1942).

RISE OF INTEREST IN SOIL MICROMORPHOLOGY

During the 1930s, C. D. Jeffries at Pennsylvania State University, C. E. Marshall at the University of Missouri, and their students showed the usefulness of petrographic work on sand fractions for soil genesis studies (Marshall 1935; Jeffries 1947). But the petrographic microscope available in the bureau's Fertilizer Division was used for bone and fertilizer studies, not for studies of soil genesis or classification. Although the bureau had petrographic competence as early as the 1920s, Hendricks and Fry used petrographic techniques in their 1930 paper, and another competent petrographer, George Faust, was hired in the late 1930s.

W. L. Kubiena, an Austrian soil scientist who pioneered soil micromorphology, spent a year at Iowa State College. Several of his students published papers on soil micromorphology, but later they moved on to other kinds of work. The first application of the petrographic microscope to soil science in the bureau came in 1942 when Faust assisted C. C. Nikiforoff in a study of the hardpan and claypan (what is now called duripan) in the San Joaquin soil. The publication of this work (Nikiforoff and Alexander 1942) included camera lucida drawings and photomicrographs.

John Cady, who had used the petrographic microscope for his thesis work at Cornell University on weathering in Podzols, joined the laboratory in 1946. He and Roger Humbert assisted C. C. Nikiforoff in a study of fragipans in Maryland, participating in the fieldwork, making mineral counts to establish the homogeneity of parent materials, and at the suggestion of Hendricks, making thin sections to study processes of pan formation. This work started not only petrographic and thin section work on soils in the bureau but the tradition of joint fieldwork by field and laboratory personnel.

Geologists developed the first thin section techniques. They used Bakelite with acetone as the impregnating medium, which left holes in the section where the acetone had evaporated. Better techniques using various plastics became available soon after.

LABORATORY SUPPORT TO THE SOIL SURVEY

The records from 1862 to 1952 are incomplete, and history can only be reconstructed from chance comments and observations. It is known that the bureau's laboratories made numerous texture checks, some selenium studies, and some salinity determinations for the Soil Survey, but aside from assistance in the selection of sampling sites, there seems to have been little close cooperation with the Soil Survey Division by either the Soil Chemistry or the Soil Physics Division until the early 1940s.

In all fairness, it must be noted that the soil surveyors of the day also had very little appreciation for the role laboratories could play in the soil survey program. M. H. Lapham, a soil surveyor and later an inspector, gives a detailed and delightful account of his experiences in the soil survey between 1899 and the early 1940s in *Crisscross Trails* (Lapham 1949), but he never mentions the work of the laboratories (except that he helped run mechanical analyses during his first few winters in the bureau).

It must be remembered, however, that the perceived objectives of the soil survey varied considerably during this time. Early soil survey reports, for example, contained rather detailed recommendations on soil and farm management and raised the ire of the USDA's Cooperative Extension Service. At other times, the objectives of the Soil Survey were very narrow. In a paper titled, "The Soil Survey—Present and Future," presented at the 1922 meeting of the American Soil Survey Association (ASSA), P. S. Lovejoy, for example, stated that "one point of view regards soil surveys [is] to do for soils what Gray, for instance, did for our native plants. From this standpoint the chief function is to discover, describe, map. . . ." He went on to say, "that from this standpoint the soil surveyor does not concern himself with the utilities, or possible utilities, or potential utilities of the soil types which he maps" (Lovejoy 1923).

In the discussion of the paper, C. F. Marbut, chief of the Soil Survey Division, stated rather emphatically that this was not one view but that "We who are doing the soil survey work know that to be the point of view." Marbut also seemed to believe that everything worth knowing about a soil could be deduced from field studies of the soil profile by what he called the "soil profile method" (Lovejoy 1923). Discussing S. D. Conner's paper on soil acidity (Conner 1923), he asked the author whether a difference in acidity between two soils was reflected in differences in the soil profile. Mr. Conner, a chemist, replied that he "didn't know anything about the profile," upon which Marbut concluded that "then it is not proved that the profile method is not satisfactory" (Conner 1923).

EFFORTS TO PROMOTE THE USE OF LABORATORY DATA IN SOIL SURVEYS

The Committee on Correlative Laboratory Work (1920–1936)

A Committee on Correlative Laboratory Work was established at the first meeting of the American Soil Survey Association in 1920. Apparently, scientists at the agricultural experiment stations did not fully share Marbut's views, although no charges to the committee were ever found, and their annual reports do not shed much light on the issue. The first report of the committee was in the form of a paper, "What Correlative Laboratory Work Is It Desirable To Do," presented by M. M. McCool at the second meeting of the association in 1921 (McCool 1921). He proposed a freezing method to distinguish various forms of soil water, and similar to Whitney, a method to measure dissolution rates of primary minerals. He presented a similar paper, "Laboratory Methods of Value to Soil Survey," at the 1922 meeting (McCool 1922).

The committee's 1922 report (Conroy 1923) presented the results of a survey of the types and extent of laboratory work carried on by 40 states engaged in soil survey, divided into chemical, physical, and physico-chemical determinations. The findings showed that a great many of the analyses reported were total chemical and mechanical analyses, but unfortunately the report does not indicate whether any of the work was actually done for the purposes of soil survey. Later reports of the committee were largely wish lists for better methods of mechanical analysis and for methods to determine such things as soil reaction, lime requirement, consistence, and "character of the clay." But no discussion ever appeared about whether or how any of these methods should be used for defining soil types.

In 1927, H. H. Krusekopf and Richard Bradfield of the University of Missouri, although not members of the committee, cooperated on two papers both titled "What Information Is Necessary for a Complete Description of a Soil?" Krusekopf addressed the field and Bradfield the laboratory aspects of the issue (Krusekopf 1927; Bradfield 1927). Bradfield, a physical chemist who had just published a classic bulletin, *Chemical Nature of Colloidal Clay* (Bradfield 1923), was more interested in the effect of clay on soil properties than on their composition. He thought past studies were too oriented toward "immediate practical applications," an approach that had taken soil scientists "about as far as we can go." Consequently, he urged investigations of a more fundamental character to explain "the behavior of soils in terms of the laws of chemistry and physics." He objected to a classification of soils based solely on field observations because he thought field observations were by necessity qualitative and subjective and unable to perceive important relations that are apparent only through accurate laboratory analyses.

Bradfield continued, "The history of the soil is written indelibly in the soil profile itself. . . . All we need to know is there if we can learn to read it." He concluded that "soils are complex but not hopelessly so" (emphasis in original) and that most of the apparent complexity is due to a few uncontrolled factors. In his paper, Bradfield listed laboratory characteristics that were considered important by the famous European soil scientists of the day and proposed a list of the absolute minimum of factors that must be considered (see Table 3.1). Although he did not specify which elements should be included in total analysis, one can assume that they would be the elements determined in mineralogical studies at that time. Bradfield closed the paper by apologizing for presenting such a formidable list. He explained that he wanted to invoke discussion "for [he wanted] very much to know what we should know about a soil in order to describe it."

The Committee on Correlative Laboratory Work never formally endorsed Bradfield's paper.

The Committee on Laboratory Studies Additional to Soil Classification (1937–1942)

In 1937, the American Soil Survey Association merged into the newly established Soil Science Society of America, which was established the preceding year. The following year, "a group of soil investigators assembled to discuss informally the possibilities of supplementing the

Table 3.1.   Minimum list of soil characteristics of outstanding importance

Physical	Chemical	Physico-Chemical
Mechanical composition	Acidity Active or pH value Reserve or titrate acidity	Carbon content
Structure	Amount and nature of the exchangeable bases	Silica/sesquioxide ratio of clay fraction
Shrinkage or swelling coefficient	Degree of saturation Flocculation value	Total N, P, and K content Total analysis

Source: Bradfield 1927.

present soil survey activities with certain laboratory studies" and recommended that the new society's president appoint a committee to investigate the idea (Baver et al. 1938). The president agreed and appointed a seven-member committee, about half of whom seemed to have a primary interest in soil survey and half a primary interest in soil chemistry, soil physics, or soil fertility. L. D. Baver, a soil physicist, became chairman and Bradfield was one of the members. Although one of the charges to the committee was to "strengthen the work of the United States Division of Soils and the United States Soil Conservation Service," neither the bureau's Soil Survey Division and Soil Chemistry Division nor the Soil Conservation Service seem to have been represented on the committee.

The committee first made a survey of the kinds of analyses that the agricultural experiment stations considered useful (see Table 3.2). The results must have disappointed them: only 24 experiment stations responded, and the only analysis on which most of them (20) agreed was the need for pH data. Fourteen of the 24 wanted available phosphorus, organic matter, nitrogen, and available potassium. Apparently, mechanical analysis was accepted as a given, since none of the experiment stations mentioned it.

In spite of this rather inauspicious beginning, the committee submitted a report in 1938 identifying two categories of desirable data; those "to be used primarily for soil classification" (see Table 3.3) and those "to be used primarily in soil management and land use recommendations" (see Table 3.4). The list is quite similar to the one Bradfield published in 1927, excepting soil water retention, for which methods had not been developed in 1927. The committee concluded that "if it were possible to obtain this information on a group of soils, much progress would be made in understanding the importance of technical data in soil classification as well as in the differentiation of soils on the basis of the data." They noted the need for standardization of methods, and in a rather obvious reference to the connotative code used by the Soil Conservation Service in soil mapping at that time, they warned that "the use of supplementary laboratory data in soil classification presupposes a differentiation of soils in well-defined classes that can be recognized pedologically and agriculturally in the field." ". . . Too much detail in field data," they continued, "will nullify any attempt to feasibly make use of the results of laboratory studies" (Baver et al. 1938).

Table 3.2. Number of U.S. states desiring certain supplementary data for soil classification

Data Desired	Number of States
pH	20
Available P	14
Organic matter	14
Nitrogen	13
Available K	12
Exchange capacity	9
Greenhouse responses	7
Exchangeable Ca	6
Volume weight	6
Salt content	5
Field responses	5
Permeability	5
Aggregation	5
Colloidal properties	4
Moisture equivalence	4
Color	4
Total analysis	3
Exchangeable Mg	3
Plasticity	2
Carbonates	2
Exchangeable H	1

Source: Baver et al. 1938.

The bureau's Soil Chemistry Division apparently ignored this report, continuing to perform total analyses until the retirement of its chief chemist in 1942. The bureau's Soil Survey Division and the Soil Conservation Service also ignored the report.

It was not ignored, however, by a young soil scientist who, as a graduate student under Hans Jenny, joined Bradfield's department at the University of Missouri shortly after Bradfield left the department. He accepted Bradfield's list of chemical and physical characteristics and his earlier admonition not to base the classification of soils solely on field observations. That graduate student was Guy D. Smith.

We believe that this 1938 report, although largely forgotten today,

Table 3.3.   Supplementary laboratory data for soil classification

Physical characteristics	Chemical characteristics
Mechanical analysis	pH
Structural analysis	Carbonates
Volume weight	
Capillary and noncapillary	
pores	
Extent and stability of	
aggregates	
Water relationships	Base exchange capacity,
Infiltration	exchangeable bases and base
Moisture equivalent	saturation: Ca, Mg, K, Na
Moisture sorption curve	
Soil consistency	Organic matter
Soil color	Nitrogen
	Phosphorus (at 2 pH values)
	Salt content (of western soils)
	Analysis of colloidal fraction:
	Exchange capacity
	Complete chemical analysis
	Mineralogical analysis

Source: Baver et al. 1938.

Table 3.4.   Supplementary laboratory data to be used primarily in soil management recommendations

Mechanical analysis	Available potassium
Volume weight	Available calcium
pH	Salt content (West)
Organic matter	Greenhouse responses
Nitrogen	Field responses
Available phosphorus	Infiltration, air and water capacity, color

Source: Baver et al. 1938.

must be considered a major milestone in the development of soil science. With time, more and more of the analyses urged by the committee were performed at soils departments of the agricultural experiment stations. After 1952, the soil survey laboratories essentially adopted the recommendations of the committee in their soil characterization program. In addition, most of the kinds of data recommended by the committee became taxonomic differentiae in *Soil Taxonomy* (Soil Survey Staff 1975).

Despite this report's significance, Bradfield, whom the senior author of this chapter knew 20 years later at Cornell University, seemed to have no further interest in soil classification.

THE LABORATORIES AFTER THE USDA'S 1952 REORGANIZATION[1]

REORGANIZATION CHANGES THE LABORATORIES

To make a cleaner separation between research and action agencies, the USDA underwent a major reorganization in 1952. As part of this reorganization, the soil surveys of the Soil Conservation Service and the Bureau of Plant Industry, Soils, and Agricultural Engineering were combined and placed in the Soil Conservation Service (SCS). At the same time, the SCS lost almost its entire research branch to the new Agricultural Research Service (ARS). Somewhat as an afterthought, however, an amendment to the reorganization directives moved those parts of the Soil Physics and Chemistry Division of the Bureau of Plant Industry, Soils and Agricultural Engineering that dealt with soil characterization and mineralogy to the SCS, with the rather narrowly defined charge of conducting investigations in support of the soil survey. Although the laboratory maintained its capability in soil mineralogy, the narrow interpretation of the meaning of "investigations" in the laboratories' new charge precluded basic research, such as that which led to the discovery of the crystallinity of clays.

Under the new organization, the SCS was divided into four "Deputy Areas," one of which was the soil survey, under the leadership of Charles E. Kellogg. Within this area was the Soil Survey Investigations Division, headed by Guy D. Smith, who had come from the bureau's Soil Survey Division. Under Smith, the leadership for the soil survey laboratories fell to Lyle T. Alexander, who moved from the bureau's Chemistry and Physics Division.

The function of the soil laboratories changed dramatically. In contrast to its counterpart in the SCS, the bureau's Soil Survey Division had used rigorous taxonomic definitions of kinds of soils, mostly defined as soil series. These rigorous definitions were now to be applied to all the SCS soil survey activities (and by extension, to soil surveys in other agencies, such as the USDA's Forest Service and the Department of the Interior's Bureau of Land Management and Bureau of Indian Affairs). One of the main tasks of the laboratories became to serve the implementation of this policy. The other major task of the laboratories was to assist Guy Smith in developing a new quantitative soil classification system, of which the "1st Approximation" had been informally distributed in early 1952.

Laboratories assisted the Soil Survey in developing and implementing the new soil classification system. George Phibbs in the Beltsville, Maryland, laboratory, ca. 1960. (Courtesy of the Natural Resources Conservation Service, U.S. Department of Agriculture)

The Three Regional Soil Survey Laboratories (1952–1975)

To serve the soil survey in all parts of the country, laboratories were established in Beltsville, Maryland, for the eastern states; Mandan, North Dakota, for the Midwest and the Great Plains; and in Riverside, California, for the West. After a few years, the Mandan laboratory was moved to a more central location in Lincoln, Nebraska. At first, only the Beltsville laboratory had the equipment and staff for mineralogical work, but this situation changed in the early 1960s.

All of the work at the laboratories was done either at the request of the field staffs or of the Washington staff, which generally meant Guy Smith. Small requests for a few analyses to verify field texture determinations, organic matter, or salinity levels, for example, called laboratory checks, were handled informally between state soil scientists and the closest laboratory. In contrast, requests for comprehensive analyses of whole pedons, called characterization analyses, were usually made for a number of soils from a soil survey area or for a number of soils with related classification or interpretation problems from several survey areas or from several states. Approval by the chief, at that time Lyle T. Alexander, was required.

Requests for characterization analyses were checked against available data for the soil series in question in the files of the Director of Correlation and Classification in Beltsville. If data were available, the request was usually rejected. In time, this system led to a situation in which data for extensive soil series that had been characterized early were often less comprehensive than data for some relatively scarce soils that were characterized later.

Since laboratory data were becoming a critical part of the soil classification and correlation process, uniform soil sampling policies and procedures had to be established. To that end, it became policy that someone from the laboratory participate in all sampling for soil characterization, and that detailed pedon descriptions be required for all sampled pedons. With the quantitative criteria of the new classification system, precision and uniformity in sampling became as important as precision in the laboratory; increasing the depth of the sample of a surface horizon, for example, might dilute the organic matter content sufficiently to move the horizon from mollic to ochric. But the main benefit of these policies was to bridge the gap that had existed between "field" and "laboratory" people.

Each of the laboratories had one or two panel trucks, as driving was

Guy D. Smith looks for "Tonhäutchen" in the B horizon of a Clarion loam, Kandiyohi County, Minnesota, September 1953. (Courtesy of the Natural Resources Conservation Service, U.S. Department of Agriculture)

the only accepted mode of transportation. Sampling trips, sometimes more than 1000 miles—for example, from Lincoln, Nebraska, to Florida or from Riverside, in southern California, to northern Idaho—frequently consumed a weekend traveling to the site and a weekend returning.

In establishing uniform sampling procedures, considerable discussion ensued about how to produce data that could be considered representative of soil taxa. Taking enough samples for a statistically valid sample of a soil series was, and still is, clearly impractical. Consequently, a policy evolved where samples were taken from two typical profiles that were morphologically indistinguishable and at least 8 kilometers apart. This policy was certainly better than just sampling one profile, and the search for the "matching profile" often turned out to be an eye opener for the survey party leader. But the logical step of going back and sampling a third or fourth pedon, if the original match was poor, was rarely if ever taken.

For some time, considerable interest existed in developing field kits to measure properties that were critical for classification and interpretation. Electrical conductivity or resistivity measurements to estimate salinity in the field had been used since the 1890s and had led to the development of the Bureau Cup, a simple cup that was filled with moist soil and mounted on a battery-driven Wheatstone Bridge. By the 1960s, more sophisticated criteria for evaluating salinity and sodicity had been developed by the ARS Salinity Laboratory in Riverside (U.S. Salinity Laboratory Staff 1964).

To enable soil scientists to estimate these properties of great practical importance and to help them estimate properties needed primarily for soil classification, such as carbonate content, cation exchange capacity (CEC), and base saturation, the Lincoln laboratory developed a very versatile field kit (Holmgren and Nelson 1977). At about the same time, the Riverside laboratory developed simpler procedures specifically for determining salinity and sodium adsorption ratio (SAR). A number of people devised other field tests and procedures, which were calibrated locally using reference samples analyzed by the soil survey laboratories or by state agricultural experiment stations.

These kits were very useful in the 1960s, but they were slow. As the turnaround time of the laboratories became better and soil scientists learned to interpolate chemical data from more numerous characterization sites, the importance of these field kits diminished, and they are rarely used now. During these early years, the kits were also used widely in international work but because they used medical syringes to extract and titrate, they began to arouse the suspicion of zealous customs officials with the coming of the drug culture.

For some time, several SCS state offices maintained one or two soil scientists at agricultural experiment station laboratories. They performed mostly texture checks and were phased out by the early 1960s, although several of the western state offices maintained the capability to determine salinity.

Throughout this period, the researchers of the three regional soil survey laboratories were primarily committed to work in direct support of soil survey activities. There were some other projects, however. For example, the laboratories provided significant support for the large geomorphology projects in New Mexico and North Carolina that were supported by the Soil Survey Investigations Division. For another project, the laboratories collected soil samples and analyzed them for strontium 90 as part of a global Atomic Energy Commission (AEC) project. Laboratory personnel also occasionally functioned as consultants to help field scientists and agricultural experiment station personnel assess and study soil-related agricultural problems. The global AEC project and occasional consulting roles foreshadowed much greater activity in support of other federal agencies and international projects later in the century.

The National Soil Survey Laboratory (1975–present)

By the early 1970s, the time for small laboratories had passed. The coming of microprocessors and minicomputers made sophisticated laboratory equipment more available, allowing greater automation in performing analyses and in processing and storing data, but the equipment was expensive. At about the same time, air travel for passengers and luggage became relatively economical. Thus advancements in electronics and in the economics of air travel exerted twin forces for consolidating all soil survey laboratory operations in one central location. Although support for the change was not unanimous, the stepwise consolidation of the three regional laboratories into one National Soil Survey Laboratory (NSSL) in Lincoln, Nebraska, took place during 1975 and 1976.

Formation of the NSSL brought together all research soil scientists of the SCS except the geomorphologists. By this time, research soil scientists were somewhat less than 0.2 percent of the SCS soil science staff, a percentage that continues to decline today. The researchers joined about half a dozen soil scientists from the three laboratories, who were charged with running the analytical operation. Almost all of the members of both groups had begun their careers mapping soils. Just two people remained from the old bureau's Soil Survey Division, and none remained from the pre-1952 SCS soil survey.

Career technicians eventually did most of the routine analytical work with highly automated equipment. These changes speeded the work and increased precision, but there were some disadvantages. The research soil scientists lost some of the knowledge and interest that come from local experience and from doing part of their own analytical work. Their personal contact with soil survey and experiment station scientists diminished initially, especially in the West and the Northeast. They had less opportunity to observe the analyses of samples they had collected. Often, they had to work with many soils that were little more than a name and a classification to them. Expectations were that computerized databases and the opportunities for Geographic Information Systems (GIS) and statistical analysis would largely compensate for these disadvantages.

The new laboratory and staff inherited the four main responsibilities of the system they replaced: 1) to produce data both to calibrate field estimates made by soil scientists and to support the classification of soils; 2) to produce a characterization database for study of soil relationships (e.g., taxonomy, estimates of properties, interpretations); 3) to supplement pedologic and geomorphic studies for the purpose of improving predictions of soil patterns and properties (e.g., geomorphic projects, active soil surveys, cooperative studies with agricultural experiment stations); and 4) to correlate soil properties to known soil performance (first for soil survey interpretations, then with increasing emphasis on data that related to environmental, sustainability, and soil utilization issues).

A spirited debate developed about the role of the new laboratory, but the field orientation of the staff and outside forces dictated the direction of services. In the beginning, the first three responsibilities outlined above dominated the laboratory's work. Equipment was modified to accommodate air travel for sampling trips, and researchers spent much time in the field. Over the next 20 years, the responsibilities of the laboratory broadened considerably as the soil survey and the SCS, which became the Natural Resources Conservation Service (NRCS) in 1994, took on more national and international responsibilities.

In 1987, the NSSL became part of the new National Soil Survey Center (NSSC) in Lincoln. The center combined in one location essentially all of the soil survey staffs with technical responsibilities, including the soil classification, correlation, interpretation, and geomorphology groups who had been distributed among the national headquarters in Washington, D.C., and the four regional soil survey offices. Ironically, what had existed in Beltsville before the 1952 reorganization was essentially re-created in Lincoln in 1987.

An acceleration of what had been a gradual shift from taxonomy and pedology investigations toward soil survey applications such as soil quality and the environment began in 1987. Robert Grossman, research soil scientist at the National Soil Survey Center, continued to pioneer procedures for field and laboratory data use in soil survey interpretations and in recording use-dependence of soil properties and soil quality. However, much work went into preparing the pedon database, data dictionaries, and data applications for use in emerging electronic media, including the National Soil Information System (NASIS) and the Internet. The NSSC continued to fly people to study areas for soil sampling, but others collected at least half the incoming samples. The researchers continued to publish and to distribute unpublished in-house reports at a rate of several dozen per year, while pedon characterization and laboratory checks continued at more than 5,000 samples and 100,000 analyses per year.

The national soil survey laboratory had reached a production level that enabled it both to support soil survey operations and to participate in major national studies in cooperation with other federal agencies, notably the Agricultural Research Service (ARS), the Environmental Protection Agency (EPA), the Food and Drug Administration (FDA), and with scientists at the state agricultural experiment stations. Projects relating to environmental, sustainability, and land utilization issues continued to draw the focus away from taxonomy, soil correlation, and basic studies of soil morphology and genesis.

These cooperative studies have included: 1) National Cooperative Soil Survey (NCSS) cooperative studies to improve soil survey quality; 2) measurement of soil lead and cadmium background levels with the EPA and the FDA; 3) a database for development of new soil erodibility factors for the Water Erosion Prediction Project (WEPP) with the ARS, the Department of Energy (DOE), and others; 4) characterization of sites to establish the effects of soil management (Conservation Reserve Program [CRP] sites) and to assess soil quality (experimental monitoring programs); 5) studies, conferences and publications related to carbon sequestration and greenhouse gas concerns; 6) pedon characterization at numerous ARS, agricultural experiment station, and other research locations and data for EPA environmental monitoring; and 7) assistance to modelers who wished to use the pedon database in their predictions and/or simulations of natural phenomena (e.g., EPIC, WEPP, and RUSLE models).

THE SEVEN APPROXIMATIONS AND SOIL TAXONOMY

The development of *Soil Taxonomy* was a major activity of the soil survey laboratories.[2] Guy Smith distributed the "1st Approximation" in the summer of 1952. The *7th Approximation* was prepared for the 1960 Congress of the International Soil Science Society held in Madison, Wisconsin (Soil Survey Staff 1960). *Soil Taxonomy*, Agriculture Handbook Number 436, was published in 1975 (Soil Survey Staff 1975).

The search for criteria for defining taxa became a major task for the laboratories. Under the direction of Smith, who initiated work to produce needed data, laboratory staffs supplied new data where data were lacking, compiled data sets, and tested new or provisional criteria. Although Smith used data from all possible sources, a major part of the information that went into creating *Soil Taxonomy* and much of the testing of taxa that followed was based on data created by the soil survey laboratories. Of the 130 pedons in appendix four of *Soil Taxonomy*, all but four were analyzed by the soil survey laboratories.

A major effort went into developing the definitions of diagnostic horizons and diagnostic soil characteristics, including much work on mineralogy and micromorphology. Many of these definitions were based on new and initially untested concepts that many soil scientists did not readily accept. It took some effort, for example, to convince Kellogg that there were two contrasting kinds of "podzolic B horizons," the argillic and the spodic horizon. Some of the definitions were also strongly influenced by the kind of data that were available at the time. The use of base saturation by sum of cations for the distinction of Ultisols from Alfisols, for example, was dictated by the fact that initially the definitions of the two taxa were focused on soils in the eastern United States, where few data by other methods were available.

Developmental work by laboratory staffs declined in 1972, the year *Soil Taxonomy* was ready for publication and Guy Smith retired. Three years later, *Soil Taxonomy* was published and the NSSL replaced the laboratories in Beltsville, Lincoln, and Riverside.

COOPERATION WITH UNIVERSITIES AND AGRICULTURAL EXPERIMENT STATIONS

From the 1950s until roughly 1990, state support for laboratories at the land-grant universities, and at one or two other universities, grew steadily. Laboratories at 18 or more universities in the eastern states contributed significant quantities of characterization data between 1952 and 1985, at least three of them operating through most of the 33-year period. The number of participating universities was smaller in the West, but the contribution was still significant from several states. Of those land-grant universities that did not characterize pedons, many did research or some level of laboratory checks to help calibrate estimates made by field soil scientists. The NSSL also exchanged samples with universities for interlaboratory standardization and at times participated in regional and international laboratory comparisons.

State support declined rapidly in the late 1980s and early 1990s, as initial soil mapping was completed and as budgets in many states became smaller. By the early 1990s, the land-grant universities appeared to have in the neighborhood of 10,000 pedons with enough data completed to warrant adding to a pedon database (based on informal discussions of university participants with one of the authors). NCSS standard methods produced most of the data, and documented methods were used to produce most of the rest.

International Activities of the Soil Survey Laboratory

The NSSL became more internationally involved when the SCS contracted with the United States Agency for International Development (USAID) in the late 1970s to assist developing countries with improving their soil survey work. The SCS created the Soil Management Support Services (SMSS) to carry out the USAID contract, and the NSSL became an important part of the project, furthering the development of the taxonomic system as an agent for technology transfer outside the United States. Knowingly or not, the project goal was similar to one expressed by Kellogg in 1949: "We need sets of predictions so that every farmer in the world can develop the optimum system for sustained production on the basis of all the accumulated research and experience that applies to the soil types on his farm, and in terms of the appropriate understanding" (Kellogg 1949). The NSSL staff participated in international committees that tested new diagnostic features, new taxonomic classes, and determinative analyses and improved definitions for several soil orders, notably Oxisols, Ultisols, Andisols, and Spodosols. The project involved a large number of soil survey cooperators from the universities, SCS field scientists, and experts from many parts of the world. Foreign and domestic interlaboratory comparisons enabled international comparisons of soils, just as Hilgard had urged the states to do before there was a federal soil survey. Today, personnel at the NSSC still participate in a variety of international technology exchanges and projects, and the laboratory has one of the largest, perhaps even the largest, global soil characterization databases in existence.

NEW LABORATORY METHODS

From the beginning, the new soil survey laboratories created by the 1952 reorganization broke completely with the traditions of the Bureau of Soils. They began "soil characterization" along lines that had been suggested in 1938 by the Soil Science Society of America (SSSA) Committee on Laboratory Studies Additional to Soil Classification. Guy Smith, head of the Soil Survey Investigations Division of which the laboratories were a part, had an interest in these ideas of quantitative soil classification and was a major factor encouraging their adoption, as was the fact that L. T. Alexander, chief of the soil survey laboratories, was chairman of this SSSA committee for some years. We do not know who first used the term "characterization."

The laboratories' small staff size and a large demand for their services forced them to take full advantage of labor-saving equipment and encouraged the development of more efficient methods. The small number of pedons that could be characterized in any one year, on the other hand, dictated that data that had been assembled over long periods of time, nearly a half century, must not be influenced by methodological changes. Inasmuch as the results of almost all soil analyses depend strongly on even small details in methods, opportunities for innovations were restricted to phases of analyses that followed dispersion and fractionation, as in particle size distribution analysis (PSDA) or extraction in other analyses.

Consequently, the laboratory still uses shakers similar to those used almost 50 years ago, as well as very similar extraction procedures. An exception might be the introduction of a mechanical soil extractor, which the laboratory staff developed in the mid-1970s as a substitute for the vacuum extraction that they (and everybody else) had used previously (Holmgren et al. 1977). This method, however, resulted in more rigorously controlled extractions that gave better and more reproducible results than the older methods, especially in fine-textured soils. When unavoidably small changes were made, elaborate tests were conducted to ensure that results did not change.

Physical Methods[3]

In 1952, standard characterization consisted of PSDA, bulk density, moisture retention analyses, and various chemical and mineralogical analyses of the several horizons. Kilmer and Alexander's paper was the basis for the methods for PSDA, which pretty much became a standard for the U.S. Soil Survey (Kilmer and Alexander 1949). The ARS shop in Beltsville built standard shakers for dispersion and separation of sand fractions, as well as special pipette tables and brass stirrers, still in use today, and shipped them to the various soil survey laboratories.

Methods for bulk density and moisture retention at first followed the published methods of the day but were soon modified to fit the particular needs of the soil survey. In particular, methods that had been developed for the plow-layers of cultivated soils proved unsatisfactory for subsoil horizons. Water retained by the soil against a pressure of one-tenth bar (10 kpa) or even one-third bar (33 kpa)[4] in crushed and sieved samples of B horizons, for example, often exceeded the total pore space of undisturbed samples of the horizons. It was also often impossible to take meaningful core samples in gravelly soils and in woodland soils with large roots. This led to Cliff Simonson, Benny Brasher, and Bob Grossman of the soil survey laboratories developing the Saran-coated clod method (Brasher et al. 1966). Since this method allowed the measurement of bulk density of clods that had been equilibrated at several moisture tensions, the Coefficient of Linear Extensibility (COLE) was soon added to the standard laboratory calculations (Grossman et al. 1968).

Standard procedures at that time paid little attention to particles larger than 2 mm. They were disregarded or at best described semiquantitatively. The proportions of particles larger than 2 mm were calculated from the often small samples that happened to get to the laboratory, where they were separated rather subjectively by the lowliest member of the laboratory staff. Such data, of course, could not be used to estimate particle size distribution of the whole soil, which engineers determined and required, or to calculate the water-holding capacity or nutrient content of the volume of soil accessible to plants. Consequently, sampling procedures were improved to provide for sieving and weighing in the field of particles that were greater than 19 mm and on a sample that was large enough (often 25 to 50 kg) to provide statistically valid data and for determining 2 to 5 mm and 5 to 19 mm fractions in the laboratory. In soils on soft sedimentary rocks, weights for particles that were greater than 2 mm were corrected in a subsample for particles that slaked in water.

Chemical Methods

Chemical characteristics were determined by methods that closely followed those described in USDA Circular 757, *Methods of Soil Analysis for Soil Fertility Investigations* (Peech et al. 1947). Cations were extracted by ammonium acetate; organic carbon was determined by wet combustion; and total nitrogen was determined by Kjeldahl. Over time more analyses were added to the standard repertoire, including extractable iron by dithionite-citrate, which was adapted from a procedure for cleaning clays for mineralogical analysis, and potassium-chloride extractable aluminum in order to compute "permanent charge."

Regional preferences by the agricultural experiment stations led to regional differences, however. The Beltsville laboratory routinely determined exchange acidity by Mehlich's triethanol-amine (TEA) method and calculated CEC (at pH 8.2) from the sum of cations plus TEA acidity. The Lincoln laboratory determined CEC by the ammonium acetate method (at pH 7.0) and measured exchange acidity only in acid soils. The Riverside laboratory used the sodium acetate method (at pH 8.2)to determine CEC. In the West, differences in CEC methods were not only the result of regional preferences, but also the need to conform, for saline and alkaline soils, to the methods of the ARS's United States Salinity Laboratory. Lincoln and Riverside also performed analyses of saturation extracts by Salinity Laboratory methods.

Equipment in the early days following the reorganization was simple, but methods for quantitative analysis were laborious and required painstaking work. Initially, the classical wet analysis methods were used to determine all cations, but flame photometers and colorimeters became standard equipment in the early 1950s. The first automatic analytical balances and mechanical calculators showed up at about the same time. A centrifuge, a few ovens, a suction apparatus, a muffle furnace, and a lot of glassware pretty much completed what were needed for a well-equipped laboratory.

The technology for performing chemical analyses changed rapidly after the formation of the NSSL in 1975. George Holmgren, in consultation with Leo Juve, developed the concept of a mechanical extractor that made the suction apparatus obsolete for extracting solutions from soil (Holmgren et al. 1977). The extractor was engineered and built commercially, and many laboratories in the United States, as well as laboratories in dozens of foreign countries, currently use it. By the end of the 1970s, the wet chemistry era was drawing to a close, plastic was replacing glassware, microprocessors were just coming on the scene, and the new soil survey laboratory had enough resources to utilize both new ideas and new technologies.

As the NSSL tested new methods for taxonomic improvements and for environmentally related analyses, the list of standard chemical procedures grew. The laboratory tested and adopted methods that employed an anode-stripping voltameter, a succession of atomic absorption instruments, direct current plasma and induced current plasma instruments, and a variety of dissolution techniques. The NSSL performed one of the first applications of a liquid chromatograph for production soil anion analysis.

Sample preparation, however, where the goal of maintaining the original particle size distribution militated against mechanical grinding, remained resistant to change. A number of machines have been designed and tested, but technicians still prepare each sample with the care indicated by the type of material, to prevent the grinding of soft rock into sand.

The first compilation of methods used in the laboratory was published as volume one of the Soil Survey Investigation Reports (SSIR) series in 1967 (Soil Conservation Service 1972). The compilation of standard methods had grown to nearly 700 pages in the NSSC's 1996 *Soil Survey Laboratory Methods Manual* (Soil Survey Laboratory Staff 1996), and was formally adopted as the source of standard analysis methods for the NCSS at the 1997 Soil Survey Conference. We think this is the first time, in a century of soil survey laboratory work, that the survey has officially adopted a standard set of laboratory methods, yet most of the pedon data produced by the SCS and cooperating universities over the last half century has been well standardized.

As long as there were three laboratories, interlaboratory comparisons were made systematically, but there has never been a formal system of interlaboratory comparisons between the NSSL and universities cooperating in the NCSS. Nevertheless, the more active laboratories and the majority of all laboratories providing soil survey data cooperated in interlaboratory comparisons between 1975 and 1990, the period of maximum laboratory activity among the university cooperators.

Mineralogy and Micromorphology

By the time the soil survey laboratories were established, most of what are now standard procedures to identify individual clay minerals had been worked out and standardized, to a degree, thanks in large part to M. L. Jackson and his students at the University of Wisconsin. In the early 1950s, Lynn Whittig, a new Ph.D. graduate from Wisconsin, joined the Beltsville laboratory and introduced the methods that are still used today, although in modified form. Only the Beltsville laboratory had equipment and capability in mineralogy in the 1950s and early 1960s, and the methods for separating and treating each specimen were cumbersome and expensive. Consequently, mineralogical work was done on only a few samples.

Around 1964, the Lincoln and the Riverside laboratories obtained x-ray equipment, making procedures less laborious. By the time the NSSL was established, in 1975, the introduction of automatic sample changers and more sophisticated DTA equipment reduced the costs of mineralogical analyses considerably. Warren Lynn, of the NSSL, introduced the use of resins to saturate clays with magnesium and potassium, further reducing time requirements. As a result of these advances, clay mineralogical analysis as part of pedon characterization in most characterization projects became possible.

The NSSL performed some electron microscopy for a time. More recently, total analysis by relatively efficient plasma technology has made something of a comeback, improving quantitative estimates of clay mineralogy. In general, however, the mineralogical work at the soil survey laboratories kept to its mandate to support the Soil Survey, and have done few highly detailed mineralogical analyses. The emphasis has been to predict and explain soil behavior, for which clay mineralogy is used as one tool in conjunction with inferences from such things as cation exchange capacity, coefficient of linear extensibility, Atterberg limits, stickiness, and water retention.

Mineral counts using a petrographic microscope continued to be made on a selected basis, but are now done mostly by technicians. With some exceptions, the emphasis continues to be with assisting in the classification of soils or determining quantitative distributions of marker minerals in the stratigraphic column, which can be of help in understanding the distribution of soils in the landscape and, consequently, in improving the efficiency of soil mapping. The study of micromorphology in thin section continues, most notably in the work of

W. D. Nettleton. But in contrast to earlier years, commercial producers now almost exclusively make the thin sections.

DATA MANAGEMENT, PUBLICATION, AND TRAINING

Initially, data were distributed informally in one of two standard formats, one for acid and neutral soils that did not contain salts, and one for saline soils that could accommodate data on the composition of saturation extracts (see Figure 3.1). Inasmuch as most soil scientists in the early days following the 1952 reorganization had not worked much with laboratory data, and quite possibly as a holdover from the procedures of the Bureau of Soils research laboratories, the initial Soil Survey policy forbid distribution of characterization data without a rather detailed interpretation of the data by one of the scientists in the laboratories. Mimeographed copies of both the data sheets and their accompanying interpretation were widely distributed to the SCS technical centers and state offices, who were in turn responsible for sending copies to NCSS cooperators.

As many of the research scientists were rather reluctant writers, and the reproduction methods of the time demanded that all the data had to be assembled before the mimeograph masters could be cut, a long delay often occurred in getting the data out. The field staff strongly resented these delays; when the reports finally arrived, they had often forgotten the question the data were supposed to answer. Thermal reproduction techniques that became available in the early 1960s allowed the laboratory staff to send partial handwritten data sheets to those who needed them most, but these data sheets were fragile and hard to read. Not until electrostatic copy machines became generally available in the mid-1960s did the situation improve. At about the same time, new copy-machine technology, which reduced the size of typewritten documents, allowed the creation of data sheets with more spaces for data (see Figure 3.2).

In 1967, the Soil Survey Investigations Division established the SSIR series to publish data and pedon descriptions, usually organized by state. Report number one of the SSIR published the methods used in the laboratory, which were identified by a code that became standard on all data sheets (Soil Conservation Service 1972). Forty SSIR reports covering most states were published by 1980. A few reports contained technical papers, and in the early 1970s, the new data code of the first report was used to prepare the 130 data sets and descriptions included in *Soil Taxonomy* (Soil Survey Staff 1975).

SOIL SURVEY LABORATORY Riverside, California

Greenfield coarse
SOIL TYPE sandy loam LOCATION Riverside County, California

SOIL NOS. S57Calif-33-16   LAB. NOS. 57564 - 57570

DEPTH INCHES	HORIZON	VERY COARSE SAND 2-1	COARSE SAND 1-0.5	MEDIUM SAND 0.5-0.25	FINE SAND 0.25-0.10	VERY FINE SAND 0.10-0.05	SILT 0.05-0.002	CLAY < 0.002	0.2-0.02	0.02-0.002	1/	TEXTURAL CLASS
0-1	Ap1	21.1	22.8	12.2	16.7	6.3	11.6	9.3	21.6	4.7	12	cosl
1-8	Ap2	18.8	24.9	12.7	17.1	6.3	14.4	5.8	22.1	7.4	2	lcos
8-18	A11	23.3	23.0	11.8	15.9	5.9	14.3	5.8	20.3	8.0	11	lcos
18-32	A12	15.0	23.3	12.9	19.4	7.3	16.0	6.1	25.1	8.4	14	lcos
32-41	B21m	18.7	21.0	11.4	15.4	5.8	13.8	13.9	19.6	8.0	16	)cosl
41-47	B22m	19.0	26.3	12.3	14.1	4.4	10.5	13.4	14.8	7.0	15	cosl
47-64	C	16.9	30.0	13.5	15.3	4.6	7.7	12.0	15.2	4.4	17	cosl

pH			ORGANIC MATTER				Elec.				MOISTURE TENSIONS		
SATU-RATED PASTE 1:1	1:5	1:10	ORGANIC CARBON %	NITRO-GEN %	C/N	CO_3 meq/l	Conduc. mmhos/cm. at 25°C.	CaCO3 equiv-alent %	HCO3 meq/l	1/10 ATMOS. %	1/3 ATMOS. %	15 ATMOS. %
6.5	7.2		1.95	0.154	12.7	n.d.	2.3	0	n.d.	15.6	10.3	4.8
6.4	6.8		0.55	0.044	12.5	0.0	0.4	0	1.5	12.3	8.3	2.5
5.8	6.4		0.29	0.027	10.7	0.0	0.2	0	0.7	10.3	7.2	2.6
6.1	6.6		0.16	0.033		0.0	0.1	0	0.7	10.3	7.5	2.6
6.2	6.6		0.08			0.0	0.3	0	1.0			5.4
6.3	6.3		0.09			0.0	0.3	0	1.1			5.6
6.4	6.5		0.05			0.0	0.2	0	0.9			4.8

CATION EXCHANGE CAPACITY (-(Na)	EXTRACTABLE CATIONS milliequivalents per 100g. soil			Exch.	Exch.	BASE SAT. %	SATURATION EXTRACT SOLUBLE milliequivalents per liter					MOISTURE AT SATU-RATION %
	Ca	Mg	H	Na	K		Na	K	Ca	Mg	Cl	
9.7	7.6	1.8	2.4	0.3	0.6	81	2.6	12.4	7.0	3.4	3.0	36.1
5.2	3.1	1.0	2.0	0.2	0.4	70	0.8	0.8	1.4	1.1	1.8	21.9
5.4	2.5	0.7	2.0	0.5	0.3	67	0.8	0.1	0.8	0.7	0.7	19.3
4.6	2.6	1.0	1.4	0.4	0.2	75	0.7	0.1	0.6	0.4	0.7	18.3
7.5	4.8	2.1	1.8	0.4	0.2	81	1.4	0.1	0.9	0.9	1.0	21.1
7.9	5.0	3.0	1.6	0.4	0.2	84	1.5	0.1	1.0	0.6	0.7	22.7
7.1	4.3	2.7	0.7	0.6	0.1	92	1.1	0.1	0.7	0.4	1.0	24.7

1/ Percent of the whole soil

Figure 3.1 Data sheet for a saline soil from the late 1950s. (Courtesy of Klaus W. Flach)

SOIL FAMILY: Natric Palexeralf, fine, montmorillonitic, thermic.
SOIL Bonsall sandy loam   SOIL No. S64Calif-37-3   LOCATION San Diego County, California
SOIL SURVEY LABORATORY   Riverside, California   LAB. Nos. 6430 - 6439

Depth (in.)	Horizon	Sand (2-0.05)	Silt (0.05-0.002)	Clay (<0.002)	Very coarse (2-1)	Coarse (1-0.5)	Medium (0.5-0.25)	Fine (0.25-0.1)	Very fine (0.1-0.05)	Int. (0.05-0.02)	Int. II (0.02-0.002)	(2-0.1)	Clay <.002	Clay <.002	>2	2-19	19-76
0- 6	A1	68.0	23.3	8.7	11.0	15.0	9.8	19.4	12.8	13.3	10.0	36.2	55.2	8.4	8.1	4	4	0
6-10	A2	62.9	24.7	12.4	9.3	13.6	9.0	18.6	12.4	13.9	10.8	36.2	50.5	11.0	10.7	3	3	0
10-14	B1	42.4	18.7	38.9	6.0	8.3	5.9	13.0	9.2	10.4	8.3	27.0	33.2	40.5	38.9	4	4	0
14-27	B21t	40.6	21.1	38.3	2.8	6.1	5.1	15.0	11.6	11.9	9.2	32.5	29.0	36.3	35.7	2	2	0
27-38	B22tca	48.2	23.3	28.5	2.0	4.1	5.5	20.8	15.8	13.8	9.5	42.3	32.4	29.0	28.1	3	3	0
38-48	B31	64.3	17.2	18.5	1.6	7.4	10.3	28.7	16.3	9.8	7.4	42.4	48.0	18.0	17.9	1	1	0
48-60	B32	69.8	10.8	19.4	3.3	14.1	16.4	27.9	8.1	7.2	3.6	29.0	61.7	19.6	19.1	2	2	0
60-89	C1	52.2	26.5	21.3	5.5	8.9	8.6	18.5	10.7	16.7	9.8	37.9	41.5	21.6	20.2	7	7	0
89-110	IIC2	56.5	29.3	14.2	12.8	12.1	6.9	13.8	10.9	12.6	16.7	30.8	45.6	20.4	12.0	31	31	0
110-120+	IIC3	50.5	32.5	17.0	3.3	10.8	8.0	16.2	12.2	12.7	19.8	33.8	38.3	18.4	17.2	7	7	0

Depth (in.)	Organic carbon Pct.	Nitrogen Pct.	C/N	Ext. Iron as Fe Pct.	Carbonate as CaCO3 Pct.	1/3 bar g/cc	Oven dry g/cc	C' g/cc	1/3 bar Pct.	15 bar Pct.	Extensibility COLEP in./in.	Extensibility COLE in./in.	Saturated Paste	H2O 1:1	H2O 1:10
0- 6	0.70	0.054	13	0.7		1.51	1.54	0.98	10.6	4.2	0.006	0.006	6.5	6.6	
6-10	0.28	0.033	8	0.8		1.76	1.78	0.98	9.5	4.5	0.003	0.003	6.4	6.8	
10-14	0.47	0.047	10	1.6		1.60	1.85	0.98	20.1	13.9	0.049	0.048	6.5	6.8	
14-27	0.21			1.0		1.73	1.99	0.99	17.7	13.1	0.047	0.047	7.6	8.0	
27-38	0.11			0.9	3					12.6			7.8	8.1	
38-48	0.02			1.3		1.60	1.78	0.99	20.2	10.6	0.036	0.036	7.7	8.0	
48-60	0.02			1.2						9.4			7.6	8.0	
60-89	0.02			1.2		1.67	1.76	0.95	15.4	11.7	0.018	0.017	7.7	8.0	
89-110	<0.01			1.4						13.3			7.9	8.3	
110-120+	<0.01			1.3		1.59	1.69	0.96	20.8	15.0	0.021	0.020	7.7	8.1	

Depth (in.)	Ca	Mg	Na	K	Sum of bases	Ext. Acidity	Cation Exch. Capacity NaOAc Sum	Ca	Mg	Na	K	CO3	HCO3	Cl	SO4	Electrical conductivity mmho/cm
0- 6	4.3	1.7	0.3	0.2	6.5	1.8	8.7	8.3	2.7	2.2	1.6	0.2	-	4.8	0.6	0.57
6-10	4.7	2.4	0.6	0.1	7.8	1.8	10.2	9.6	0.7	1.0	2.4	0.1	-	1.8	0.6	0.34
10-14	11.2	8.8	2.1	0.2	22.3	3.1	25.6	25.4	0.9	0.6	5.0	0.1	-	1.6	3.0	0.66
14-27	13.8	8.2	3.5	0.2	25.7	0.6	24.7	26.3	2.1	2.9	15.3	0.2	-	2.2	14.3	2.16
27-38	20.2	10.1	6.7	0.2	37.2	0.2	26.8	37.4	5.4	6.4	33.8	0.2	-	2.2	38.5	4.80
38-48	10.2	7.5	6.3	0.1	24.1	0.6	23.0	24.7	5.5	6.6	39.8	0.1	-	1.5	46.4	5.65
48-60	8.5	6.2	5.1	0.1	19.9	0.8	20.1	20.7	3.5	3.5	30.2	0.1	-	1.0	33.9	4.07
60-89	10.6	7.4	5.9	0.1	24.0	0.8	24.6	24.8	2.1	2.9	27.5	0.1	-	0.9	29.2	3.56
89-110	18.1	8.6	7.1	0.1	33.7	1.3	36.7	35.0	1.0	0.8	13.2	0.1	-	2.6	11.7	1.79
110-120+	14.1	8.6	5.9	0.1	28.7	1.3	31.0	30.0	1.0	0.8	12.8	0.1	-	1.0	13.1	1.74

Depth (in.)	Water at Saturation Pct.	Exchangeable Na Pct.	SAR	Base Sat. NaOAc CEC Pct.	Base Sat. Sum Cations Pct.	Mont.	Verm.	Hydro- biotite Mica	Mica	Kaol.	Kaol. 7A3	Kaol. (DTA) in ground whole soil (DTA)	7A2
0- 6	25.8	2	1	78	73	-	t	xx	xxxx	xx		15	1
6-10	19.8	5	3	83	77	xx	t	xx	x	xx			
10-14	44.7	7	6	88	86	xx	x	t	t	xx			
14-27	50.8	11	10	98	101	xxx	t	t	t	xxx		26	14
27-38	49.6	19	14	99	133								
38-48	38.7	20	16	97	98	xxx	-	t	t	xx		20	10
48-60	33.9	20	16	96	94								
60-89	39.4	20	17	97	93							11	10
89-110	47.1	14	14	95	90								
110-120+	47.2	17	14	96	91							5	7

1/ From characterization sample;
   determined by shaking overnight
   in a sodium hexametaphosphate
   solution.
* Analysis of ground whole soil;
  results expressed on <2-mm basis
** Analysis of ground whole soil;
   results expressed on whole-soil
   basis.
- = looked for but not found
t = trace
x = small
xx = moderate
xxx = abundant
xxxx = dominant

Figure 3.2 Data sheet for a saline soil from the early 1970s. (Courtesy of Klaus W. Flach)

The ready availability of computers in the late 1960s further changed the laboratories' procedures. The Lincoln laboratory was first to contract with a commercial data processor to do the laborious PSDA calculations; shortly after the Riverside laboratory contracted for access to the computer facilities of the University of California at Riverside. The laboratory used a provision in a 50-year-old cooperative agreement between the Bureau of Soils and the University of California specifying "that the costs of livery be shared equally" to obtain in-house rates for computer use. Laboratory staff members Benny Brasher and Klaus Flach took a course in Fortran and eventually, after many trials and tribulations, automated all of the laboratory calculations, creating as a fringe benefit the first laboratory database. Procedures were cumbersome by today's standards, of course. Instrument readings were manually entered on data sheets, hand carried to the computer center, key punched on IBM cards, and processed in batch mode. Eventually, the huge savings made available by use of highly automated instruments and recording equipment played an important role in the decision to consolidate the laboratories in 1975.

Brasher reworked the Fortran programs from the Riverside laboratory for the new NSSL, creating an operable system within one year, much of which is still used today. For about 20 years, the NSSL system remained one of the most complex and effective produced within the SCS. Each laboratory section leader programmed microprocessors for the initial data handling of the section; the most notable of these was Fred Kaisaki's programming of an 8-bit machine to handle the bookkeeping and initial calculations for the entire chemistry section.

As the laboratories entered the 1990s, electronic communications began to replace hard-copy publications of data and descriptions, ending in the creation of a large electronic database of more than 15,000 pedons by mid-decade. By that time, a comprehensive database was needed for running computer models such as the Erosion Productivity Impact Calculator (EPIC) to meet certain requirements of the 1985 Farm Bill. Texas A&M University was contracted to enter essentially all pedon data assembled by the SCS since 1952 that was not already in an electronic database. Today, the pedon database contains approximately 20,000 pedons, and about half have complete characterization to depths of 1.5 to 2 meters. No longer needed to disseminate pedon data and descriptions, the SSIR series now primarily publishes methods and special studies.

The testing of the *7th Approximation* in the 1960s and the adoption of *Soil Taxonomy* in the 1970s required that all NCSS soil scientists develop competence in the use of laboratory data. Consequently, the three regional laboratories and later the NSSL developed one-week and two-week training sessions in which groups of soil scientists gained hands-on experience in the laboratory techniques and intensive training in the use of laboratory data. For several years as many as 10 percent of NCSS field soil scientists attended these sessions each year. The SCS also continued a policy of contracting with universities to conduct Soil Survey Institutes, intensive six-week refresher courses for soil scientists, which did much to increase the competence of soil scientists in the use of scientific information. Finally, in May 1995 Rebecca Burt of the NSSL developed a comprehensive *Soil Survey Laboratory Information Manual*, published as SSIR number 45 (Soil Survey Laboratory Staff 1995).

As the competence of the laboratory data users and the quantity of data increased, the necessity for and the possibility of sending interpretive reports with every release of data declined. Whereas researchers of the 1950s and 1960s spent much of their time writing interpretive reports, the focus has now shifted toward putting the data, and the tools for their use, into electronic media that can quickly convey the information to skilled practitioners.

THE ROAD AHEAD

Two major trends are now affecting the use of and demand for laboratory data. One relates to land-use issues, the other to the information age—what some call the democratization of science. More specifically, the two trends are 1) the increasing economic impact of land-use decisions made with the help of soil surveys, and 2) the growing availability of soil survey information, including laboratory data, to any scientist, business, or household with a computer and modem. These trends are increasing the demand for varied laboratory analyses, producing challenges to traditional sampling and characterization paradigms and changing the roles of researchers associated with the soil survey laboratory.

With growing attention to the environment, to long-term sustainability of quality resources, and to profit margins in a global economy, the demand for accurate and precise soil survey applications is leading to increased backstopping by laboratory data. These new applications are produced to meet the needs of a particular set of final customers, and thus impact people more directly than previous applications that paid more attention to taxonomy and soil genesis.

There is some danger, however, that traditional pedological and taxonomic skills may be lost because of the emphasis these new trends place on numbers and simulation models. Will the future see skills for manipulating data replace morphological descriptive capabilities, or will fuzzy logic and other new approaches incorporate key qualitative field observations into tools that create and quantify soil survey interpretations? Only time will tell, and the outcome will perhaps hinge upon decisions made close to the NSSC laboratory.

The increasing availability of information-age tools contributes to the democratization of science because it spreads the opportunities for understanding and testing of data and other tools of science. *Soil Taxonomy* democratized soil classification by enabling all competent soil scientists to make decisions that before had been reserved for the few at the top of the scientific or bureaucratic ladder. Electronic networks do the same for laboratory data and their uses. Entrepreneurial uses of soils information from many sources have created challenges to traditional methods of soil characterization and sampling.

What lies ahead as we enter the second century of the Soil Survey? What developments will the democratization of science and growing attention to choices in the use of the land bring? We believe that the beginning of the Soil Survey's second century is as pivotal to pedological science as was the beginning of the first. But we also believe that the Soil Survey will be able to address the issues of the new century only if the purveyors of traditional skills find their proper role in the entrepreneurial environment of the new century.

ACKNOWLEDGMENTS

Although we emphasize the contributions of the soil survey laboratories in this chapter, we acknowledge with appreciation the enormous amount of data, ideas, and other information supplied by colleagues at U.S. and international universities and by soil survey organizations in other countries, as well as the ideas, criteria, and testing by the SCS correlation staffs.

NOTES

1. The remainder of the article is based on the personal knowledge and experience of the authors, except where references are specifically cited.
2. Although we emphasize the contributions of the Soil Survey laboratories to the development of *Soil Taxonomy* in this paper, we acknowledge with appreciation the enormous amount of data, ideas, and other information supplied by colleagues at U.S. and international universities and by soil survey organizations in other countries, as well as the ideas, criteria, and testing by SCS correlation staffs.
3. In this section and throughout the chapter, methods are identified by terms that are commonly used in the soil science literature. Excellent general descriptions of the methods for field-oriented soil scientists are available in *Soil Survey Laboratory Information Manual* (Soil Survey Laboratory Staff 1995), and detailed descriptions for the laboratory-oriented scientist can be found in *Soil Survey Laboratory Methods Manual* (Soil Survey Laboratory Staff 1996).
4. Throughout this chapter, traditional units that are no longer accepted by scientific publications are followed by SI (Systeme International d'Unites) units (American Society of Agronomy 1998).

REFERENCES

American Society of Agronomy. 1998. *Publication Handbook and Style Manual*. Madison, Wis.: American Society of Agronomy.

Baver L. D., F. J. Alway, Richard Bradfield, W. H. Pierre, F. C. Shaw, R. S. Smith, and M. B. Sturgis. 1938. Report of committee on laboratory studies supplementary to soil survey. *Soil Science Society of America Proceedings* 3:351–354.

Bradfield, R. 1923. *The Chemical Nature of a Colloidal Clay*. Bulletin 60. Missouri Agricultural Experiment Station.

Bradfield, R. 1927. What information is necessary for a complete description of a soil? B. Laboratory aspects. *The American Soil Survey Association Bulletin* 8:104–111.

Brasher, B. P., D. P. Franzmeier, V. T. Volassis, and S. E. Davidson. 1966. Use of Saran resin to coat natural soil clods for bulk density and water retention measurements. *Soil Science* 101:108.

Brown, I. C., and J. Thorp. 1942. *Morphology and Composition of Some Soils of the Miami Family and the Miami Catena*. U.S. Department of Agriculture Technical Bulletin No. 834. U.S. Department of Agriculture, Washington, D.C.

Byers, H. G., M. S. Anderson, and R. Bradfield. 1938a. General chemistry of the soil. In *Soils and Men*, pp. 911–928. Yearbook of Agriculture 1938. Washington, D.C.: Government Printing Office.

Byers, H. G., C. E. Kellogg, M. S. Anderson, and J. Thorp. 1938b. Formation of soils. In *Soils and Men,* pp. 949–978. Yearbook of Agriculture 1938. Washington, D.C.: Government Printing Office.

Conner, S. D. 1923. Acidity tests as an aid in soil survey work: Laboratory viewpoint. *American Soil Survey Association Bulletin* 4:23–28.

Conroy, G. W. 1923. Report of the Committee on Correlative Laboratory Work. *American Soil Survey Association Bulletin* 1:177–110.

Grossman, R. B., B. R. Brasher, D. P. Franzmeier, and J. L. Walker. 1968. Linear extensibility as calculated from natural-clod bulk density measurements. *Soil Science Society of America Proceedings* 32:570–573.

Hendricks, S. B., and W. H. Fry. 1930. The results of x-ray and microscopical examination of soil colloids. *American Soil Survey Association Bulletin* 9:194–195.

Holmgren, G. S., R. L. Juve, and R. C. Geschwender. 1977. A mechanically controlled variable rate leaching device. *Soil Science Society of America Journal* 41:1207–1208.

Holmgren, G. S., and R. E. Nelson. 1977. A field procedure for base saturation using KCl-triethanolamine. pH 8.1. *Soil Science Society of American Journal* 41:824–827.

Jeffries, C. D. 1947. Some developments in the field of soil mineralogy. *Soil Science Society of America Proceedings* 11:16–18.

Kellogg, C. E. 1949. *The Soils That Support Us: An Introduction to the Study of Soils and Their Use by Men.* New York: Macmillan.

Kilmer, V. J., and L. T. Alexander. 1949. Methods of making mechanical analyses of soils. *Soil Science* 68:15–24.

King, F. J. 1905. *Investigations in soil management. II. Relations of crop yields to the amount of water-soluble plant-food materials recovered from soil.* U.S. Department of Agriculture Bureau of Soils Bulletin. U.S. Department of Agriculture, Washington, D.C.

Krusekopf, H. H. 1927. What information is necessary for a complete description of a soil? A. Field aspects. *American Soil Survey Association Bulletin* 8:89–104.

Lapham, M. H. 1949. *Crisscross Trails: Narrative of a Soil Surveyor.* Berkeley, Calif.: Willis E. Berg.

Lovejoy, P. S. 1923. The soil survey—Present and future. *American Soil Survey Association Bulletin* 4:53–59.

Marbut, C. F. 1935. *Soils of the United States.* Atlas of American Agriculture, Part III. U.S. Department of Agriculture, Washington, D.C.

Marshall, C. E. 1935. Mineralogical methods for the study of silts and clays. *Zeitschrift. Kristallographie* 90:8–34.

McCaughey, W. J., and W. H. Fry. 1913. *The Microscopic Determination of Soil-Forming Minerals.* U.S. Department of Agriculture Bulletin. U.S. Department of Agriculture, Washington, D.C.

McCool, M. M. 1921. What correlative laboratory work is it desirable to do? *American Soil Survey Association Bulletin* 1:32–37.

McCool, M. M. 1922. Laboratory methods of value to soil survey. *American Soil Survey Association Bulletin* 3:32–37.

Nikiforoff, C. C., and L. T. Alexander. 1942. The hardpan and the claypan in a San Joaquin soil. *Soil Science* 53:157–172.

Peech, M., L. T. Alexander, L. A. Dean, and R. F. Reed. 1947. *Methods of Soils Analysis for Soil Fertility Investigations.* U.S. Department of Agriculture Circular 757. U.S. Department of Agriculture, Washington, D.C.

Robinson, W. G., and R. S. Holmes. 1924. *The Chemical Composition of Soil Colloids.* U.S. Department of Agriculture Bureau of Soils, Bulletin 1311. U.S. Department of Agriculture, Washington, D.C.

Russel, E. J. 1911. The recent work of the American Soil Bureau. *Journal of Agricultural Sciences* 1:327–346.

Soil Conservation Service. 1972. *Soil Survey Laboratory Methods and Procedures for Collecting Soil Samples.* Soil Survey Investigation Report Number 1. Soil Conservation Service, U.S. Department of Agriculture, Washington, D.C.

Soil Survey Laboratory Staff. 1995. *Soil Survey Laboratory Information Manual.* Soil Survey Investigations Report Number 45, Ver. 1.0. Natural Resources Conservation Service, U.S. Department of Agriculture, Washington, D.C.

Soil Survey Laboratory Staff. 1996. *Soil Survey Laboratory Methods Manual.* Soil Survey Investigations Manual 42, Ver. 3.0. Natural Resources Conservation Service, U.S. Department of Agriculture, Washington, D.C.

Soil Survey Staff. 1960. *Soil Classification: A Comprehensive System, 7th Approximation.* Soil Conservation Service, U.S. Department of Agriculture, Washington, D.C.

Soil Survey Staff. 1975. *Soil Taxonomy: A Basic System of Soil Classification for Making and Interpreting Soil Surveys.* Soil Conservation Service Agriculture Handbook Number 436. U.S. Department of Agriculture, Washington, D.C.

U.S. Salinity Laboratory Staff. 1964 *Diagnosis and Improvement of Saline and Alkali Soils.* Agriculture Handbook Number 60. U.S. Department of Agriculture, Washington, D.C.

Weber, G. A. 1928. *The Bureau of Chemistry and Soils.* Baltimore, Md.: Johns Hopkins University Press.

Whitney, M., and F. K. Cameron. 1903. *The Chemistry of the Soil as Related to Crop Production.* U.S. Department of Agriculture Bureau of Soils Bulletin 22. U.S. Department of Agriculture, Washington, D.C.

4

MEMOIRS OF A SOIL CORRELATOR
Joe D. Nichols

I was a soil scientist for the Soil Conservation Service (SCS) for 37 years. For 20 of these years I had correlator in my job title, including 12 years as principal soil correlator for the southern states at the South Technical Service Center (TSC) in Fort Worth, Texas. In the broader definition though, I did correlation work almost every day of my career, since correlation was a function of soil surveys.

A SHORT HISTORY OF CORRELATION

Since my work as a soil scientist with the SCS did not begin until 1956, a brief description of developments in the practice of soil correlation and interpretation in the U.S. Soil Survey before that time will provide a starting point and some background for my own story.

Some form of work examination or inspection would have been necessary from the beginning of the soil survey program in 1899, but both official written procedures and inspection were scarce early in the survey (Lapham 1949). According to Dr. Roy Simonson (1987), the Bureau of Soils issued field books each year beginning in 1902, but the first field book to offer much in the way of instruction was the 319-page, 1906 version. For the first five years of the Soil Survey, each survey stood on its own, with no attempt to correlate soil series among them. The 1903 field book, which mentions the need to relate soil series across survey lines, was perhaps where some correlation began.

The first designation of the correlator position came when the position of inspector was established in 1909. Bureau of Soils Bulletin 96 (Marbut et al. 1913) established organized field inspection. Dr. Curtis Marbut served as chairman of this field inspection, and Dr. Hugh H.

Bennett, J. E. Lapham, and Macy H. Lapham served as inspectors for the South, North, and West divisions, respectively (Lapham 1949).

The next update in instructions came with the 1914 field book, which sufficed as official written instructions until the 1937 *Soil Survey Manual* (Kellogg 1937). Dr. Charles E. Kellogg, principal soil scientist of the Soil Survey Division in the U.S. Department of Agriculture's (USDA) Bureau of Chemistry and Soils, wrote the 136-page-long manual. Intended for the use of soil surveyors in the field, the manual gives good instructions on the making of soil surveys, the preparation of soil survey reports, and the classification and correlation of soils. A section toward the back of the book on inspection and correlation provided an example of the report of inspection.

Dr. Kellogg treated the subject of inspection and correlation separately, even though they were usually done concurrently. He listed items for the first inspection and the final inspection and included mention of other inspections. He then discussed correlation under a separate subtitle. His wording of its meaning and purpose is worth repeating:

> The correlation of soils refers to the procedure for the determination of the final names to be given the mapping units shown on the published maps. This is accomplished by the correlation committee, consisting of the district inspectors, one of whom was chairman, from a study of the type descriptions and samples submitted by the chief of party and the final correlation memorandum of the inspecting scientist in relation to standard samples and descriptions of established soil units. (Kellogg 1937)

The correlation committee leaders were called regional inspectors. A map in the manual showed the inspection districts. The 1937 manual also lists and explains soil complexes, intricate mixtures of two or more soils that could not be separated at the mapping scale. Some counties had used complexes in published surveys before this manual was published, while some counties' surveys lacked soil complexes even after the 1937 manual became available.

The 1951 *Soil Survey Manual* was published as Agricultural Handbook Number 18 (Soil Survey Staff 1951). Although the publication listed Soil Survey Staff as the author, Dr. Kellogg led this group and, I understood, was responsible for much of the new manual. The book built on the 1937 version by adding detail, such as pictures of soil structure with specific size limits for each type. The 1937 manual had listed types such as blocky or prismatic structure but without size limits. The 1951 manual introduced the title "soil correlator" to replace "soil inspector." A replacement for the 1951 manual was not begun for more than 20 years and was not published until 1993.

The SCS was formed in 1935 from the former Soil Erosion Service. Because their farm-planning conservation program required soil maps, they began to make their own, rather than use those of the Soil Survey Division of the Bureau of Chemistry and Soils; the Bureau did not have enough people to meet the SCS's needs and their map scale was too small. With two groups doing the same kind of work, you might expect that there would be trouble. In this case, you would be right, although the trouble apparently was not enough of a problem to cause the secretary of agriculture to reorganize the USDA's soil mapping program until 1952.

The SCS produced its own manual, a 20-page, pocket-sized booklet, *The Procedure for Making Soil Conservation Surveys*, by Glenn L. Fuller (1936), head of the Section of Conservation Surveys. The manual instructed mappers on procedures for soil surveys, including guidelines for mapping erosion, land use, and slope. The soils were to be identified following the system used by the Bureau of Chemistry and Soils, and I assume the soil scientists used the bureau manual.

The SCS had its own inspection staff of regional soil scientists to do quality control. Richard Marshall was the regional soil scientist for the Gulf Coast region, which included Oklahoma. The authors of the SCS manual note, however, that the interbureau Soil Correlation Committee, which consisted of representatives from the bureau and the SCS, would make the final correlation of soil types.

E. A. Norton, principal soil scientist in the Physical Surveys Division of the SCS, and his staff authored a new 40-page *Soil Conservation Survey Handbook* in 1939 that went into more detail about erosion, land use, and slope than the 1936 manual (Norton 1939). It also provided the soil map symbols and instructions on the collecting of soil samples and the writing of the soil survey report. The handbook also reiterated the information contained in Fuller's 1936 manual about classification and the Soil Correlation Committee's involvement in final correlations.

In his book, *Soil Conservation*, Dr. Hugh Hammond Bennett (1939) devoted an entire chapter to soil conservation surveys, providing mapping information and symbolization for land use, erosion, and slopes. Dr. Bennett sounded like the old correlator when he wrote, "In order to avoid confusion through multiplicity of detail, it is necessary to group the numerous variations entering into the physical complex. For instance, it is necessary to establish definite slope groups for particular kinds of soils, in order to express the correct meaning of the dominant slope factor in relation to the practical use of the land and to avoid every slight change in gradient." He did confuse the issue a bit when he wrote on the next page that the actual procedure is to superimpose three classification systems on the map—soil, slope, and erosion.

The bureau made their own surveys, mainly published at one inch per mile, although two inches and even more to the mile were used in later years. The SCS maps were generally four inches per mile. According to Dr. Simonson (1987), a 1942 order from the secretary of agriculture stated that the bureau would prepare all legends and conduct the fieldwork reviews. The rules were amended in 1945 to allow the SCS control of its simplified farm-planning soil surveys.

Soil survey production was not high during World War II. In some cases, use of the simplified soil mapping system met the need for conservation surveys in Oklahoma. The simplified soil surveys were apparently used more frequently in other states. After the war the long-standing friction between the SCS and the bureau over soil surveys reached a peak (Gardner 1998; Cline 1977). The secretary of agriculture made several attempts to correct the situation. Finally, on 14 October 1952, as part of the major reorganization of USDA's science staff into the Agricultural Research Service, Secretary of Agriculture Brannan transferred all personnel and funds of the Division of Soil Survey to the Soil Conservation Service and designated the SCS as the single agency for soil surveys in the USDA (Gardner 1998).

The combination of the surveys was not made without a great deal of cooperation on both sides. A little over one year later, the SCS regional offices were abolished, according to Dr. Gardner (1998), as part of an ongoing attempt by state interests to take over the popular SCS. The takeover failed, as had several others before, but the resulting shakeup benefited the Soil Survey, I believe, because while SCS headquarters officials concerned themselves with the "other" problem—the attempted state takeover—rank and file soil scientists had time to work out their differences. Dr. Charles Kellogg, of the old bureau, became head of the Soil Survey with the title of assistant administrator. In a featured speech at the 1957 National Work Planning Conference of the Nation Cooperative Soil Survey that focused largely on soil interpretations, Dr. Kellogg announced a goal of correlating 50 soil surveys in 1958.

HISTORY OF SOIL INTERPRETATIONS

Although a thorough history of soil interpretations would be more extensive than a history of soil correlation, the following brief synopsis follows the development of soil interpretations only as it relates to soil correlation and inspection. The first soil surveys began in 1899 and were intended for use by agricultural interests, with tobacco high on the list. M. H. King appeared as head of soil management in the 1901 report, *Field Operations of the Bureau of Soils* (Bureau of Soils 1901). I have no documentary evidence of correlation of soil management or interpretations, but interpretations in the soil survey reports increased with each period. They certainly would have received some scrutiny from supervisors and cooperators. In fact, an experiment station representative wrote the section dealing with soil management in many early reports.

The 1937 *Soil Survey Manual* (Kellogg 1937) included a good section on data gathering during the course of the survey. The Soil Survey party chief could receive assistance for this task from other soil scientists and the supervisor. The soil interpretations were not listed among the items for correlation nor were they mentioned in the suggested format for the inspection report. However, a statement in the manual noted that all parts of the maps, legends, and reports must be carefully checked for accuracy before being released for publication, indicating some sort of quality control of interpretations. The 1938 USDA Yearbook of Agriculture, *Soils and Men* (USDA 1938), contained a large amount of material on soil use, and it remained a reference for at least 20 years for those using or making soil surveys. Its sections on the formation and classification of soils also made it an important reference for anyone involved in the study of soils.

The 1951 *Soil Survey Manual* (Soil Survey Staff 1951) provided much more material on interpretation, including guidelines on data gathering and a form for crop yield data. An adjective rating was provided to use in inspection reports to indicate the reliability of yields and management classes. But inspection and correlation were still listed separately, and soil management was not included as part of the final correlation report. Still, a reader of the 1951 manual would certainly be impressed with the importance of soil interpretations to the whole soil survey. Dr. Kellogg established the position of soil correlator for interpretations in 1953. The title was changed to assistant principal soil correlator for interpretations at a later date.

BEGINNING AS A SOIL SCIENTIST

In February 1956, I reported for training to Party Chief Earl Nance at Buffalo, Oklahoma. I had a B.S. and an M.S. degree in soil science from Oklahoma A&M College (later Oklahoma State University). I had also been a student trainee-employee for two summers. I began mapping under his close supervision in a few days. I had an old pickup truck, a soils auger to bore holes, and a sharpshooter shovel to dig holes. We used aerial photographs for maps and sketched our soils area lines on the photograph. We had a soils color book to compare and name soil colors, a bottle of dilute acid to check for free lime or calcium carbonate, and a pH kit to determine the acidity when no lime was present. In the spring, I got a newer pickup with an electric power probe that pushed a 1-inch tube into the ground and removed a soil core, a big step forward.

The *Soil Survey Manual* (Soil Survey Staff 1951) was our bible for making soil surveys. It contained instructions on how to describe soils, horizon by horizon, in precise terms and order, and it explained map symbols, plotting soil boundaries and soil drainage, soil stoniness and other phase distinctions such as texture, slope, coarse fragments, and others. I still have that first manual. I carried it in the pickup for reference until it was so worn and soiled that I was given a new book. We also used some Soils Memoranda issued from the state and national level that included instructions for certain jobs.

Our job was to remap an older survey, bring it up-to-date, and complete the unmapped portion. The SCS had made a number of simplified soil conservation surveys during World War II, and perhaps earlier in some parts of the state. These simplified surveys were based on soil texture, soil permeability, and soil depth and also included slope and erosion. Their purpose had been for immediate use in the farm-level conservation plans. The Harper County survey on which I worked, however, was a soil series type-standard soil survey, as were most of the soil surveys in Oklahoma before World War II made soil scientists scarce. For this type of survey, we identified the soil series and phases, such as texture and slope, and related them to a soil mapping unit. The aerial limits of the mapping unit and its symbol would be placed on the aerial photograph.

The landscape, or the soil mapping unit, usually contains a dominant soil such as Carey and a small amount of other, usually similar soils; a soil scientist must be able to identify the dominant soil series on the landscape. The dominant soil and the similar soils, along with a few dissimilar soils, occur together in the soil area or soil landscape and are called the soil mapping unit. The soils that are not of the dominant series are called inclusions. In some cases, because of the scale of the maps and the intricate pattern of mapping units, two dominant soils might occur. These were named as complexes, such as Quinlan-Woodward complex, combining the names of both series.

We finished the survey in the late fall. During the time we made the survey, we had just one visit from the soil survey supervisor, Ray Marshall, who was also the acting state soil scientist. I could see the importance of the party chief position. They were well-trained, and operated much on their own with a little higher guidance.

MY FIRST CORRELATOR

When Earl Nance, our party chief, told me about an upcoming soil correlation field study, we rechecked our soil profile and map unit descriptions and revisited some of the typical soil series sites, as well as checked to make sure each map joined with the adjoining map. I could tell the week-long field study with the correlator was serious business. Edward H. Templin, senior soil correlator for Oklahoma and Kansas, showed up on Monday with Dr. Fenton Gray. I had taken introductory and advanced soil morphology and genesis under Dr. Gray; I did not realize at that time that he would be looking over my shoulder the rest of my career in his job as an Oklahoma State Experiment Station soil survey representative, which he held in conjunction with his job as soils professor at Oklahoma A&M.

I was learning that the soil survey was a cooperative venture with the Experiment Station and with several other agencies interested in soil surveys of certain areas. The senior soil correlator, Mr. Templin, had been with the soil survey at Texas A&M University in 1923 when the experiment stations furnished part of the team of field soil scientists. In his later position, his office was in the College Agronomy office at Oklahoma A&M, where I had seen him while a student. The SCS later moved him to the state office on the edge of the campus. I also remembered him from a field trip he had gone on with Dr. Gray's class to help us understand soils.

The field study week was cold, with some snow blowing early in the week. Edward Templin was thorough and knowledgeable. He would write the field correlation report in his office and then send it to the principal soil correlator's office for the Great Plains. There, Dr. Andrew Aandahl, perhaps aided by one of his assistants, would look over the report and issue the intermediate correlation for the county. The report then went to Dr. Roy Simonson, director of soil classification and correlation in Washington, D.C., who would issue the final report, from which the soil survey manuscript would eventually be written for publication.

The report included a listing of the soil map units and their symbols. In most cases, the soil map unit used the name of the soil series followed by the name of the dominant texture of the topsoil, called the type, which was then followed by the dominant soil gradient or slope phase and perhaps other descriptive terms—Carey silt loam, 1 to 3 percent slopes, for example. The field symbol was correlated to the publication symbol—Ca, in the case of Carey. The correlation report also listed the soil series with their classification—Carey is an example of a series name, which was in the Reddish Chestnut great soil group in 1957. Soil sample data from the SCS laboratory and the State Highway laboratory were also checked and correlated.

FROM PARTY CHIEF TO SOIL SPECIALIST

I moved to the SCS area office in Woodward, Oklahoma, in the fall of 1956. I worked on the general soil map section of the Harper County soil survey report and also made simplified soil surveys for four surrounding counties on request. In the summer of 1957, I moved to Roger Mills County, Oklahoma, where I joined Party Chief Dent Burgess on a survey already under way. Odos Henson, who worked closely with me to gain some training, joined us later. We completed that survey about two years later, in the fall of 1959. By then, Edward Templin had moved to the Great Plains principal soil correlator's office, and Henry Otsuki had replaced him as senior correlator for Oklahoma and Kansas.

As the Roger Mills County work finished, I moved to Osage County, the largest county in the state, as party chief. I was born and had started to school in Osage County, so this was almost like going home. We had barely settled in when Henry Otsuki, state soil scientist Louis Derr, and soil survey supervisor Fred Dries showed up for the initial field review to get me started, accompanied by state range conservationist Clarence Kingrey.

The Osage County survey included a large acreage of rangeland, which was to be surveyed at a different intensity, using a procedure that had been tried in Nebraska. The smallest delineation in the range area would be an animal grazing unit, or the amount of acres a cow would need to graze for a year, about 10 to 20 acres. I would then convert some of the range site surveys made by the range conservationist to soil surveys. The manual only allowed for detailed and reconnaissance surveys; the rangeland survey we were trying fell between the two. The procedure proved useful, and we would use more of these in later years.

Certain amounts and kinds of grasses, combinations that served as brief definitions of range sites, grew on certain kinds of soil, making it possible to coordinate soil and range sites. A range site converted fairly easily into a soil mapping unit with one soil name. Complexes were a problem, however, because the mapping unit was interpreted in the conversion instead of the soil. Although a site with deep soil might be best identified by one soil and a site with very shallow soil by another, if the two soils occurred within a complex mapping unit, an average depth for the complex, such as shallow, was used to describe the soils on both range sites. The capability class units for farmland were done the same way, so I accepted this system as being the right way.

After we got a good start in the survey, the regional range conservationist, Arnold Heerwagen, visited the county with a big retinue of people from the state office. Heerwagen had a new concept for converting the range surveys into soil surveys. He advocated coordinating the range site with the soil types and not the mapping unit. In other words, complex soil mapping units would also be complex range sites. The concept sounded logical to those concerned, and the change was made, although it required some measurements and effort to implement. In the process, we were moving toward correlation of interpretations. A similar conceptual change in the interpretation of capability classes for crop and pastureland came several years later. The Major Land Resource Areas (MLRA) eventually were used to coordinate the interpretations for capability class units. Soil phases were grouped according to capability class units, with treatment alternatives listed so conservationists could offer farmers alternatives for land use (USDA 1951).

Detailed studies of the kinds and amounts of grasses on range soils showed that breaks in vegetation did not exactly match breaks between soils, especially the 20-inch-to-hard-rock soil break. This did not come as a big surprise because soil mappers were quite aware that when interpretations were made across several boundaries, some of them would not fit exactly. Natural conditions change gradually, and when interpretations are made abruptly there is bound to be some mismatch. At the scale of soil mapping used for farm and ranch planning, the slight mismatches usually were not important, and the other benefits of mapping soils with this system outweighed the benefits of accuracy from a more flexible classification system.

In late February, Carl Fisher, a western Oklahoma party chief, and I attended a two-week soil correlation school at Lincoln, Nebraska, where Andrew Aandahl, principal soil correlator for the Great Plains states and his staff were stationed. About 30 people from the region attended the school. I would see many of them numerous times in future years and even work with some. Edward Templin, the first correlator I had ever seen and one of Dr. Aandahl's assistants, ran the school. Our training involved going through a lot of the *Soil Survey Manual* with Dr. Aandahl and his staff, as well as going into soil correlation in detail, including soil interpretations. Dr. Aandahl wrapped up our time at the school by telling us to keep like soils together; he was less concerned about the hair-splitting decisions we made than that the obvious be right.

Dr. Roy W. Simonson, director of soil classification and correlation, came to the school from the headquarters in Washington, D.C., He told us about the new soil classification system under development. This new system had been under some kind of development since the 1940s and had gone through six approximations. It had been published as *Soil Classification, a Comprehensive System, the 7th Approximation* (Soil Survey Staff 1960) a few months before our school was held. Dr. Simonson gave us the rundown on what the new system entailed and on what it would do to mapping, describing, and correlating soils. One of the big changes involved the concept of soil series. The old system used a modal, or central, concept of a series with somewhat fuzzy boundaries between the series. The new system used discrete boundaries between categories, with soil series as the lowest category level.

We spent a great deal of time with it in our future years with the soil survey; it came to be known to us as "The Seventh Approximation," "The Seventh," or just "the brown book."

Because we had received Dr. Simonson's training, Carl Fisher and I were privileged to attend and help when Louis Derr and his staff made the first classification of the Oklahoma soil using "The Seventh." (Classification of soils had to be reviewed each time a taxonomy change was made.)

I began work in the southeastern part of Osage County, where the city of Tulsa was spilling over from its home county. The area was mainly in general farming. As I mapped on the aerial photographs, I tested the mapping units and described the typical soils in the field. Standard procedure required 10 soil descriptions to characterize each unit by the end of the survey. I also noted the other soils or inclusions occurring in small amounts within the mapping units. My typical method was to go to a farm and talk to the farmer about what I was doing, then take my pickup truck into the pasture areas and, where crops were not growing, make soil borings with the hydraulic soil probe. Where I could not drive, I walked with a soil auger that allowed me to bore into the soil and check the soil profile. I tried to describe at least one soil per day by digging a small pit with my sharpshooter shovel then continuing to bedrock or to about 70 inches with the hand auger.

People frequently asked me how often I dug holes. Soil scientists did not dig holes to discover what soil was there but to confirm what they predicted from experience. My judgment was to dig enough holes to confirm my predictions of the soil in the area but not so many that I took too much time and did not meet expected production. I dug at least one hole and usually several in each soil area. Our instructions were to see each mapping unit throughout its extent. That did not mean we had to be on top of every part of the line, but it did mean that we had to get closer than a distant hill to view it. To get an idea of the composition of the unit, especially complex units, we ran soil transects, a series of usually 10 holes equally spaced across the area. This was a case when the soil coring machine on the pickup really saved time and strain on our backs.

When I found an area of soils not on my legend of previously

Joe D. Nichols as a party chief, with pickup and soil-coring machine. (Courtesy of Joe D. Nichols)

mapped units, I described it well, assigned a symbol, and test mapped the unit until about 500 acres were located. When I found a good typical location to represent the soil series, I made a careful soil profile description from a small pit and saved a sample of each horizon in a small paper bag designed for that purpose. I then sent the new description to the state soil scientist for approval. He either wrote a letter temporarily approving the unit until the next progress review or sent the soil specialist out to our work site for a day or two to look over the new unit and our other work.

I talked to farmers, gathering information from them about the kinds of crops they planted and their yields. I would record this information in my soil handbook, along with the level of management the farmer used. We also recorded roads, schools, churches, farm ponds, and other landmarks to assist people in locating their property on the maps. All of the information about a soil mapping unit went into the soil handbook. The soil handbooks (not always a "book"—some party chiefs used a file box and file cards) were the record of our work in the county. The soil manuscript would be written from this data after it was adjusted by the final correlation.

In the office, I sketched soil lines around each soil area on the aerial photograph, first with a pencil, using a scale of 1:20,000, or 3.168 inches per mile. The smallest somewhat circular area matched the size of the eraser on the end of a regular wooden pencil and encompassed about 3.5 to 4.0 acres. This met the needs of the local conservationist, as well as many other users. Smaller but very important features for farm planning, such as rock outcrops in deep soil, could be denoted with a spot symbol. Later when the map had been completed and joined with adjoining maps, the delineation lines would be inked and my name would go on the map.

When enough data was available from our soil mapping and complementary information gathering, we wrote the technical soil description and the description of the mapping unit. Along with the record of our soil descriptions, soil samples, and data such as crop yields and management behavior, these descriptions would be the basis for correlation and would be used to write the final soil survey manuscript.

I tried to work in the field whenever the weather permitted and spent rainy days in the office working on my soil handbook and inking maps. Sometimes the weather would be good, but the soils were too wet to drive on in this part of the county; I misjudged the wetness of one field on 20 February 1962 and found myself stuck in the mud as John Glenn orbited the earth. There was an area of sandy surface soils with scrub-oak vegetation in the southwest part of the county that was good for mapping in wet periods.

I also made maps of individual farms or ranches when the conservationists needed them for current planning. In one such case, the range conservationist was working on a plan for a 50,000-acre ranch in the northeastern part of the county. The ranch had some bison and some longhorn cattle, as well as modern breeds of cattle. My pickup got stuck one late afternoon in the fall, and when I finally determined I could not get out, I walked the fence line with the bison on one side and the longhorns on the other. To my surprise, they didn't bother me or even act like they saw me. When I reached the ranch house, I called the range conservationist, who came out after dark to get me. I left the pickup stuck until the following week, when the Bureau of Indian Affairs soil scientist assigned to this area took me back and helped me pull it out. As soon as we had it out, we got right back to work on the survey.

Getting a pickup stuck wasn't too unusual. The SCS in Oklahoma did not buy four-wheel drive vehicles in those days—they were too expensive to buy and to maintain. But if you didn't drive off-road in difficult places, you had to walk, which slowed the mapping, so most of us just took the risk. When I did get stuck, I could usually get out by digging, letting part of the air out of the tires, using a heavy-duty jack or some boards for traction, or some combination of these methods. If I couldn't get the truck out, most of the time someone came to get me, although once I just hitchhiked a ride to town. The traveling state mechanics often complained that the soil scientists' pickups looked like junk vehicles, what with the boards, etc. that we carried to deal with the inevitable. We, of course, preferred to think they looked like safari vehicles.

After three and one-half years I had the legend and soil handbook in good shape. I was selected to replace John Allen, the soil specialist from western Oklahoma, who was going to Kansas as the state correlator. I was moved to the area office at Pauls Valley instead of Clinton, where Allen had been located, with responsibility for one-third of the state, divided into three administrative areas.

I moved my family to Pauls Valley in July 1963, our fifth location in eight and one-half years. As a Soil Specialist, I assisted 12 soil survey part chiefs and reported directly to Louis Derr, the state soil scientist. I was responsible for approving new mapping units and for making both progress field reviews and short trips to assist with problems as they arose. Seven of the locations were too far to drive to and come back in a day, so the word "travel" entered our vocabulary. The job involved some unofficial correlation work—as I've already mentioned, all soil scientists were involved in correlation—but Henry Otsuki, the state correlator, would make the comprehensive and final reviews, as well as any initial reviews we needed. The June 1964 supplement of the 7th Approximation gave us more information for our mapping work, and we were adapting soil series to this system with new descriptions, as fast as possible.

In the winter of 1964, I went to a special school established at Cornell University by Dr. Charles Kellogg and his staff to bring selected soil scientists up-to-date on recent advances in soil science. To this "Okie," Ithaca, New York was nearly in the Ice Age! I arrived at Tompkins County airport in the snow and left a month later in the snow.

The Cornell University Agronomy staff kept us extremely busy. There were about 30 soil scientists from around the United States, some of whom had been at the correlation school I had attended in Lincoln four years earlier. The Agronomy Department chairman, Dr. Marlin Cline, headed the school. Cline, who had been an adjunct correlator with the soil survey, had much to do with the survey's development; he had worked on both soil classification and soil interpretations. His teaching would be extremely important to me in the future.

ACHIEVING THE TITLE OF SOIL CORRELATOR

After three and one-half years, Louis Derr, Oklahoma State soil scientist, moved to the headquarters in Washington, D.C., to become assistant operations director for the soil survey. Henry Otsuki was selected to replace him, and I was selected to replace Henry as state correlator. After years of unofficial correlation experience, I now had correlator in my title. We moved once again in the fall of 1965, this time to Stillwater, Oklahoma, where the state office was located. Because my wife and I had attended college there, Stillwater was also like going home.

In my new job, I found that one thing that I would never have enough of was time. My responsibilities included initial reviews, comprehensive progress reviews, and final reviews, all in the field. Following the field reviews, I had to write each review in the proper format and provide them to the appropriate people. The initial reviews and comprehensive progress reviews went to the field soil scientists and their supervisor, the area conservationist. The final reviews and field correlations had to be signed by the state soil scientist and then sent through the state conservationist to the principal soil correlator's office. As a state soil correlator, I worked as a state specialist alongside engineers of various sorts, agronomists, range conservationists, woodland conservationists, biologists, and other specialists, who all had their parts in the interpretation of soil surveys.

During my tenure in Stillwater, the new soil classification system, based on the 7th Approximation, was being phased in and all of the official soil series descriptions had to be rewritten. This work occupied a lot of my time, but it was an important investment. Correlations cannot be made without good soil descriptions and good mapping unit descriptions. The job was made more difficult because some of the sites where earlier official series descriptions had been made had not been seen in years; a few had been absorbed by urban areas or had become inaccessible for some reason. When we revised a series, soil scientists using those series in state and in adjoining states had an opportunity to review them before they became official.

The SCS reorganized in 1964. I discovered over the years that reorganizations were a common occurrence for administrative or "boss-type" people, and they tended to affect all of us. The SCS had not had regional offices since 1953. Except for engineering units, which had been left at the old regional office locations, they tried to get along with only a few specialists beyond the state office level, whom they assigned to a number of scattered locations. Dr. Kellogg had managed to keep the principal soil correlators during this period of deemphasis of technical services. In 1964, however, these specialists were brought together and new specialists were selected to staff four new TSCs. These centers evenly divided the workload of technical assistance to the states. Before the reorganization, there had been five principal correlator offices, but these were realigned and combined with the four new TSCs. What I call TSCs had at least four names over the next 20 years, but to simplify things I will just use the acronym that we used for the longest time.

The new principal correlator's office at Fort Worth, Texas, served the southern states: Oklahoma, Texas, Arkansas, Louisiana, Mississippi, Alabama, Tennessee, Georgia, Florida, and North and South Carolina, with Kentucky then Puerto Rico added later. The western office was located at Portland, Oregon, the midwestern office at Lincoln, Nebraska, and the northeastern office at Upper Darby, Pennsylvania. Edward Templin and Dr. James DeMent moved from Lincoln to the Fort Worth office. Also in the Fort Worth office were William Bender, assistant principal correlator for interpretations, and Keith Young, the manuscript soil scientist.

One of the my first assignments as state soil correlator was to travel to Fort Worth with Henry Otsuki and Fenton Gray to review the field correlation of Sequoyah County at the principal correlator's office. Edward Templin, the Great Plains principal correlator, did the reviewing, and Dr. Lindo J. Bartelli, the south principal soil correlator, sat in part time.

Edward Templin, like many of the correlators and state soil scientists, had a large World War II military survival folding knife, which he used to dig in the soil while on field reviews. These knives were purchased or were obtained from the SCS laboratory. They were cheap in the years following the war, but by the 1960s they were no longer available in army surplus stores. I mentioned to Louis Derr, when he was on a visit to our office, that I coveted one of those knives. In a few weeks, one arrived in the mail from Washington, still covered in its original cosmoline, the army protective grease. I used that knife on field and sampling trips until I retired, and brought it out of retirement several times for a volunteer sample trip and later for some organization field trips that I still like to attend. I recently gave the knife to the National Soil Survey Center at Lincoln, Nebraska, for the new soil museum.

The principal soil correlator's review of the field correlation involved checking the material produced by the survey party the way it had been checked at the field correlation. The soil characteristics in the descriptions were compared with defined and named soils, and the map units and available interpretations were checked. We had small samples of each soil horizon, and some of these were checked against the soil color book and perhaps against vial samples from the official type locations that were kept in a large file cabinet.

I made one more trip to Fort Worth with Henry Otsuki before Christmas to attend the South State Soil Scientists Meeting. This was an interesting gathering of the principal soil correlator staff, several or all of the staff of the state soil scientists, some of the Washington staff, and few members of other principal soil correlator staffs. The meeting's purpose was coordination. With *Soil Taxonomy* still under development, it was heavy on soil classification. This was a chance to meet our counterparts from other states. We carried what soil laboratory data we had in briefcases to share with others and to test new ideas. The conversations carried over after work into happy hour or whatever, and we solved some difficult problems during these times. We often went to a special place to eat dinner, and I remember in particular a favorite place that flew in fresh seafood and had a specialty of all the boiled, peel-it-yourself shrimp you could eat.

MY FIRST FULL YEAR AS A CORRELATOR

In January 1966, we started the correlation on Cherokee County, where Peter Warth was the party chief. Henry Otsuki had already done some work on this county. He, soil specialist Fred Dries, and I spent a week there for a field study. Later that month, following the Cherokee County work, I went to Caddo County, in the southwestern part of the state, to begin that field correlation.

Caddo County was in my old area, and I had worked with party chief Harold Moffatt before. Caddo was Harold's first completed county, and I was a fairly new state correlator, so we would learn together. I had an advantage in being a correlator at a time when a new soil classification system was still under development; I saw the first changes and had the opportunity to test them. For the field scientist, however, this was a rather tough time. They were working with a changing system. They could not afford to predict too much. Their job required them to describe and map the soil with the instructions on hand. Their map units had to be designed to be mappable and to allow the needed interpretations.

My work in Caddo was followed by a correlation in Delaware County in March. Like Moffatt, this was also party chief Everett Cole's first completed county. Delaware County was partly in the Cherokee Prairies MLRA, but most of it was in the Ozark Highlands MLRA. The MLRAs were designated divisions on a national small-scale map that included soils and land use. The divisions were based on variations in soil characteristics and land-use patterns. USDA agencies used these maps for inventories, reports, conservation work, and predictions. The more than 150 MLRAs were combined into 20 Land Resource Regions (LRAs) for the continental United States (USDA 1951).

I had mapped soils in the Cherokee Prairies, but the Ozarks were new to me. Although the area was interesting and the forested hills certainly scenic, I began to understand why both of my grandfathers left these cherty and partly stony, often steep, soils to farm in the center of the state. My father had told us that his family had never sharpened a hoe while farming in the Ozark region because the chert gravel dulled it so fast. I was sure this was exaggerated somewhat.

The acceleration of the soil survey and the fact that few surveys had been completed in the 1940s and early 1950s meant not only revising existing series but establishing new series to fill gaps where there were no surveys and where new series were needed to fit the more precise *Soil Taxonomy*. When we mapped a sizable acreage of a new soil, we wrote the proposal and sent the soil pedon description with 10 supporting pedons to the principal correlator's office. We also circulated copies to other soil scientists in Oklahoma and in adjoining states. When the principal soil correlator's office approved the series, they sent a copy to the director of soil classification at SCS headquarters. The principal soil correlator had been given the authority for final correlation of surveys, but the director retained final series approval.

In Delaware County, the party chief had found something quite odd to us. There appeared to be a developed (argillic) horizon on the bottomlands of larger creeks. We looked until we were sure, then the Arkansas state correlator, Oliver "Ben" Carter, came to check the "joins" where Delaware County bordered that state. It took awhile, but we finally signed off on the developed soil in the bottomland. The soil occurred on floodplains, but because the floodwater had only a small amount of sediment, new soil material was not being added. As a result, the soils on the stream bottoms were stable. The incident provided us a good lesson—in science there are often exceptions.

Back in Stillwater, I prepared and taught a session in Dr. Fenton Gray's soil morphology class, which gave me the chance to put ideas down on paper and explain them to students. Teaching seemed easy until I tried it!

During this interlude in the office, there were more soil series descriptions to write and review. I also prepared for the following week's field trip. At about this time, I was informed that an edict had come down from Washington, D.C. Henceforth, the term party leader would be used instead of party chief—another old bureau term was lost. I noticed later that they changed the title of the SCS "boss" from administrator to chief—probably a coincidence.

I traveled to the south central part of the state to begin the final field review of Pontotoc County. The county included the Rolling Red Prairies MLRA on the west side and the Cherokee Prairies MLRA on the east side, as well as some of the Cross Timbers MLRA that had scrub oak and grass vegetation. The southern part of the county included some of the Grand Prairie MLRA, denoting very clayey soils that moved around so much with shrinking and swelling that the developed B, or argillic, horizons did not form. In the new system, they were called Vertisols. Most of the soils in the county, however, had mollic epipedons, which were soils of dark color, with high levels of organic matter and adequate levels of base elements. Obviously, the county was not simple, and fortunately, party leader Vinson Bogard was experienced.

In a week, I was back at the office and found a new version of the key for *Soil Classification*. The changes were not big, mainly at the subgroup level, which is just above the bottom level. I checked out my notes on Pontotoc County and prepared to go back to Caddo County to continue work on the field review there, checking the highway engineering samples against the sites and descriptions and working on the legends.

On my next work trip, I went to the principal soil correlator's office in Fort Worth to go over our classification of Oklahoma soils with principal soil correlator Dr. Lindo J. Bartelli. While there, I also reviewed a soil manuscript from Georgia to help out. Reviewing the manuscript helped me by giving me an opportunity to ask questions and learn more about soil manuscripts.

Travel was pretty much a constant in my work as a state soil correlator. In the next few weeks, I would go back to Delaware County to work on the final field review there, then on to a progress field review of Grady County with Earl Nance and Armer Fielder. In between, I completed a paper that I was to give on the upcoming field tour in conjunction with the Soil Science Society of America and American Society of Agronomy Meetings. I took one more trip before the meetings to work on the correlation of Jefferson County.

The Soil Science Society of America and the American Society of Agronomy Meetings took place in Stillwater the following week. These meetings were big for a small city like Stillwater, and Oklahoma State University opened a new dormitory to help house the group. The soils tour was a success, though I did not get to ride on one of the tour buses. I drove directly to the Summit soil site in Osage County to dress up the pit and prepare for the group of soil scientists from around the United States and a few other countries who were coming to hear my talk. They stood around the pit while I gave my paper on the data and description of the soil. They then filed down the steps, which were cut into the soil, looked at the pit, and exited on steps cut up the other side. Field soil scientists rarely got to examine soils in large, handy pits such as those prepared for field tours. There were many influential soil scientists in attendance who asked some questions and made a few comments. After my paper, the group was treated to a traditional outdoor barbecue lunch at the headquarters of the large ranch where the pit had been dug, whose owner happened also to be the chairman of the board of an oil company.

Later at the meetings, as I was walking down the hall in the student union building, I spotted Dr. Charles Kellogg, head of the Soil Survey, walking by himself. I overcame a little apprehension about speaking to such an important person and chatted with him for a while. He was very cordial, and I believe that much of what made the Soil Survey work was the result of his diligent, somewhat patient, and gentlemanly attitude.

Back at the office, I worked on the correlation of Cherokee and Delaware Counties. We were combining the counties for correlation and for publication. Combining the soil surveys, with some overlap of legends, would be done under the direction of party leader Everett Cole and would save money and time in publication.

Following another trip to Caddo County, I went to Fort Worth for management training, which I did periodically over the years, each time the management people came out with a new system. As I remember, this particular system was management by objectives. I suppose we soil scientists have no right to complain, since we ourselves were revising the soil classification system. But it certainly seemed management systems had a shorter life than soil classification systems.

In those days, the principal correlator's office was understaffed, and they were utilizing state staff people for some jobs, such as my earlier work on manuscript reviews. David Slusher, state soil scientist of Louisiana, came to Pontotoc County for the principal soil correlator's office and went over the pale great group soils with us. These were deeply weathered soils that show characteristics of being on very old, stable landscapes. The pale great group soils were a new concept for classifying soils that had escaped glaciation, being either eroded away or receiving sediment. We succeeded in showing Dave that our soils matched the concept.

Still in 1966, my first year as a state soil correlator, I traveled to Texas in October for a soil sampling trip with Allen Newman of the Texas state soil staff to determine base saturation of soils on the old coastal plain of northeast Texas. Dr. Reuben Nelson of the Riverside SCS laboratory had a kit that he used to run base saturation in the motel rooms, which allowed us to use a transect to show the gradient of change instead of just piecemeal sampling. The test worked fine.

For the next field trip, Henry Otsuki and I picked up R. C. Carter, assistant principal soil correlator, at the Tulsa, Oklahoma, airport and went to Tahlequah, Oklahoma, for the review of the field correlation of Delaware and Cherokee Counties. Later in the month, Henry and I went to Fort Worth for the regional state soil scientists meeting, my second since becoming a state soil correlator. I took one more trip before the end of the year, to sample one soil and look for a good type location for another. After reviewing the manuscript for the Pittsburg

County survey, I took leave for the rest of the year—a full two weeks.

My first full year as a correlator had gone well. I finished most of the work on four field correlations and part of the work on three others. Classification development and changes and rewriting soil series descriptions took much of the rest of the time. A few years before, Dr. Kellogg had been hoping for 50 correlations across the whole country; we had completed almost four for Oklahoma alone. We had 34 people in the field making soil surveys, up from 17 ten years earlier, when most of the work had been request work for conservation use.

HOW DOES A CORRELATOR CORRELATE?

A correlation begins before the trip, with a review of the soil handbook in the office. Once we arrived in the county to be reviewed, we would go to the site, or location, for each soil series mapped by the soil survey party; there were usually about 30 to 40 of these. Most series occurred over a dozen or so counties, but some occurred in only a few counties, and some stretched across several states. In some cases, the county site of a series would be the same site used for the official series description. When mapped at a particular site, the series was given a phase name, which usually included slope and might also include other important features such as flooding or erosion. The primary soil series plus the inclusions of other soils in the delineation comprised the mapping unit. As part of the correlation, we would also check the soil description in the other phases or mapping units of the same series. There were usually about 80 mapping units in a county. The description determined through a correlation was the soil description for the county that went into the soil survey report.

We carried a box containing the soil maps sketched on aerial photos as we drove from one site to another, and I kept the aerial photograph oriented with the landscape as we traveled. If an area looked different from what we expected based on the map, we would stop and look. We made some random stops as well. At a stop, we would usually take a 2-inch diameter core with the soil-coring machine to a depth greater than 60 inches or to parent material. We also dug some holes with the sharpshooter shovel. We moistened the soil to feel the texture—the amount of sand, silt, and clay—and matched the color with the standard color book. We checked the soil structure—the arrangement of soil particles into blocks, granules, or prisms—and measured the size and strength. Soil consistence, or feel, was supposed to be measured at different moisture levels—moist, wet, and dry—but since we had only a single visit, this check required some experience on our part. We also observed, or measured, the soil horizons and how they graded to other horizons and other items such as gravel content and color variations. Those interested in knowing more specifically and completely what we measured, observed, estimated, or inferred are invited to examine the *Soil Survey Manual* (1993 or earlier), as the list is long and complicated. We carried the official series descriptions in a book, and after making all our observations and measurements, we compared our county site to the official descriptions. Whether we did this at the site, back in the vehicle, or back at the office depended somewhat on the weather.

The party chiefs or leaders also took correlation samples, which we compared to the description to see if they matched. We knew that when the principal correlator's personnel did the review of field correlation they would check at least some of them. We also checked the field notes for the standard 10 descriptions for each mapping unit. The parties also sampled 10 pedons to send to the State Highway Department laboratory for engineering analysis. These analyses would also go into the survey publication and so needed to be checked. In all, a county field correlation would take two to three weeks in the field, although not all at one time.

NEW CONCEPTS AND TERMINOLOGY

It may seem that I sneaked in many new concepts one at a time in my narrative, but that is the way we received them or had them explained to us as the new taxonomy replaced the old. For example, old soil descriptions were called soil profile descriptions, in reference to the face of a pit in the soil (the soil profile) from which the descriptions were taken. We realized that there was thickness as well as face dimensions when we took samples or described structure. The new taxonomy, however, changed the soil concept from a "profile" to a three-dimensional body that was large enough to encompass the soil horizon variations. Called pedons, these units were usually about one meter in each dimension, but could be up to 10 meters square and as deep as necessary. The soil mapping unit contained many of these pedons, called polypedons when occurring alongside other pedons, as well as areas of other included soil; few or no mapping units are pure to a soil classifier. A good description of the soil we classify and of pedons can be found in

Soil Taxonomy (Soil Survey Staff 1975) and in a paper by Johnson (1963).

Another concept that changed with the new taxonomy was the inclusion of soil moisture and temperature as an integral part of classification. In the past, these characteristics were mostly handled by the concept of soil zonality—certain soils occurred in certain zones. In Roger Mills County, Oklahoma, for example, the main vegetation was mid-grass, with some short-grass plains. On the west side of the county, however, was an area of soils with tall grasses and tiny oak trees from 2 to 4 feet tall. The soils had thick sand topsoils with a leached layer and reddish, sandy loam or sandy clay loam, slightly acid subsoils. These soils looked like the Red and Yellow Podzolic soils of the eastern part of Oklahoma and the southeastern part of the United States. Edward Templin correlated these soils as Reddish Brown, since they occurred in the Reddish Brown zone, but said they had some of the characteristics of the eastern soils. Under the new classification system, these soils could have been classified simply according to their own characteristics, without restrictions based on the zone where they were found. This change, eliminating zonality, would concern some of the old-timers for years.

TWO MORE YEARS AS OKLAHOMA STATE CORRELATOR

I remained the state soil correlator for Oklahoma for another two years. The first of those years, 1967, started out busy. I spent some time in February with Dr. John Stone, soil physics professor at Oklahoma State University, working with some neutron soil moisture data he and others had gathered over several years. We plotted the data over a decreasing moisture gradient, from central to western Oklahoma, and calculated it into days-dry in certain parts for the soil moisture definition in the *Soil Taxonomy*. These calculations were what we needed to define Aridic subgroups (drier intergrades) for the ustolls and ustalfs in the Oklahoma and Texas panhandles. Dr. Bartelli and the Texas soil scientists accepted the definition, and the addition eventually received approval for general use; most people concerned agreed the refinement was needed.

The following month, I found myself occupied with a very different kind of job. I went to Kiowa County to look at a site where the Museum of the Great Plains was excavating a mammoth, which is a kind of prehistoric elephant. I described the soils at the site and took some samples to send to the laboratory at Lincoln. I went back one weekend later in the year for a tour and had an opportunity to meet with the other researchers and discuss our findings. An issue of the museum's journal was devoted to our work. The whole experience was a real treat, especially because I had an interest in archaeology and geology.

In April, I began the field correlation of McCurtain County, the second largest county in the state. The county has the highest rainfall in the state and large areas of commercial forest. The county is very scenic, and I think of it as the pine forest area. We also completed several of the correlations that were begun the previous year with reviews of the field correlation by members of the principal soil correlator's staff either in the field or at the principal soil correlator's office.

Later in April, I traveled to Oklahoma City for the National Land Judging contest, and in September I flew to North Carolina to attend a tour of the coastal plains soils and geology research project, one of four such projects in the country. The SCS researchers Dr. Raymond Daniels and Dr. Erling Gamble hosted the tour. These projects were very important in developing principles for identifying soils on landscapes and for soil classification.

In October, Dr. Reuben Nelson of the Riverside laboratory was in southeastern Oklahoma to run the soil analyses for another sample trip. He had gone one step further than the previous year with his on-site testing capabilities and could now run the analyses out of the back of a station wagon. This trip spanned three counties, and we worked from early to late. We dug deep pits with a backhoe on soils that the local party leaders, state soil specialist, and I had located several weeks before. Soil scientists from the Texas, Arkansas, and Louisiana state staffs attended. Robert Reasoner, the party leader from McCurtain County, did a good job with logistics. The data from this trip and the previous year's Texas trip gave us the kind of information we needed to separate the Alfisol order from the more leached Ultisol order in these areas.

During the fall, we made a special trip to Tillman County in southwestern Oklahoma to look at the Paleustoll soils, which were old and deeply weathered. Our task was to characterize these soils and see how extensive they were in the county. On our way to the site, Henry Otsuki and I picked up Clifford Rhodes, who had been the party leader for most of the survey but had switched to a conservationist job before it was completed. Dr. Lyle Alexander, chief of Soil Survey Investigations for the SCS, and Dr. Robert Grossman, head of the Lincoln soil survey laboratory, attended as well. Texas state soil scientist Gordon McKee and Dr. James DeMent from the Fort Worth office were also present.

In December, we picked up Dr. Ray Daniels at the Tulsa Airport; he came on a visit from the soil geomorphology project in North Carolina to look at some prairie Mollisols that seemed to have a water table. He helped us set up some water-table measuring wells for party leader Claude Newland to test weekly. The classification precision of the new *Soil Taxonomy* took more knowledge than the old system, but it allowed for more precise interpretations, or use management, which was the main purpose of the survey.

I traveled to Clemson, South Carolina, in July of 1968 for my first Southern Regional Work Planning Conference. These conferences were important meetings. We invited staff of agencies that cooperated with the soil survey, such as the U.S. Forest Service soil scientists and experiment station soil scientists, who taught soil classification at the land-grant colleges. Committees at these conferences often worked out new ideas, which were then sent to the National Work Planning Conference for further refinement or blessing. Much of what would become new policy and procedure began in these conferences.

Later that summer I attended radiological monitor instructors training at the University of Oklahoma campus in Norman, then was loaned to the principal correlator's office to review a field correlation of Meigs County in eastern Tennessee.

In September, we returned to Tillman County for the Paleustoll sampling trip, for which we had prepared the previous year. I had been there earlier to select sites. Drs. Lindo Bartelli and James DeMent of the Fort Worth office attended, as did Texas state soil scientist Gordon McKee and two of his staff, Robert Elder and his assistant, Westal Fuchs. Leland Gile and Dr. John Hawley, soil scientist and geologist from the Desert Research Project, another of the four projects of which the North Carolina coastal plains study was a part, also came to help us. We had three soil pits in Oklahoma and three in Texas. We found well-developed, blocky structure in the lower B horizon at a depth of about 60 inches. The interior of the blocks was noncalcareous, and thick coats of calcium carbonate were on the coatings of the blocks. These characteristics indicated that the soils were leached at an earlier time when the rainfall was higher, then calcium carbonate was added from dust that leached down into the horizon as the climate dried out.

In conjunction with this trip, we dug a pit in the adjoining county for Dr. Bartelli to look at a proposed new series, Indiahoma. The soil was high in shrinking and swelling clay, and occurred as two soils in a repeating pattern on a ridge and swale landscape. One occurred on the 6-inch-high ridge and the other on the slightly lower swale. The ridge and swales were 5 feet wide at the top of the slope and exceeded the combined 23 feet allowed by *Soil Taxonomy* at the bottom of the slope. Here and in some other parts of the world, this landscape was known as gilgai, along with a number of other terms. Since the ridge soils were reddish brown and the swale soils dark brown, the local farmers called the soils candy-striped land. They were even vegetated by different kinds of grasses. The new *Soil Taxonomy* allowed for soils in a repeating pattern to go into one series, which was a new concept. Since the ridge soils and the swale soils were in different classifications, I proposed classifying the dominant swale soil for the series and including a pedon description of each soil in the series description to make the range of characteristics easier to write. At first Dr. Bartelli liked the two-soil idea, but he called the next week and said to keep the new series but to cut it back to one soil. I think he was afraid that Dr. Roy Simonson would think we were losing our minds. Dr. Richard Arnold, who later became head of the soil survey, was also concerned about the repeating soil pattern concept, but the best answer for how to handle these soils may be in the future (Arnold 1964).

Henry Otsuki, Fred Dries, and I attended the state soil scientists meeting at Fort Worth again in December. The meeting emphasized soil interpretations. Dr. Bartelli and his staff provided an entry form with the most used soil interpretations, and the state soil scientists agreed to test the form with the newly developed guides. Use of the entry form marked the beginning of coordinated soil interpretations. The concept was a big step forward, and like all big steps, it caused concern for a number of people.

We finished out the year with a review of field correlation in the field with R. C. Carter of the principal soil correlator's office, then in late January I went to the Kellogg Center on the University of Oklahoma campus for a one-month course titled Program Planning Leadership. The course included a lot of economics and information on working with nonagricultural people. I did not realize how quickly I would put this information to use. While I was still at the Kellogg Center, I received an offer to become the state soil scientist for Colorado. I accepted. We agreed that I would report as soon as I completed the course, and that I would return to Oklahoma in the spring for the principal soil correlator's office review of the field correlation of McCurtain County.

I was state soil scientist in Colorado for two and one-half years. The state soil scientist served as the program staff soil scientist and worked under the state conservationist to manage the soils program in the state. I worked with the area conservationists and the other program staff on a variety of jobs. I did some soil correlation work, but did more with soil interpretations in response to the rapid development that resulted from many people moving into the state. While I was there, Robert Dansdill served as assistant state soil scientist and Arvad Cline as state correlator. Both were highly skilled soil scientists. We had a good program that was highly varied; there was no time to get bored.

TO FORT WORTH TO BECOME A CORRELATOR AGAIN

By 1971, Keith Young had moved to Washington to be on the national interpretations staff, and I was offered the job in Fort Worth as assistant principal correlator for interpretations. The principal soil correltor's (PSC) office assured me that I would still make my share of correlations. The decision was difficult for my family and me. We liked Colorado, and my work at the state office was going fine. But SCS tradition suggested it was unwise to turn down offers; we moved to Fort Worth in July.

In addition to Dr. Lindo Bartelli as principal soil correlator, the other staff included Dr. James DeMent, James Coover, and Thomas Yeager, who worked on soil classification and correlation, and Westal Fuchs, who worked on soil survey manuscripts. I arrived in the middle of a project to computerize the soil survey interpretations sheet so that soil manuscript tables could be computer-printed. After a late August trip to the Alabama state soils meeting to talk about the programs at the Fort Worth TSC, I flew back to Dallas and took a cab to the building where a test of the new computerized system was to be demonstrated. A group from the headquarters in Washington was there, including William Johnson, deputy administrator for soil survey, and A. A.

Klingebiel, director of interpretations, Homer Taff, our TSC assistant director, and Dr. Lindo Bartelli and Westal Fuchs from the principal soil correlator's office.

The demonstration went well. We got permission to enter the soil interpretation forms for two counties into the computer and to print out the tables for two soil manuscripts. This also was a success. We waited a few months for approval on that project, after which Keith Young came to see us with a computerized form that Iowa State University would process and store for us. We would be able to use this centralized system to coordinate interpretations and print out tables needed for soil manuscripts. To ensure interpretations could be made uniformly, we needed a guide. The instructions were ready to complete the form, but they needed to be tested and improved over the next few years. Through this process, interpretations had been worked into the correlation phase.

As assistant principal soil correlator, I went to the field with staff members of other principal soil correlator offices to see the soils as we reviewed the written material. There were usually some problems or questions related to soil classification or correlation, and the field visits let the principal soil correlator see more soils. In November, for example, I made a trip to Walterboro, South Carolina, for a field tour on the lower coastal plain. The survey we reviewed was in an area with few recent surveys and with several soil problems, having to do mainly with how many new soil series were needed.

Before I left Colorado, the automatic data processing staff member with whom we worked in the Portland, Oregon, technical center told me that he had a copy of a cell-type, computer-based soil map program the Forest Service had developed. I didn't have sufficient time to look into this map project before I left, but when I arrived in Fort Worth, I explained the system to Dr. Lindo Bartelli, who gave me unqualified support to pursue it. Henry Otsuki had just embarked on a project to make a new general soil map to update the Oklahoma County, Oklahoma, survey because of urban development. He came to Fort Worth, where we showed him how to overlay the detailed maps with a grid sheet and put the dominant soil on the sheet for a 40-acre grid. We could then produce and print generalized soils maps that kept the detailed legend. We could print any soil interpretation map that could be made from the database. Henry tried both 160- and 40-acre grids, and was pleased with the results.

Don Hazlewood, from our TSC computer section, Harold Tallman, from the cartographic unit, and I developed a production system for these maps. Dr. Lindo Bartelli and I wrote a paper on the system that I gave at the Soil and Water Conservation Society meeting at Portland, Oregon, that year, which later was published in their journal (Nichols and Bartelli 1974). I also wrote a paper on how the gridding affected soil maps for the Soil Science Society of America journal (Nichols 1975). In the next few years, what had come to be called the Map Information and Display System (MIADS) would cover all of Oklahoma and over 300 counties across the South. It was also used some in other parts of the country. The system, as we saw it, was only for general soils maps, made by generalizing more detailed maps. The SCS was working on an improved computerized mapping system that would use soil lines instead of square cells, but MIADS helped us make computerized maps until the better system was developed. When the work began to take too much of our time in the TCS, the cartographic unit contracted out the code entering and computer work to a Dallas company.

In October, I picked up Benjamin Matzek of the classification and correlation staff in Washington, D.C., at the airport and drove to Lubbock, on the south plains of Texas, below the panhandle. Ben and those of us at the southern TSC thought the visit was probably the result of doubts that Dr. Roy Simonson had about Paleustolls in this somewhat dry area. The landscape was a large, smooth plain, with few incised streams and some poorly developed drainage patterns with broad shallow depressions, called playas, that have water in them part of the year. The area had been named the Southern High Plains MLRA. Texas state correlator Earl Blakley, Texas Technical University soils professor Dr. B. L. Allen, and area soil scientist Dan Blackstock took us for a tour. It did not take many stops, looking at deeply weathered, reddish subsoil soils that had little reduction in clay content to great depths, to convince Ben that we had Paleustolls.

Dr. Lindo Bartelli and I traveled to Washington, D.C., to the principal soil correlator's conference in late November of 1971. The Washington staff and two or three people from the principal correlator staffs were also there. We would bring each other up-to-date on what was going on, try to solve some problems, and do some planning. At the same time, we received guidelines and instructions from the Washington staff.

A little over a year later, in January 1973, we attended the National Technical Work Planning Conference in Charleston, South Carolina.

The big news at the conference was the accelerated plan for publishing soil surveys. Eighty soil surveys, a record number, had been sent to the printers the previous year. *Soil Taxonomy* was in galley proofs, and a procedure for amending the document was about to be issued. Headquarters also announced that a handbook of soils for engineering uses would be issued. In response to continuing complaints, the conference also examined instructions, or guides, for the soil interpretation, and a committee met on what kinds of soil maps were needed. The old system had only two kinds: detailed and reconnaissance maps. Many believed we needed five levels, in part because of the widespread perception that reconnaissance surveys were extensive and did not furnish enough detail for planning.

The technical staff of the soil survey often debated the use of standards versus guidelines. Standards were to be followed exactly; guidelines offered a good approximation and required some judgment. Soil interpretations were made by comparing the soil series and phases to the guide or instructions. The important slope breaks for interpretations did not always occur at the same slope gradient as the mapping, so when the guide was compared with the mapped slopes, there was not a 100-percent match. The same was true of other soil characteristics as well. As I've mentioned several times throughout this chapter, the mapping units were not pure—they often included other soils.

Readers interested in pursuing this subject further should consult Dr. Richard Guthrie's article that was published in *Soil Survey Horizons*, the fall 1982 issue (Guthrie 1982), as well as two letters of comment on the article and the answer from Guthrie in the spring 1983 issue. Readers may also wish to look at papers by Dr. Berman Hudson (1992) and Dr. George G. S. Holmgren (1988). I recall Dr. Lindo Bartelli saying that one group of soil scientists followed *Soil Taxonomy* like they followed their Bible, which I took to mean that they were following it too closely without any judgment. Then again, Dr. Bartelli had been raised in Michigan, and the folks we worked with out of the Fort Worth office were southerners. It was a lesson in communication.

In May, I attended the Canadian Soil Survey Work Planning Conference in Saskatoon, Canada. I found their problems and opportunities to be like ours. They were using our *Soil Taxonomy* but modifying it to meet their needs. I traveled by car to Edmonton with Dr. Julian Dumanski of the Canadian Soil Survey staff to see what the Canadians had done with a low intensity, small-scale soil mapping job on forestland in the foothills of the Rocky Mountains. This kind of mapping seemed like an economical use of time that might be appropriate for areas that we thought of as having low-intensity use. Unfortunately, survey users in these low-intensity areas sometimes thought we meant their uses were less important, which posed a different sort of problem for establishing this kind of survey.

Later in 1973, Dr. Lindo Bartelli transferred to headquarters as director of soil interpretations. James Coover replaced him at the southern TSC as principal soil correlator. One of the pressing jobs he inherited was getting the states oriented to using the computer form for the soil survey record, which had quite complicated instructions. James Coover assigned Tom Yager to that job, and Tom did a good job until he retired and went to South Africa to work on soil interpretations with their soil survey. I never learned whether he tried to get the South Africans to use a computer form.

In May, I made a trip to eastern Oklahoma with the South TSC range conservationist and forester to look at the question of whether some soils were rangeland or woodland sites. This trip was not the first time this problem had arisen. We, the soil scientists at the TSC, thought that either site could occur on a particular soil and had constructed the soil interpretation record to accept both. We did not solve the problem.

Dr. Lindo Bartelli kept a folder of letters from Dr. Roy Simonson that contained judgments on correlations. We also had several Soils Memoranda on correlation, but the Simonson letters carried the weight of a judge's decision. When we had a special problem, "Bart" would pull these letters out and look at them for guidance. Dr. James DeMent asked James Coover whether it would be better if we all had a copy of these letters, to which Coover agreed. After that, we all had our own copies. Dr. DeMent got his letters out to solve a problem that arose while he was working with Jimmie Frie, state correlator from Oklahoma. Frie said that if he had copies of those letters, he could have done the job right the first time. As a result, copies of the letters were sent to each state correlator's office. Eventually, the National Soil Survey Handbook (Soil Survey Staff 1983) would replace the Soils Memoranda, providing us with more precise and maybe better guidance. The price we paid, however, was that the handbook section on correlation was much larger.

September 1973 was the first of two meetings at New Orleans on how to improve the procedure for the soil survey publication and maps.

William Vaught, director of the Fort Worth TSC, was chairman of the task force. James Coover and Dr. James DeMent attended the meetings to work on maps, and I attended to work on the soil survey report. The task force included people from all over the United States. Over the course of two meetings, and a lot of writing and talking, we recommended some dramatic changes. The SCS administrator accepted many of them, which left us facing the implementation process.

Over the following year, in addition to the regular work of the principal soil correlator's office, Horace Leithead, regional range conservationist, and I went to Lubbock, Texas, to join Earl Blakely, Texas state soil correlator, and Rudy Peterson, area range conservationist, to examine soils and range sites with the New Mexico state soils staff. In some areas, the soils were near the boundary between Aridisols, or desert soils, and nondesert soils. We had a good week; problems that were hard to solve by correspondence were often easier to solve on site.

In 1975, we acquired two Linolex word-processing machines for use in typing the soil survey reports in our office. These were very early word-processing systems. We soon also hired an editor to edit soil survey manuscripts in our office instead of in Washington. This was part of the task force's recommendations on manuscripts. Coover made another change that saved time. He told the state soils staffs that if their field correlation was in good condition, the principal soil correlator's office would not retype it as a final correlation, but would just put a letter on the original document, accepting it as final. The idea was a good one and worked in some cases. But Larry Ratliff, assistant principal soil correlator at that time, reminded me years later that we had not utilized this program well because we had been too strict on minor errors.

THE JOB OF PRINCIPAL SOIL CORRELATOR

James Coover retired in the fall of 1975, and I was named principal correlator for the southern TSC early in 1976. However, I was listed as head of the soils staff to match the head of the engineering staff, head of the cartographic staff, and other titles on the TSC listing. William Johnson still addressed my letters to principal soil correlator, and that remained the title in the Soils Memoranda. The soils staff had grown with the extra responsibility shared with Washington. The soil correlators were Dr. James DeMent, Gerald Latshaw, Franklin (Ted) Miller, and Talbert Gerald, with Dr. Gordon Decker added soon after. I was fortunate in having Juanita Ray stay on as head secretary. She had been head secretary for Dr. Lindo Bartelli and James Coover and knew the system well. James Benson served as editor for soil survey manuscripts, and we had three word-processor operators, three secretaries, and a file clerk. We soon added another word processor and another clerk, bringing the total to 17 people. I needed a short-term and a long-term plan to keep things moving.

As head of the TSC soils staff, I attended meetings with my counterparts in the states and other TSCs to help the technical center better assist the states. We had a new director at the Fort Worth TSC, and cooperation was becoming a buzzword. Each correlator, including me, made three to six comprehensive reviews in the field per year, to speed up completion of the maps and manuscripts after correlation. We would do about the same number of reviews of field correlations, either in the field or in our "correlation room."

We held a soil correlation school for about 30 state staff and party leaders each year, and every other year, in even years, we had a Southern Regional Soil Survey Work Planning Conference that rotated from state to state. I served as chairman of the organizing committee, while

Joe D. Nichols as principal soil correlator. (Courtesy of Joe D. Nichols)

the state soil scientist and experiment station soil survey leader in the state where the meeting was held alternated as meeting chairman. In alternate years from the work planning conferences, we held a state soil scientists meeting, which was attended by the state soil scientists from the region and by one or two of their staff. The experiment station soil survey leaders were also invited and as many as half of them usually attended, even though the meetings were held during the academic year. Changes were arriving at a fast pace during the 1970s and coordination was needed.

State soils staffs also had a planning conference or meeting each year to go over the year's progress and plan for upcoming years. The meeting involved the state soils staff, the state conservationist, the experiment station soils staff, and any other cooperators. We tried to have one of our staff at each of those meetings to help us do a better job of planning and to see where our assistance could be best used. The Washington, D.C., office held a meeting every fall for principal correlators and one or two of our staff to bring us up-to-date with national developments and to plan for the next year.

The National Work Planning Conference was held in odd-numbered years, which I attended with one or two of my staff. Travel money was always a problem, and we had to negotiate each time with the TSC director, with no guarantee that the money would not be cut during the year either by the director or by headquarters. Travel money available to the state staffs also affected our ability to work closely with them.

The National Cooperative Soil Survey (NCSS) operated on goodwill and cooperation. I tried to visit with each of the state soil scientists, the state soil survey leaders, the Forest Service regional soil scientist, and the SCS headquarters principal soil staff each year. When I could not manage a personal visit, I called, and I invited any of them to write or call at any time if they saw an opportunity or had a problem. We tried to see the state conservationist or a representative whenever we traveled to one of our states. I needed to be able to call on any of these people for help.

One reason the survey worked well was the involvement of the soil survey leaders, who were also soil science professors at state land-grant universities. The professors could furnish technical assistance and reach into other parts of the agronomy departments at their colleges to talk to other professors about yields, fertilizers, clay mineralogy, and other topics important to the survey. I went each year to the meeting of the agronomy department heads from the land-grant universities in our region. I attended these meetings with one of the experiment station soil survey leaders from the region and the headquarters staff, who furnished part of the agronomy departments' money. I was invited because I chaired the Soil Survey Work Planning Conference, one of their official committees. The agronomy department heads were always interested in the progress and direction of the soil survey; several of them had held a job working with the soil survey earlier in their careers.

Trips to the field on reviews were a welcome change from the fast pace at the office, where something always needed to be done. On these reviews, the state correlator and I would travel out in the state and look at a survey with a party leader and his crew. I always saw this as a return to where the important work was being done. The state soil scientist and the experiment station soil survey leader often joined us. Earlier I remarked on how important the party leaders were and how little supervision they actually received. Their job was describing the relationship between closely occurring soils without much variation in climate. Ours was to see that the characterization of local soils was correct, and that the big picture worked. It did not take long to train a good party leader to be a correlator, but some never had the opportunity because the survey hired only one correlator for every fifteen or more party leaders. Others never took the opportunity because they liked the party leader job more, and many did not want the travel a correlator position required.

We tried to meet in our correlation room on Fridays and talk about the review trips coming up in the next week or two. We went over the legend for each county to see if any of the correlators knew of a problem with any of the soil series. The idea was that the correlator going to the county would have the combined experience of all of us. Of course, some correlators knew more about some parts of the South than they did about other parts, but we tried to train everyone for the entire area. I went over the correlations with each correlator, as had my two predecessors before me. This helped with my training also and assured a last check before the signature went on the document.

In April of 1976, William Johnson and Dr. John McClelland came from Washington to give us a program appraisal. The inspection was valuable, as it gave us the chance to see what headquarters wanted and to understand how we could better serve the states. But it was not without incident. When I picked them up, it was the first time I had driven to the fairly new airport. I was a little late, so I parked in the "close-in" parking and met them at the gate. I asked them to wait on the curb with their luggage while I went for the car. I reached the parking area quickly, but the car was gone. I began a systematic search and with some relief, found I was on the wrong floor of the parking garage. But when I met them at the curb, where they had been waiting for much longer than they expected, I had to tell them that I couldn't find the car. Not a terribly auspicious beginning for a new principal soil correlator's first program appraisal!

Our Southern Work Planning Conference took place that same month in Jackson, Mississippi. We had committees on soil potentials, kinds of soils maps, soils yields, and waste disposal on soils. Dr. R. H. Griffin from NASA gave a talk on using MIADS, the cell-type computer mapping system, to make interpretive soil maps in Mississippi. I imagined that NASA's computers were better than ours.

By 1976, the soil geomorphology research projects were being closed down. The scientists who had worked on them were being placed at TSCs, one scientist to each, to provide direct assistance to the states. We had gained much new knowledge for mapping and classification from these projects. The desert project and the coastal plain project especially would continue to be used as training and tour sites.

Dr. James DeMent and I attended the National Work Planning Conference with our TSC director Vernon Martin in January 1977. To me this seemed to be the most important conference I had yet attended, mostly due, I suspect, to the fact that the *Soil Taxonomy* was in print. Committee work included modernizing soil surveys and improving soil survey techniques, soils and fertilizer responses, and waste treatment of soils. We heard reports on the soil surveys in Belgium, Canada, France, and the Netherlands. The U.S. Forest Service, U.S. Geological Survey, the Bureau of Land Management, and the Bureau of Reclamation also reported on their surveys. William Johnson's address provided the details of the SCS soil survey's recent accomplishments and where we were headed. He reported on a reorganization in Washington, but said it would not affect the field much. The new organization would have fewer deputy chiefs, so Johnson was to become a supervisor not only of the soil survey but also of plant scientists and others. Dr. Klaus Flach became the new assistant administrator for the soil survey. He would be in charge of the biennial work planning conference and directly in charge of the soil survey. I interpreted this as a downgrade of the soil survey; William Johnson remained Dr. Klaus Flach's boss, but he eventually would be replaced by a nonsoil scientist. I also learned that Dr. Lindo Bartelli had retired and been replaced by Dr. Donald McCormack.

Vernon Martin took me and our TSC agronomist T. V. Jamieson to Mississippi to determine why the state technical staff could not make the Universal Soil Loss Equation (USLE) work. We met with a large group of Mississippi SCS staff, who were joined by Agricultural Research Service (ARS) scientists from the Oxford Research Station. With several adjustments, mainly to slope length, we got the equation working, and they were able to begin using the modern system.

In June, I attended the first International Soil Classification tour in Brazil. Dr. John McClelland, director of the Soil Classification and Correlation Division, and I represented the U.S. SCS. Dr. Guy Smith, known as the editor of *Soil Taxonomy* and a retired member of the SCS, was the guest of honor. Dr. Roy Simonson made the trip as well; he had traveled over part of this area years before. Dr. Richard Arnold from Cornell University, later director of the soil survey, also attended. Dr. Rene Tavernier, head of the Belgian soil survey (and influential in the development of *Soil Taxonomy*) and a number of other Europeans were also present; in all, about 70 to 80 people made all or part of the trip.

This trip was the beginning of an effort to make our soil classification system fit around the world. Dr. John McClelland knew the *Soil Taxonomy* as well as anyone, since he had finished writing it after Dr. Guy Smith retired. Dr. Hari Eswaran, also on the tour, was a young professor from Belgium. He was later hired to come to the United States to head our foreign operation when the SCS received money from the Agency for International Development for an international assistance project. Known as Soil Management Support Services (SMSS), the project issued several documents to assist the soils programs in the rest of the world and sponsored a number of tours and committees that would assist in fitting the *Soil Taxonomy* to the rest of the world. The Brazilian tour was on low-activity clay soils, or soils that had been highly weathered. The southern United States had these soils, as did tropical areas of the world. The committee for these soils continued for several years. Other committees were added for additional soil classification problems.

It was an honor to gather with such a large and distinguished group. Dr. John McClelland and I roomed together in towns where we had to share a room. After our first field trip, he asked me for a rundown of what I had seen. I told him that it seemed odd that less than half of the people on the trip had gone into the soil pit. He, too, had noticed this. Many of the people on the tour accepted the soil description of the local soil scientist and looked at the printed soils laboratory data. The U.S. soil scientists liked to get into the pit and make their own description and sometimes argue about what they saw.

We covered much of southern Brazil by bus, plane, and train. We had churasco, or Brazilian barbecue. One meal was particularly special because all, or nearly all, of the cow had been churascoed—you are probably not interested in more details. Coffee was served in very small cups and beer in larger-than-normal (by U.S. standards) bottles. We spent some time chasing down cigars for Dr. Guy Smith, who loved good cigars.

Packing clothes and supplies for an extended tour where the location changed frequently was not easy. My secretary and I photocopied the important keys to *Soil Taxonomy* and the amendments and other frequently used material into a single booklet, measuring about four-by-nine inches and less than a one-half inch thick. I carried this in my pocket, saving a lot of weight, and most of the tour participants wanted one. At the end of the tour, I gave the copy we had prepared for the trip to the chairman of the group, Dr. Frank Moorman. The SMSS began issuing these booklets in 1983, I believe. Their first edition was one-half-inch thick; by the seventh edition in 1996, the booklet was nearly one and one-half inches thick, reflecting the rapid additions to *Soil Taxonomy*.

In October, I had the pleasure of picking up Franklin Newhall at the Dallas-Ft. Worth airport for a trip to look at soils and soil moisture regimes. Frank had written the Newhall soil moisture model that calculated soil moisture regimes for *Soil Taxonomy*, but he had never seen the soil and plants in the drier part of this country. Dr. Ron Paetzold, a soil physicist from Lincoln, accompanied us as we traveled from the high plains of Texas, which receive about 17 inches of annual precipitation, to eastern Oklahoma, which receives about 40 inches of annual precipitation. Across this range, we saw the plant cover change from short-grass to tall-grass prairie. Frank was pleased to see that his work reflected local conditions.

We stopped for lunch near Waurika, Oklahoma, at a drive-in café adjacent to a cow pasture. A working cowboy had tied his horse to the fence and climbed through the fence to eat in the café. The cowboy was dressed in jeans and chaps, and wore spurs on his cowboy boots and a cowboy hat. Nothing fancy, just practical range dress. Frank wore a dress shirt, slacks with suspenders, and a small-brimmed hat that was typical of eastern city dress. As they glanced at each other, I think they recognized in themselves the contrasts between the East and the West.

Around the same time, we began testing a computer-assisted scheduling program with the acronym CASPUS. This program was designed to help schedule surveys so that people were where they needed to be to begin mapping and to ensure that maps were delivered on time. Reviews could be scheduled on time and manuscripts and completed maps could be ready to go to the printer simultaneously. The system worked well but it had some problems, since people are good at creating problems. A committee of nonsoils people later determined that CASPUS did not work. I thought that it had helped us considerably, but first efforts are rarely the final solution.

In 1978, Carter Steers joined our staff from the Cartographic Unit. He had been a member of the committee on orders of soil and was one of the early participants in work on soil transects for Order 3 soil surveys. He was a correlator but also assumed the inventorying and monitoring responsibility of the soils unit as a part-time assignment. Soil scientists have traditionally been heavily involved in inventories, and many of the state soil scientists led their state programs. We lost responsibility for the inventory program, however, when national headquarters in Washington assigned a full-time position to the program and located it on another TSC staff.

Over the years, several work-planning conference committees had developed the concept of orders of soil surveys, and in 1978, it became policy. Our regular detailed soil survey became Order 2. Order 1 was added as a more intensive survey. Surveys slightly less intensive than the regular Order 2 survey, with minimum delineations of about 4 to 40 acres, were named Order 3. Less intensive surveys, with minimum delineations of up to about 250 acres, were named Order 4. Order 5 designated the old exploratory survey without much detail.

Using this new policy, Dr. James DeMent pointed out a map unit in a Florida survey that was flooded by tidal waters and was an association of soils. He noted that it was really an Order 4 survey, based on the way it was made and the size of the delineations. Florida state soil scientist Robert Johnson preferred to call it an Order 3. When I inquired why, he explained that Order 3 surveys could be interpreted in the tables in the soil survey manuscript, but that Order 4 surveys were only interpreted in the map unit description. We went along with his request, but the incident made me realize that our publication policy was affecting our correlation policy.

The concept of ordering soil surveys accelerated the already active statistical measurement of soil delineations for content of soil or soils within the delineations. These statistical studies were a field in themselves, and some of the comments they produced caused considerable anxiety among field mappers. One of the problems was the precision allowed by *Soil Taxonomy*. Another problem was the advance of soil laboratory methods and the number of analyses that could be made. I was of two minds about the advances. I believed the practical purity of soil mapping units was based on the estimated properties made by the soil mapper, and that the important feature was whether a dominant part of the mapping unit was interpreted the same way from survey to survey (Miller and Nichols 1979; Byrd 1991). At the same time, I thought that the statistical checks were correct because otherwise we had difficulty explaining mapping purity to users, and if we had enough actual data, our inclusions of other soils would be a higher percent than our handbook allowed.

MY NEXT TEN YEARS AS A CORRELATOR: A SUMMARY

Having provided considerable detail about my first two years as the principal soil correlator for the South, the remainder of this chapter touches only on the highlights of my more recent experiences. In January 1978, I left Fort Worth in the snow for the tropical climate of Kenya, where Maynard Scilley and Robert Sketchley were making a small-scale soil survey of two large areas. I went to provide technical assistance for a month, mainly on soil classification. We stayed in some rather primitive motels and even had a tent camp near Kitui. I liked the tent camp particularly. When I read Robert Ruark's and Ernest Hemingway's articles on Kenya as a young man, I never dreamed I would actually sleep under canvas on the plains of Kenya. While we were in Nairobi, I stayed where they had stayed, at the New Stanley Hotel, and I even drank beer at the "long bar," as they did. I returned through England for a visit with Anthony Smyth, a soil scientist with the British Overseas Ministries Program who had visited us earlier in Colorado when he was headquartered in Rome with the United Nations Food and Agriculture Organization (FAO) soils staff. For hours his staff asked questions on the *Soil Taxonomy*.

In June 1978, Dr. James DeMent retired from the Fort Worth office, leaving correlators Earl Blakley, James Brasfield, Dr. Gordon Decker, Talbert Gerald, Dr. Richard Guthrie, and Gerald Latshaw. In November 1979, Keith Young and I held an interagency trip with the U.S. Fish and Wildlife Service, the U.S. Army Corps of Engineers, and the SCS to Louisiana and South Carolina to develop wetland concepts. We did not know then how big this project would get, but we gathered enough facts for Keith to begin serious work on defining wet or hydric soils. Our editing and word-processing crews were working well, and the soil survey backlog was getting smaller. Our use of the word-processing machines, editors at the TSCs, and using soil scientists to compile soil maps was a success. Dr. Donald McCormack (1980) reported a record 133 soil surveys published in 1979.

By 1979, the soil correlations issued by our office were no longer being checked in Washington. Franklin Miller, Maurice Stout, Richard Kover, and I were the principal soil correlators, and we tried to keep uniformity among the different regions of the country through uniform policy and direction. We would discuss and discuss what that policy and direction should be, but when we went to Washington with a recommendation, we all supported the idea. Dr. John McClelland, who retired in about 1980, completed a large amount of work and was a real leader in coordinating the work of regional soil correlators. But gradually over the years more responsibility was given to the principal soil correlators. Once, under a reorganization that did not include a national correlator at the national office, the principal soil correlators asked Dr. Klaus Flach to add the position, which he did. Earlier, the principal soil correlators had a letter that Vernon Martin, South TSC director, received from Washington that authorized us to call our own meetings, with the idea that we would work to keep the regions coordinated. The letter was soon canceled with the comment that no other group within the SCS could call their own meetings.

The SCS underwent another reorganization in 1980. Dr. Richard Arnold was moved from classification and correlation to director of soil survey. Dr. Klaus Flach was the assistant deputy chief to Dr. Ralph Mc-

Cracken, who had come to the SCS from the ARS to replace the retired William Johnson as deputy chief. He had mapped soils early in his career. The director level that was occupied by Dr. Arnold and reported to Dr. McCracken was one more step down for the soil survey; the 1982 list of Washington staff included 50 directors, making the new organization quite a comedown for soils, which did not have its own deputy chief. The former director positions, which had headed divisions under the old system, became national leaders; for example, National Leader for Soil Classification and Correlation.

Another big change, at least to those involved, followed in 1982 with the consolidation of the four regional TSC cartographic units at Fort Worth, which was accomplished with a reduction in force. The total number of people did not seem as many as were there before, although I never checked the final figures. We at the Fort Worth TSC were still able to walk down the hall and resolve some soil map problems; the other three TSC soils units had to attempt such resolutions by mail or by telephone. Further changes to come suggested that the technical emphasis in soil survey took a back seat to other needs. As long as the TSC remained, they would continue to be able to support the states. Some time after I retired, even the TSCs were abolished, but that is beyond the scope of this chapter.

In January 1982, Dr. Guy Smith, long retired from the SCS, came to Lubbock, Texas, which was one of several sites around the country where he answered questions on the rationale of *Soil Taxonomy*. Several SCS and university scientists attended to ask questions about soil from this part of the country. From these discussion sessions about the concepts and rationales of *Soil Taxonomy*, the SCS produced a book for those interested in the concepts of our soil classification system and in how we built such a documented system (Smith 1986).

I had commented at one time in a headquarters staff conference that I thought the survey areas with both Order 2 and Order 3 surveys should be interpreted in separate tables. I was voted down on that concept because of the trouble and cost. For awhile I had my way, in part, since the Order 3 map units had a symbol all in capital letters, in effect recording a separate interpretation for these units. I lost this when many of the surveys replaced letter symbols with numbers for soil mapping unit symbols on the soil survey maps.

At the 1983 National Work Planning Conference, however, I received a committee assignment to look into whether different orders of mapping should have different kinds of interpretations. I was allowed to choose the members of my committee, and we made some headway. But interpretations continued to be made, as they had been for years, for the taxonomic unit, or the pedon, at the phases of series level, with that rating applied to the map unit according to which soil or soils dominated in the unit. We could also map at any of the five orders of soil survey or at phases of any of the four taxonomy classes above series. We did not have a good way of interpreting the taxonomic classes above the series level. In other words, our mapping allowed for more precision than we had a system to interpret. I had my chance to change this, though, and did not improve the situation much, so I maintained silence on the subject. I thought we, the principal soil correlators, spent too much time on day-to-day correlation and publication activities and not enough time developing a more advanced survey and soil survey publication. I received a Christmas card and note from Louis Derr several years after he had retired, in which he suggested that we had not spent enough time explaining soil surveys to the public. I agreed, but recognized that we were limited in time, just as he was when he had been working, by the push for producing soil surveys. It would even get worse as the years went by.

Our office held a special tour in the spring of 1985 to look at Vertisols and related soils in the Mississippi River alluvial area, which was attended by soil scientists from Arkansas, Louisiana, and Mississippi. These tours were really a very good way to do soil correlations since we all looked at the same pedon in a pit and, after resolving any disagreements, decided what the soil should be. Many of the correlation problems were centered on soils whose characteristics fell near to the classification boundaries. Dr. Marlin Cline had told us at the Cornell correlators' training session that two pedons of soil adjacent to each other on the boundaries of a system could be more like each other than like the central pedon in their class, which makes field mapping and interpretation difficult. This was the same concept referred to in our training by Dr. Andrew Aandahl, when he told us about the hair-splitting decisions we would have to make.

I thought we never spent enough time determining how wide the allowed range of series characteristics should be. The need for precision probably deterred our interpretations. The width of our subgroup characteristics affected the number of our soil series and likely also received inadequate attention. The soils series did represent landscape entities, but the decision on whether to divide one series into two series involved considerable judgment. A study I made of the number of series per taxonomic subgroup indicated that the unwritten criteria varied across the country.

Another difficult job for principal soil correlators was updating older surveys to modern standards. We were often tempted to throw the old survey away and make a new survey. One of the principal soil correlators summarized a significant part of the problem when he noted, "Soil scientists are not very tolerant of other soil scientists' work." The problems associated with updating soil maps and manuscripts will only magnify in the future.

With passage of the 1985 Farm Bill came changes in the way we did a number of jobs. The field office people found themselves working with all farmers instead of just those who volunteered to cooperate. Two of our soil interpretations, highly erodible land and wetlands, became standards; soils either were, or they were not, of these two types, with no weasel space. A soil survey decision on this interpretation could cause a farmer to lose subsidy money or to be disallowed from developing an area, depending on the interpretation. DeWayne Williams, assistant principal soil correlator, took over most of our responsibility for wetland delineations, which had grown beyond just the SCS.

The 1986 budget for the USDA did not look favorable for the SCS. The presidential request amounted to about a 25-percent cut. The final budget was not that bad, but seemed to serve notice of tight times ahead. Our TSC staff was not affected; we had 19 people and had been given responsibility for the quality of soil maps. Previously, we had been checking them with the cartographic unit in Fort Worth, but now had final approval authority. We traditionally held a state soil scientists meeting for our region every other year to try to keep the state staffs and the experiment station leaders up-to-date on the rapid changes. Associate TSC director Glen Black informed me during one of the years following these budget tightening efforts that we would have to cancel our scheduled state soil scientists meeting because of travel cuts. I responded, "Do you mean to tell me the state conservationists of the South have met six times this year and we can't meet once in two years?" When I started counting off the conservationists' meetings to him, he began signaling stop with his hands and cleared our meeting. He knew I was exaggerating by one or two meetings, but I knew he understood our need to communicate with the state staffs.

Changes continued to come. At the 1987 National Work Planning Conference, participants heard a report from a new committee that had been charged with studying the soil survey. The report used such words as centralization, reorganization, and definitive responsibility for each organizational level and mentioned the states that would become responsible for correlations. There were no soil scientists, either from the SCS or the experiment stations, on the committee. I wondered why the committee had been charged with studying the soil survey but not with the engineering or conservation or plant sciences programs. Was it because the leader of the soil survey no longer held a deputy chief position? It was interesting that nonsoil scientists thought that they knew how much quality control soils needed, but did not dare question quality control ideas of the engineers. The committee commented that some of our surveys were overdesigned. I wondered if they understood how much effort we had put into proper design when some states had wanted to make all soil surveys at the same intensity. Further, some soils scientists did not like to map the less extensive Order 3 survey. As a state soil scientist once remarked, "Some soil scientists resist all efforts to keep them from walking all over the land and digging all of the holes they want to dig."

I thought that the SCS would not tamper with the principal correlator concept, which along with the precursor inspector position, had been around for 78 years and predated the SCS by 52 years—I was wrong. I went on my last field review in the summer of 1987 to Sullivan County, Tennessee. My first field review for the South TSC had also been to Tennessee, 19 years before, when I was on loan to the principal soil correlator's office from Oklahoma.

The Washington office sent some people to Fort Worth in September to describe the reorganization to us: The principal correlators were gone. The editors, some of whom had been moved to Fort Worth in an earlier effort to decentralize, would be centralized again. The correlators who were willing to be centralized could move. All of the quality assurance work—this new term had been substituted for the older term quality control—would be done out of Lincoln, Nebraska. In the end, Washington backed out on most of the moves, and the realignment, as they called it, went smoothly. In a few years, Lincoln lost control of correlation to the states. I continued to work for a few more years in my new position as head of the soil interpretation staff at the South TSC, but I will end the story of my career as a correlator here. I think, from the way the reassignments progressed, that I was the last principal soil correlator.

I will, as William Johnson did in 1977, end with simply a description of the reorganization and leave the explanation and interpretation of changes to someone less involved than myself. Reorganizations seem to be a common occurrence and may even perform some useful function. When I finished typing the draft of this chapter on 18 March 1998, Richard Cheney, former secretary of defense, was being interviewed on the television news. He stated that Washington could make a train wreck sound like progress.

I would like to thank the SCS, and those I worked with, for the help and good-natured assistance I received through the years. I feel fortunate to have worked in the soil survey program during the rapid expansion of mapping and the development of the soil classification and interpretation systems. The soil survey has proven to be very resilient for more than 100 years, but there remains much work to do to keep the soil survey abreast of rapidly developing science. As the soil survey enters its second hundred years, challenges are not lacking.

REFERENCES

Arnold, R. W. 1964. Cyclic variations and the pedon. *Soil Science Society of America Proceedings* 28:801–804.

Bennett, Hugh Hammond. 1939. *Soil Conservation*. New York: McGraw-Hill.

Bureau of Soils. 1901. *Field Operations of the Bureau of Soils*. U.S. Department of Agriculture, Washington, D.C.

Byrd, Hubert. 1991. Speaking out on soil surveys. Letters to the Editor, *Soil Survey Horizons* 23(4):126–127.

Cline, Marlin G. 1977. Historical highlights in soil genesis, morphology and classification. *Soil Science Society of America Journal* 41:250–254.

Fuller, Glenn L. 1936. *Procedure for Making Soil Conservation Surveys: Outline No. 4*. Soil Conservation Service, U.S. Department of Agriculture, Washington, D.C.

Gardner, David Rice. 1998. *The National Cooperative Soil Survey of the United States*. Historical Notes Number 7. Soil Survey Division, Natural Resources Conservation Service, U.S. Department of Agriculture, Washington, D.C.

Guthrie, R. L. 1982. The relationship between soil taxonomy and soil mapping. *Soil Survey Horizons* 23(3):5–9.

Holmgren, George G. S. 1988. The point representation of soil. *Soil Science Society of America Journal* 52:712–716.

Hudson, Berman D. 1992. The soil survey as a paradigm–based science. *Soil Science Society of America Journal* 56:836–841.

Johnson, William M. 1963. The pedon and the polypedon. *Soil Science Society of America Proceedings* 27(2):212–231.

Kellogg, Charles E. 1937. *Soil Survey Manual*. U.S. Department of Agriculture Miscellaneous Publication Number 274. U.S. Department of Agriculture, Washington, D.C.

Lapham, Macy H. 1949. *Crisscross Trails: Narrative of a Soil Surveyor*. Berkeley, Calif.: Willis E. Berg.

Marbut, C. F., H. H. Bennett, J. E. Lapham, and M. E. Lapham. 1913. *Soils of the United States*. U.S. Department of Agriculture Bureau of Soils Bulletin 96. U.S. Department of Agriculture, Washington, D.C.

McCormack, Donald E. 1980. Record number of soil surveys published. *Soil Survey Horizons* 21:23–24.

Miller, F. Ted, and Joe D. Nichols. 1979. Soils data. In *Planning the Use and Management of Land*, edited by M. T. Beatty, G. W. Peterson, and L. D. Swindale. Madison, Wis.: ASA-CSSA-SSSA.

Nichols, J. D. 1975. Characteristics of computerized soil maps. *Soil Science Society of America Proceedings* 39(5):927–932.

Nichols, J. D., and L. J. Bartelli. 1974. Computer-generated interpretive soil maps. *Journal of Soil and Water Conservation* 29(5):232–235.

Norton, E. A. 1939. *Soil Conservation Survey Handbook*. U.S. Department of Agriculture Miscellaneous Publication Number 352. U.S. Department of Agriculture, Washington, D.C.

Smith, Guy D. 1986. *The Guy Smith Interviews: Rationale for Concepts in Soil Taxonomy*. SMSS Technical Monograph 11. Soil Management Support Services, Soil Conservation Service, U.S. Department of Agriculture, Washington, D.C. and Department of Agronomy, Cornell University, Ithaca, N.Y.

Simonson, Roy W. 1987. *Historical Aspects of Soil Survey and Soil Classification*. Madison, Wis.: Soil Science Society of America.

Soil Survey Staff. 1951. *Soil Survey Manual*. U.S. Department of Agriculture Handbook Number 18. U.S. Department of Agriculture, Washington, D.C.

Soil Survey Staff. 1960. *Soil Classification: A Comprehensive System, 7th Approximation*. Soil Conservation Service, U.S. Department of Agriculture, Washington, D.C.

Soil Survey Staff. 1975. *Soil Taxonomy: A Basic System of Soil Classification for Making and Interpreting Soil Surveys*. U.S. Department of Agriculture Handbook Number 436. U.S. Department of Agriculture, Washington, D.C.

Soil Survey Staff. 1983. *National Soil Handbook*. Soil Conservation Service, U.S. Department of Agriculture, Washington, D.C.

U.S. Department of Agriculture (USDA) Staff. 1938. *Soils and Men*. Yearbook of Agriculture 1938. Washington, D.C.: Government Printing Office.

U.S. Department of Agriculture (USDA) Staff. 1951. *Land Resource Regions and Major Land Resource Areas of the U.S.* U.S. Department of Agriculture Handbook Number 296. Soil Conservation Service, U.S. Department of Agriculture, Washington, D.C.

5

OPENING OPPORTUNITIES: WOMEN IN SOIL SCIENCE AND THE SOIL SURVEY

Maxine J. Levin

Women have been involved in soil science and soil survey since the interest in this most basic of natural resources emerged. Women employed by the early U.S. Soil Survey were largely restricted to office activities, but soon they ventured into fieldwork. Albeit brief, the first appointment of a woman to a field party came about in 1901, only two years after the soil survey began in earnest. It would be another 45 years, however, before the Soil Conservation Service (SCS) would appoint a woman as a field soil scientist.

The delay in women joining the ranks of field scientists has significance for telling the story of women's place in the soil survey. Soil science is first and foremost a field-based science. Soil mapping begins by studying the landscape and building a conceptual model of how the topography, geology, plants, climate, water, and animals interact to predict soil characteristics. Mapping then continues in the field where the soil scientists validate their predictions of soil types by digging, describing, and sampling the soils and vegetation. Benchmark samples are analyzed in the lab to validate conceptual models. Finally, the information is consolidated onto aerial photo maps, associated computer databases, and manuscripts for publication. Until women gained an equal place as scientists in the field, they remained in supporting roles in the soil survey.

THE PIONEERS (1895–1965)

Women's roles in the earliest years of the soil survey appear to have been limited to clerical work, copyediting of manuscripts and cartographic drafting of maps, although women with appropriate academic training soon began to work in the laboratories (American Society of Agronomy n.d., 4). Janette Steuart and Sorena Haygood, who maintained the laboratory and field records for the U.S. Department of Agriculture's (USDA) Division of Agricultural Soils, were among the first women to work for the soil survey. According to Macy H. Lapham's account, Steuart was hired 4 January 1895, making her the first woman appointed to the Division of Soils, then part of the U.S. Weather Bureau, located in the USDA. Both Steuart and Haygood were career employees; Steuart retired in the late 1920s with nearly 30 years of service and Haygood retired sometime later (Lapham 1945, 344).

The soil survey remained resistant to women in field parties. Julia R. Pearce, one of the earliest female pioneers in soil survey, joined the soil survey as a member of the field party at Hanford, California, in June 1901, but never had the opportunity to join her party in the field. Pearce had been one of only two 1901 graduates in agriculture from the University of California at Berkeley (UCB). In his commencement address, Secretary of Agriculture "Tama Jim" Wilson lamented at the small size of the graduating class and emphasized that the Department of Agriculture needed candidates trained for technical positions. After the speech, Pearce sought Wilson out and told him "she was ready and willing to come to the relief of the Department." Pearce found herself almost immediately appointed as an assistant to Macy Lapham's all-male field party in Hanford (Lapham 1949, 28–29).

While Secretary Wilson was sympathetic to the idea of women in the workplace, Lapham was uncomfortable with having women in the field with an all-male crew (Helms 1992; Baker 1976). On the day that Pearce arrived in Hanford, it was said that Lapham sent a telegram that stated, "Miss Pearce is here, what in hell shall I do with her?" In the end, he put her to work copying maps. In 1903, she was transferred to the Bureau of Soils in Washington as an assistant in soil survey and later transferred to the Bureau of Plant Industry as a laboratory assistant (Lapham 1949; Bureau of the Census 1911, 630).

From Lapham's memoirs (Lapham 1949), it appears fieldwork was out of the question for women before the 1940s. Women found places in the field in unofficial capacities, however. Mary Baldwin, the wife of soil inspector Mark Baldwin (employed by the Soil Survey 1912–1944), mapped with her husband in northern Wisconsin and the Boundary Waters of Minnesota during the early 1920s. They worked during the summer months, camping and using a small boat to go from island to island. Mary would drop Mark off on one side of the island, he would map on foot, and she would wait for him with the boat on the other side of the island. While she waited, she might search for survey markers or make observations on her own of the general area. Mary recalled that there were times when she wished that she could have tried mapping on her own. As it was, she accompanied her husband everywhere during his remote mapping experiences, transcribing or taking field notes for him and assisting with the sampling (Baldwin 1992).

Mary Baldwin might still have had difficulty finding employment with the USDA, even if she had the background and training to map soils. At the time, married couples were discouraged from working for the same federal agency, and the USDA already employed Mary's husband. A clause in a 1932 appropriations law even stated that married persons living with a federally employed spouse would be dismissed first in the case of government reductions in force and that preference should be given to others for new appointments. Although this legislation was repealed in 1937, it limited married women's employment by the federal government at that time (Baker 1976) and perhaps set a precedent for the future.

During the decades of the 1930s, 1940s, and 1950s, women contributed to the soil survey through editing, writing erosion history, and conducting lab work. Lillian H. Weiland—the first female employee of the newly established Soil Erosion Service and secretary to Hugh Hammond Bennett—put together a "Bibliography on Soil Conservation Compiled in the Office of the Chief of SCS" in 1935 (Helms 1992). The bibliography consolidated ideas for soil erosion control technology for the new agency (Weiland 1935). In 1937, Lois Olson and Dr. Arthur Hall spoke on studies in erosion history as part of a series of research seminars for SCS staff. Some of today's thinking on interpretations of the soil survey and field practices to control erosion can be attributed to this series of lectures. Olson, a geographer by training, headed the SCS's Erosion History Section (*Service News* 1940).

Charlotte Whitford (Coulton), a graduate of Ohio State University with a masters degree in botany, joined the SCS as a secretary with a field soils staff in Zanesville, Ohio, in the mid-1930s. J. Gordon Steele, an old classmate, soon recruited Whitford to work as an assistant soil technologist in Washington, D.C., on a series of reports on soil erosion. She later worked as an editor on soil surveys and eventually became head of the SCS publication staff. She retired in the 1980s with almost 50 years of service (Helms 1992).

Dorothy Nickerson, a soil color technologist for the USDA from the late 1920s through the 1940s, was instrumental in developing the soil color standards for soil survey. Nickerson had been an assistant manager of the Munsell Color Company before joining the USDA in 1927 (Jaques Cattell Press, Vol. 4 1972). She made extensive colorimetric tests in the lab and worked with soil scientists in the field to match soils to the Munsell colors and to create a new set of color names, first introduced in preliminary form in 1941. She then worked with Thomas D. Rice, Kenneth Kelly, and Albert H. Munsell to adapt the Munsell color chart system for describing soil color in the lab and the field. The U.S. Soil Survey adopted the Munsell color charts and new color names in 1949 (Simonson 1993).

The first woman soil scientist officially assigned in the field for the SCS was Mary C. Baltz (Tyler). Mary Baltz graduated from Cornell University and joined the soil survey as a "junior soil surveyor" in 1946. Labor shortages during World War II provided the opportunity for her to work in a job that, up to that time, appeared to be reserved for men (Helms 1992). By 1951, Mary was responsible for mapping in Madison and Oneida Counties in New York, and later she was assigned the task of map measurement for the entire state. In contrast with today's electronic techniques, the work was done by cutting out the soil map delineations on copies of the field sheets. Areas with the same label were weighed together and a factor converted the weight to acres. She hired a team of women to do this conversion job in the winter months (Rice 1998b).

Erwin Rice, a retired soil scientist in New York, mapped under Mary Baltz's direction. He remembered Mary as a confident, petite woman who enjoyed mapping in the field, was comfortable with the all-male crews, and had a good sense of humor. He called her a "splitter," a soil scientist who tends to separate out concepts for new soils as opposed to lumping them together under general categories of old soil names. Mary Baltz worked for the SCS until about 1965 (Rice 1998a).

Ester Perry was a major figure in the California soil survey effort. Her 1939 Ph.D. in soil science from UCB was the first received by a woman in the United States. For her doctoral research, "Profile Studies of the More Extensive Primary Soils Derived from Granitic Rocks in California," she may have been one of the first students to use x-ray diffraction to look at clay mineralogy structure in soils (Cattell 1944; Sposito 1998). She studied under Charles Shaw at Berkeley, and Kelly, Doer, and Brown were mentors and coworkers with her in Riverside,

California, where she worked at the Subtropical Horticulture Research Center during her graduate studies (Sposito 1998). From 1928 to 1939 she worked as an associate soil technologist for the California Agricultural Experiment Station (Cattell 1944; Cattell et al. 1961).

From 1939 until she retired in 1965, Perry essentially ran the USDA Soil Survey Lab in Room 33 in the basement of Hilgard Hall at UCB. She moved up through the ranks during that time, from "junior soil technologist" (1939–1954) to "associate specialist soils" (1954–1960) to "specialist" (1960–1965). In a 1952 presentation to the Western Soil Science Society in Corvallis, Oregon, Perry praised the benefits of close collaboration between a soils lab and a field soil scientist for quick turnaround of information in support of mapping (Perry et al. 1952). After the establishment of the Beltsville Agricultural Research Center in 1952, the Berkeley Soils Lab was slowly phased out.

In soil science, or more specifically in pedology (the study of soil genesis), as with all the earth sciences, there were very few women working in the field before the 1970s. Gary Sposito, who was a student in Ester Perry's lab for a year, recalls that she was well aware of being a pioneer in her profession. As one of "Ester's boys" (as students who worked part-time in her lab were known), he thought she effectively mentored many young men and women into a soil science career. Dr. Perry maintained an all-business approach in the lab, but also remembered to bring birthday cakes for her students and provided a bed in the lab for those who might work through the night on important projects. Despite her accomplishments, Ester Perry was never promoted to associate professor or put on the tenure track. This was not unusual, however, since many women and men worked as researchers or technicians for their entire careers at the agricultural experiment station without receiving academic status (Sposito 1998; Huntington 1998; Birkeland 1998). She also was not acknowledged in USDA records as an official soil survey collaborator (Soil Conservation Service 1952).

FOUNDATIONS: BUILDING ON THE PIONEERS (1959–1975)

During the 1950s and 1960s few women ventured directly into the field of soil science. Some arrived through other disciplines, such as geology, microbiology, or one of the plant sciences. They found through their graduate studies that soil science was a key element in their research and then continued on to further studies in soil science. Some were mentored and encouraged by major professors (as Ester Perry had been by Shaw and Kelly) to pursue soil science and stick with it. As scientists and teachers, the women soil scientists who started in the 1950s and 1960s spent a good deal of their careers as mentors themselves, and many placed a high value on that aspect of their work.

Ester Perry, who bridged the gap from the pioneer era to this period of building foundations, was herself one such mentor, not only through her work with students in the Berkeley Soils Lab, but also particularly in her effort to bring equal access to soil survey field training to women in the mid-1950s. In that decade, a summer field course—Soils 105 field trip—was a requirement for a soil science major from UCB and the University of California at Davis (UCD). The course was offered each summer, and since the 1930s had convinced many a prospective soil science student that soil survey could be a lifelong interest and career path. In 1953, Eva Esterman, a soil science honors student, was the first woman to request to take the course. The UCB Soils Department offered Eva an option to graduate without participating in the field course, but she wanted to take the trip just like all of the other students. For six weeks in the field, the university arranged for Eva to have separate sleeping facilities and comfort stops. Dr. Earl Story's wife served as chaperone, accompanying the students in a separate car and with some discomfort. The academic dean at the time, Dr. Frank Haridine, considered the experiment a complete disaster and swore publicly that no women would ever again take the field course (Berc 1998; Esterman 1998; Huntington 1998).

The event triggered Ester Perry to step in and offer a Soils 105F course ("F" for female), which she planned and made available from 1956 to 1959. The trip was soil survey-oriented but because Ester had different professional contacts, the course had a somewhat different approach. Three women took the course in 1959, the last year the specialized course was offered. In 1965, the "Soils 105" course officially became co-ed and included two women students. Perry accompanied the group as a chaperone (Huntington 1998). Seven years later, the class was 50 percent women, and there was no women's chaperone.

In addition to mentoring students of soil science, many women professors and researchers made substantial contributions to our understanding of soils and soil science during the 1960s and 1970s. In the United States, Cornelia Cameron, Jane Forsyth, Jaya Iyer, Eva Ester-

Soils 105 field class, for students of the University of California at Berkeley and Davis, 1972. The author is in the front row, fourth from the left. (Courtesy of Maxine Levin)

man, Nellie Stark, and Elizabeth Klepper have been among the most prominent, although there are no doubt many others. Scientific publications usually list the first names of researchers only by their initials, which makes identifying the authors by gender difficult at best; the individuals included here were identified by their students and colleagues.

Dr. Cornelia C. Cameron completed her Ph.D. in geology (with an emphasis in geomorphology) at the University of Iowa in 1940. After 11 years teaching earth sciences, Dr. Cameron joined the U.S. Geological Survey (USGS) in 1951 and spent the next 43 years in the field. Her field career began in military geology, with terrain analysis of military sites in over 30 countries on five continents, many of them dangerous militarized zones at the time. Dr. Cameron's colleagues remember her as quite a character in the field.

In part, Dr. Cameron's reputation for eccentricity had to do with her mother. One of this country's first female Ph.D.s in botany, Dr. Cameron's mother had a strong interest in her daughter's work and accompanied Cornelia on field expeditions until she was 103 years old. The younger Dr. Cameron joked that when her mother got so old that her eyesight had deteriorated, she put a cow bell on her so she could find her if she wandered off. In a story about daughter and mother's adventurous military terrain investigations in the Caribbean area in early 1961 before the Bay of Pigs invasion, Dr. Cameron recounted that "Mother and I were a perfect pair. We told everyone that we were Canadian tourists. One time, as I was doing traverses along the slopes of one of the islands, Mother stayed in the car. I was upslope from her when I saw a truck full of guerrillas pull up. Mother simply charmed them and they drove off" (Harden 1998; USGS 1986, 1994).

Dr. Cameron was an internationally recognized authority on peat soils and their use as a soil additive and source of energy, and on the impact of peat removal on swamp and bog environments; she wrote prolifically— 110 publications—on the subject. Both the USGS and the Department of the Interior recognized her accomplishments in research and public service. Dr. Cameron received the USGS's Meritorious Service Award in 1977 and its Distinguished Service Award in 1986, and received the Department of the Interior's Public Service Recognition Award in 1990.

Dr. Jane L. Forsyth, a professor of geology at Bowling Green State University in Ohio, has contributed much to our understanding of the age relationship of soils and till to northern Ohio glacial geology. Among her peers she has been affectionately dubbed the "Queen of the Pleistocene," according to her colleague Peter Birkeland, retired from the faculty at University of Colorado, Boulder. Dr. Forsyth earned her Ph.D. from Ohio State University in 1956 and taught at University of Cincinnati, Miami University, UCB, and Ohio State University, before joining the faculty at Bowling Green State University in 1965 (Birkeland 1998; Wasserman 1992; Forsyth 1990).

Dr. Jaya Iyer, professor of soil science at the University of Wisconsin, Madison, has focused her research on relating soil properties and tree growth. Dr. Iyer earned her Ph.D. in botany from the University of Bombay, India, in 1959. The external referee for her Ph.D. research, soil scientist Dr. Sergei Wilde of the University of Wisconsin and a member of the Wisconsin Forestry Hall of Fame, encouraged her to take an interest in soils, and she completed a second Ph.D. in soil science at the University of Wisconsin, Madison in 1969. Dr. Iyer has had a highly successful career as a national expert in soils for tree nurseries, specializing in urban, Christmas tree, and forestry production (Iyer 1998).

Dr. Eva Esterman, who received her Ph.D. in soil science from UCB in 1958, nearly 20 years after Ester Perry, was only the second woman to earn a soil science Ph.D. from Berkeley. She went on to become a professor at San Francisco State University in 1960, where she taught botany and later added soil science to the curriculum. She focused her soils research on soil microbiology and biochemistry before she retired in 1982 and began raising sheep (Berc 1998; Esterman 1998).

Dr. Nellie Stark, tenured professor and forest soil ecologist at the University of Montana, Missoula, from 1970 to 1992, gave us the theory of the "biological life of a soil," which describes how soils and plants interact during development and decline phases of soil genesis. The theory explains the variation in nutrient uptake by plants based on the stage of soil genesis. Indirect nutrient cycling, which involves uptake of ions by the roots from the soil, predominates when a soil is young; direct nutrient cycling, which involves uptake of ions by the roots directly from the litter, bypassing the soil, occurs as the soil becomes older and depleted by weathering (Stark 1998a, 1998b).

Stark earned a Ph.D. in botany (ecology) from Duke University in 1962, with a minor in soils based on credits she collected at Oregon State University in 1961. Dr. Stark's research with the Desert Research Institute, Reno, Nevada, focused on soils and nutrient cycling of litter in the tropical ecosystems of Brazil and Peru. The soil chemistry lab for forestry that Dr. Stark operated at the University of Montana was well known and received and processed samples from all over the world (Stark 1998a, 1998b).

Finally, Dr. Elizabeth L. Klepper, a research leader and plant physiologist at the Columbia Plateau Conservation Research Center, Pendleton, Oregon, concentrates her research on root growth and functioning under field conditions and plant-soil water relations. Dr. Klepper holds degrees from Vanderbilt University and Duke University. She has been recognized for her accomplishments by all three professional agronomic research societies of the United States: the American Society of Agronomy (ASA), the Crop Society of America (CSA), and the Soil Science Society of America (SSSA). She was the first woman ever to receive the Fellow award from the SSSA (American Society of Agronomy n.d.).

These are but a few of the women who dedicated their lives and research to soil science and soil survey during this period. It is inspiring to consider their achievements. Over the course of about 50 years, women in soil science and soil survey moved out from under their restriction to clerical support to become influential researchers and field investigators. Both Dr. Stark's theory of the biological life of a soil and Dr. Cameron's important and risky fieldwork would have been virtually unimaginable when the soil survey began in 1899.

IN THE CLASSROOM, IN THE FIELD, AND IN THE LAB (1970–1990)

Despite great strides in the field of soil science, women were still not actively recruited into the USDA's Soil Survey Division in the 1960s. For example, in a 1962 recruitment speech to the Agronomic Education Division of the American Society of Agronomy in Ithaca, New York, Assistant Soil Survey Administrator Charles Kellogg expressed his agencies' concern about recruiting good candidates, "especially of well-trained, broadly educated young men who can develop rapidly" (Kellogg 1963). His comments were not surprising perhaps, since the professional workforce of the country at that time was still predominantly male.

The transition during the 1960s was profound, however. By the 1970s, career counseling documents were beginning to discuss ways to channel girls into nontraditional careers, and encouraging young women to enter nontraditional occupations continued as a theme into the 1980s (Women's Bureau 1960–1980). Corresponding changes occurred in the classrooms; as more and more young women began to enter previously male science and employment territories, materials and approaches to education changed to meet the needs of this more diverse student population.

In the soil survey, as well as in some of the other earth sciences professions, a woman still needed to be persistent in the 1970s to obtain a field appointment. In the SCS of the 1970s, there were fewer than 15 women in the federal employment series soil scientist (470 series) at any one time nationally, despite an acceleration in soil survey mapping and a general increase in field crews (Association of Women Soil Scientists 1960–1998).

Most of these women soil scientists thought they were the only female soil scientist in the agency. In addition, there were no formal professional organizations for women field soil scientists; the Association of Women Soil Scientists (AWSS), organized by a group of women soil scientists in the U.S. Forest Service, was not formed until the early 1980s. But slowly more career opportunities began to emerge for women. Title VII of the Civil Rights Act of 1964 and the Civil Service Reform Act of 1978 helped increase opportunity by prohibiting sex discrimination in employment and requiring diversity in the workforce (Helms 1992), and the Women in Science and Technology Equal Opportunity Act of 1980 opened up more opportunities for women to receive support in university settings.

In the 1970s and early 1980s, the SCS soil survey staff in California included five women field soil scientists (Arlene Tugel, Nancy Severy, Chris Bartlett, Lisa Holkolt, and Maxine Levin), a crowd compared to other states. Many states had only one woman working as a soil scientist in the field—Carole Jett in Nevada; Carol Wettstein in Florida; Sue Southard in Utah; Margaret Rice in Mississippi; Caryl Radatz in Minnesota; Mary Collins in Iowa; and Debbie Brasfield in Tennessee, for example. Some states had two or more women in the field, and there may have been more—records of employees for those years are spotty and have not been saved comprehensively—but in any event, women still comprised a small percentage of the total field soil scientist staff. Nevertheless, their contribution to soil survey was sizable, with millions of acres mapped, at times with some physical hardship.

Many of the women who worked in field parties during the 1970s went on to achieve greater responsibilities and position. By the late 1980s, the SCS had appointed the first woman state soil scientist, followed by others in the early 1990s, and women grew more prominent as soil survey party leaders. Carol Wettstein became the first woman state soil scientist, serving as state soil scientist in Maryland from 1988 to 1989 and as state soil scientist in Colorado from 1990 to 1995. Carole Jett served as state soil scientist in California in 1991, and Carol Franks was state soil scientist in Arizona in 1994. In 2000, Maxine Levin was appointed national program manager of the Soil Survey Division. In the 1980s, there were at least three published soil surveys for which women were the party leaders or the principal field investigators: Sacramento County, California (Arlene Tugel); city of Baltimore, Maryland (Maxine Levin); and Indian River County, Florida (Carol Wettstein).

We can anticipate seeing more women listed in soil surveys of the 1990s, as the number of women party leaders increased significantly during the decade. An all-female crew of soil scientists led by Deborah Prevost mapped the Hualapai-Havasupai Indian Reservation, Arizona, in the late 1980s and early 1990s. This soil survey was published in 1999—exactly 100 years after the establishment of the USDA Soil Survey (Prevost 1998; Prevost and Linsay 1999).

Deborah Prevost, on right, and Jennifer Foster at work on the Hualapai-Havasupai soil survey, fall 1990. (Courtesy of Natural Resources Conservation Service, U.S. Department of Agriculture)

Women have also contributed to soil surveys in many other ways that are not reflected in publications. Soil correlators and data management specialists make significant contributions to soil survey data and manuscripts, and a number of women have held these positions: Sue Southard (California), Renee Gross (Nebraska), Carmen Santiago (Puerto Rico), Panola Rivers (Pennsylvania), Kathy Swain (New Hampshire), Laurie Kiniry (Texas), Diane Shields (Delaware), Susan Davis (Maryland), Marjorie Faber (Connecticut), Tammy Cheever (Nebraska), and Deborah Anderson (North Carolina). In the last few years, women soil scientists have been instrumental in the effort to digitize the soils information that is used in the publications, including Vivian Owen (Texas), Jennifer Brookover (Sweet) (Texas), Darlene Monds (Massachusetts), Barbara Alexander (Connecticut), Caroline Alves (Vermont), Lindsay Hodgman (Maine), Caryl Radatz (Missouri), Adrian Smith (Nebraska), Amanda Moore (Oregon), Sharon Schneider (Oregon), Marcella Callahan (Arkansas), Brandi Baird (Oregon), and Jackie Pashnik (Rhode Island).

In the National Cooperative Soil Survey there are also a number of women field soil scientists who work mostly with soil survey interpretations and education, including Sue Southard (California—volcanic soils and vertisols), Lenore M. Vasilas (Maryland—hydric soils), Sheryl Kunickis (Washington, D.C.—landscape analysis), Susan Ploetz (Minnesota—resource inventory), Susan Casby-Horton (Texas—soil geomorphology), Christine Clarke (Maryland—geographic information systems), Jeannine Freyman (Virginia), Karen Kotlar (New York), Lisa Krall (Connecticut), Gay Lynn Kinter (Michigan), Donna Hinz (Nebraska), Patricia Wright-Koll (Minnesota), Jeanette Bradley (Arkansas), and Deborah Prevost (Nevada). Like agricultural extension specialists, these soil scientists act as a bridge between university research, soil survey mapping, and the public, interpreting soil surveys for practical use by both agencies and individuals, including providing on-site field investigations.

Women in the SCS, now the Natural Resources Conservation Service (NRCS), have made significant contributions to soil science in the National Soil Survey Laboratory (NSSL) and the National Soil Survey Center (NSSC), and as researchers in the Soil Quality and Watershed Sciences Institutes. Carolyn G. Olson has been a lead research scientist at the NSSC, located in Lincoln, Nebraska, since 1989. Dr. Olson's research focuses on soil geomorphology, Quaternary geology, and clay mineralogy. Olson received the honor of being made a Fellow of the Soil Science Society of America in 1996 (Olson 1998; American Society of Agronomy n.d.). Other women soil scientists at NSSL and NSSC include Rebecca Burt (soil chemical properties), Joyce Scheyer (urban soil properties), Susan Samson-Liebig (soil quality), Deborah Harms (soil physical properties), Sharon Waltman (national soil survey databases and GIS interpretations), and Carol Franks (soil biology). In the Institutes, Arlene Tugel (New Mexico), Betty McQuaid (North Carolina), and Cathy Seybold (Oregon) have been working with soil quality and watershed health indicators (Natural Resources Conservation Service 1997; Burt 1998; Scheyer 1998; Franks 1998).

Other federal agencies, such as the USGS, also provide opportunities for women in soil science research. Jennifer W. Harden, with USGS in Menlo Park, California, built on her Ph.D. research using soil chronosequencing to develop the Harden Index, which used soil horizons and carbon-dating to measure time in the alluvium sequencing. Since then, she

Deborah Prevost recording landscape surface characteristics on the
Hualapai Indian Reservation, summer 1990. (Courtesy of Natural
Resources Conservation Service, U.S. Department of Agriculture)

has worked on the effect of climate on soil, particularly as it relates to
groundwater recharge and wetland assessment. She has been a front-
runner in research on global change issues of soil carbon, carbon dioxide
emissions, and soil carbon sequestration (Harden 1998). Marith Reheis,
with the USGS in Denver, has also done significant research in using soil
properties as a paleoclimatic record for chronosequence mapping in

Rocky Mountain glacial outwash. Originally a geologist by training, she received her Ph.D. in soil science under Pete Birkeland at the University of Colorado, Boulder, in 1984 (Reheis 1998).

At NASA's Goddard Space Flight Center, Elissa Levine has been working with soils in forested ecosystems since 1987. She models soil physics and soil chemistry to assess watershed leaching, soil carbon ecosystem effects, and the effects of acid precipitation on soils and groundwater. She was recently appointed lead scientist for the NASA Global Change Master Directory. She was also selected as a Fellow of the Brandwein Institute for Science Education, an award based on her work as the principal scientist in the GLOBE program's soil investigations for teaching soil science worldwide to K–12 students (Levine 1998).

Since the 1970s, women scientists with the U.S. Forest Service have been involved with the National Cooperative Soil Survey ecological unit inventories, as well as with technical soil interpretations in the specialties of forest soil productivity, soil erodibility, fire ecology, and forest ecosystem health. Some of the prominent women involved in this effort include Gretta Boley (Washington, D.C.), Clare Johnson (Six Rivers National Forest, California), Carol Smith (SCS-USDA and Tahoe National Forest, California), Barbara Leuelling (Superior National Forest, Minnesota), Connie Carpenter (White Mountain National Forest) and Mary Beth Adams (Northeast Forest Experiment Station, West Virginia) (Association of Women Soil Scientists 1960–1998).

In 2000, three women held positions as pedology (soil genesis) professors in U.S. universities: Janice L. Boettinger, Utah State University, Logan; Christine Evans, University of New Hampshire, Durham; and Mary Collins, University of Florida, Gainesville. Dr. Boettinger is working on an extensive review of worldwide zeolite mineral occurrences in soils and the use of zeolite and clinoptilolite for waste disposal systems of animal production operations. She is also working on characterizing selected soil resources of Utah, which includes research on saline, wet soils and irrigation-induced hydric soil characteristics (Current Research Information System 1998). Dr. Evans is focusing her research in the field of describing anthropogenic (human-influenced) soils and developing terminology to describe soil properties derived from human activity (Evans 1998; University of New Hampshire 1998).

Dr. Collins' research at the University of Florida focuses on the genesis, morphology, and classification of soils; identifying and delineating hydric soils; using ground penetrating radar to study subsurface properties; and pedoarcheology (American Society of Agronomy n.d.). She is best known for her dedication to soil survey fieldwork and reaching out to other countries to spread soils technology. As part of the People to People Program, she first opened the door to doing ground penetrating radar soil investigations in China and Portugal. Dr. Collins was made a Fellow of the American Society of Agronomy in 1996 and a Fellow of the Soil Science Society of America in 1997 (Olson 1998).

A number of women who have been made Fellows of the Soil Science Society of America have provided outstanding contributions to soil science. Mary Beth Kirkham, made a Fellow in 1987, is a professor at the Kansas State University Evapotranspiration Lab. Her work has focused on heavy metal uptake by plants and soil-plant-water relations for over 20 years. Mary K. Firestone, who became a Fellow in 1995 and also received the Emil Truog Soil Science Award, is a professor of Soil Microbiology at UCB. Her research focuses on the microbial population-basis of carbon and nitrogen processing in ecosystems. Jean L. Steiner, made a Fellow in 1996, is director of the USDA Agricultural Research Service (ARS) Southern Piedmont Conservation Research Laboratory in Watkinsville, Georgia. Her research is on humid region water balance studies in complex topographies. Finally, Diane E. Stott, a 1997 American Society of Agronomy Fellow, is a soil microbiologist with ARS in West Lafayette, Indiana. Her work has focused on the effects of organic matter dynamics on soil structure and erodibility and modeling the effects of plant residue decay on erodibility (American Society of Agronomy n.d.).

The women described above are only a sampling of the many women who have contributed research, mapping, applications, and education to the field of soil science in the last 25 years. Others, to name only a few more, include Nancy Cavallaro (University of Puerto Rico—soil chemistry and tropical soil fertility), Laurie Drinkwater (Rodale Institute, Pennsylvania—sustainable agriculture), Kate Skow (University of California, Davis—soil microbiology), Laurie Osher (University of Maine, Bangor—soil science), Samantha Langley (University of Southern Maine, Gorham—soil science education), Kate Showers (Boston University—soil conservation), Jeri Berc (NRCS-USDA—soil conservation), and Katherine Newkirk (Woodshole Marine Biological Lab—global warming).

1990 AND AHEAD

Clearly, women have made significant and numerous contributions to the field of soil science and soil survey through research, mapping, applications, and education. As the numbers of women have increased in the classroom, lab, and field, changes have also taken place within the soil science discipline. The women who are graduating in soil science in the 1990s are confident and intellectually engaged and are quickly gaining recognition for their work. As an example, Eva M. Muller of Spokane, Washington, a soil survey project leader with only seven years of experience, was awarded the National Cooperative Soil Survey Soil Scientist of the Year Award in 2001. Lenore M. Vasilas, who finished her M.S. in soils in 1997, remarked about any remaining stereotypes in her work, "Oh we don't think about it... We just go ahead and do it!" (Vasilas 1998).

There is a world of difference between Julia Pearce's experience in the early 1900s and Lenore Vasilas's reality in 1998. In the educational realm, between 1987 and 1996, soil science, along with education, communication, and social science, experienced the largest percentage growth of female participation. While overall enrollment of students (B.S., M.S., and Ph.D.) in the soil sciences held relatively steady between 1987 and 1996, fluctuating between 1,200 and 1,500 students, enrollment of women in the soil sciences rose from 16.2 percent in 1987 to 32 percent in 1996. In 1996 there were 228 female B.S. graduates in soil science, almost double that of 10 years before. Ph.D. and M.S. candidates in the soil sciences in 1996 were also about one-third female, once again double from 10 years before (FAEIS 1997).

Progress in employment numbers in the SCS/NRCS appears a little less dramatic. In 1985, the SCS employed 85 women soil scientists at the federal level; in May 1998, there were 94 women soil scientists in various level positions. The progress is significant, however, given the overall reduction in the total number of soil scientists in the agency. Currently, about half of the NRCS's new soil scientists are women (Natural Resources Conservation Service 1992/1998). We can anticipate a twenty-first century that witnesses a continuing trend of more women working in the field in soil survey and private consulting firms and as teachers and researchers in the field of soil science in university and laboratory settings across the nation and the world.

ACKNOWLEDGMENTS

The author wishes to thank John Tandarich, Soil Science Society of America (SSSA) Historian; Douglas Helms, Natural Resources Conservation Service (NRCS) Historian; Gordon Huntington, University of California, Davis; and Gary Sposito, University of California, Berkeley, for their assistance. They have been investigating the history of soil science for many years and provided her with very helpful comments. She also wants to thank the 20 or so other men and women whom she interviewed informally over the phone for their thoughts and experiences.

REFERENCES

American Society of Agronomy. n.d. List of fellows database awards and biographies. Soil Science Society of America, pp. 1–99.

Association of Women Soil Scientists (AWSS). 1960–1998. Informal records provided to M. Levin by M. Faber, AWSS Historian 1988–1998.

Baker, Gladys L. 1976. Women in the U.S. Department of Agriculture. *Agricultural History* 50(1, January):190–201.

Baldwin, Mary, interview by John Tandarich and John Goodman, 12 December 1992.

Berc, Jeri, telephone interview by author, 7 August 1998.

Birkeland, Peter, telephone interview by author, 22 May 1998.

Bureau of the Census. 1911. *Official Register: Persons in the Civil, Military, and Naval Service of the United States, and List of Vessels*. Vol. 1. Washington, D.C.: Bureau of the Census, Department of Commerce and Labor.

Burt, Rebecca, telephone interview by author, May 1998.

Cattell, Jaques, ed. 1944. *American Men of Science: A Biographical Directory*. Lancaster, Penn.: Science Press.

Cattell, Jaques, Garrison Cattell, and Dorothy Hancock, eds. 1961. *American Men of Science: A Biographical Directory, the Physical and Biological Sciences*. 10th ed. Tempe, Ariz.: Jaques Cattell Press.

Current Research Information System (CRIS). 1998. Summaries of J. L. Boettinger, PSB research 7/01/93–6/30/98. (http://ext.usu.edu/agx/CRISreports/soi.html)

Esterman, Eva, telephone interview by author, 7 August 1998.

Evans, Chris, telephone interview by author, 11 May 1998.

Food and Agricultural Education Information System (FAEIS). 1997 (August). Fall 1996 enrollment for Agriculture, Renewable Natural Resources, and Forestry. (http://faeis.tamu.edu/)

Forsyth, Jane L. 1990 (December). Biography. University Files, Bowling Green State University, Bowling Green, Ohio.

Franks, Carol, telephone interview by author, May 1998.

Harden, Jennifer, telephone interview by author, 26 May 1998.

Helms, Douglas. 1992. Women in the Soil Conservation Service. *Women in Natural Resources* 14(1, September):88–93.

Huntington, Gordon, telephone interview by author, 26 May 1998.

Iyer, Jaya, telephone interview by author, 26 May 1998.

Jaques Cattell Press, ed. 1972. *American Men and Women of Science, Formerly American Men of Science, the Physical and Biological Sciences,* Vol. 4. 12th ed. New York: Jaques Cattell Press/R. R. Bowker Company.

Kellogg, Charles. 1963. Opportunities for soil scientists and agronomists in the Soil Conservation Service. *Agronomy Journal* 55:575–576.

Lapham, Macy H. 1945. The soil survey from the horse-and-buggy days to the modern age of the flying machine. *Soil Science Society of America Proceedings* 10:344.

Lapham, Macy H. 1949. *Crisscross Trails: Narrative of a Soil Surveyor.* Berkeley, Calif.: Willis E. Berg.

Levine, Elissa. 1998 (April). Resume. NASA Goddard Space Flight Center Laboratory for Terrestrial Physics, Greenbelt, Md.

Natural Resources Conservation Service (NRCS). 1992/1998. Employment databases. Personnel Division, Natural Resources Conservation Service, U.S. Department of Agriculture, Washington, D.C.

Natural Resources Conservation Service, U.S. Department of Agriculture (NRCS/USDA). 1997. Internal directories. U.S. Department of Agriculture, Washington, D.C.

Olson, Carolyn, telephone interview by author, 20 May 1998.

Perry, Ester P., G. Huntington, and F. Barrandino. 1952. Laboratory tests in soil survey and land classification. Abstract, Western Soil Science Society Meeting, Corvallis, Ore.

Prevost, Deborah, telephone interview by author, 13 July 1998.

Prevost, Deborah, and Bruce A. Linsay. 1999. *Soil Survey of Hualapai-Havasupai Area, Arizona, Parts of Coconino, Mohave, and Yavapai Counties.* Natural Resources Conservation Service, U.S. Department of Agriculture, Washington, D.C.

Reheis, Marith, interview by author, 22 May 1998.

Rice, Erwin, letter to author, 28 May 1998.

Rice, Erwin, telephone interview by author, 13 May 1998.

Scheyer, Joyce, telephone interview by author, May 1998.

Service News. 6 February 1940. U.S. Department of Agriculture, Soil Conservation Service Historical Files. Record Group 114, Records of the Natural Resources Conservation Service. National Archives at College Park, Maryland. (Lois Olson, head of the SCS Erosion History Section, published an article, Erosion: A heritage from the past, in *Agricultural History* 13(4, October 1939):161–170.)

Simonson, Roy W. 1993. Soil color standards and terms for field use—History of their development. In *Soil Color,* edited by J. M. Bigham and E. J. Ciolkosz, pp. 1–20. Madison, Wis.: Soil Science Society of America.

Soil Conservation Service (SCS). 28 October 1952. Records of Soil Survey Units 1933–1974; Historical Files; Soil Conservation Service, U.S. Department of

Agriculture; Record Group 114, Records of the Natural Resources Conservation Service. Record of collaborators and personnel in soil survey. National Archives at College Park, Maryland.

Sposito, Gary, telephone interview by author, 14 May 1998.

Stark, Nellie, telephone interview by author, 11 May 1998a.

Stark, Nellie, letter to author, 13 May 1998b.

U.S. Geological Survey (USGS). 1986. Cornelia Clermont Cameron (1911–1994) Memorial. Internal Documents, U.S. Geological Survey, Washington, D.C.

U.S. Geological Survey (USGS). 1994. Justification for nomination of Distinguished Service Award, Cornelia C. Cameron. Internal Documents, U.S. Geological Survey, Washington, D.C.

University of New Hampshire. 1998. *Research Highlights*. Spring.

Vasilas, Lenore M., telephone interview by author, 16 June 1998.

Wasserman, Mary Ann. 27 May 1992. BGSU geology prof helped trace Ohio's Ice Age roots. Bowling Green, Ohio, *Centennial Tribune*.

Weiland , Lillian H., ed. 1935. Bibliography on Soil Conservation compiled in the office of the chief of SCS. Revised by June Henderson, 1936. Historical Files; Soil Conservation Service, U.S. Department of Agriculture; Record Group 114. National Archives at College Park, Maryland.

Women's Bureau. 1960–1980. Observations; Records of the Women's Bureau 1892–1972, 1995; Record Group 86.3. National Archives at College Park, Maryland.

6

Contributions of African-Americans and the 1890 Land-Grant Universities to Soil Science and the Soil Survey

M. Dewayne Mays, Horace Smith, and Douglas Helms

The U.S. Soil Survey offered few opportunities for minorities during its first 60 years. Since the passage of civil rights legislation in the 1960s, however, the soil survey has become more inclusive. The employment profile of the soil survey in the 1990s, while still predominantly white and male, now includes minorities, particularly African-Americans, in both the federal service and among cooperators at universities and other institutions. Access to education and employment has improved, and the atmosphere in both universities and the soil survey is more open and encouraging than in earlier times. As with early leaders of the soil survey, particular professors and universities have helped guide students toward an interest in soils and the soil survey; the informal networks that emerged have persisted—some for nearly 50 years.

African-Americans did not begin to enter the Soil Conservation Service (SCS) in significant numbers until after the passage of the Civil Rights Act of 1964, when the agency developed an active recruiting campaign (Helms 1991). However, the historically black land-grant institutions, commonly called the 1890 universities, began practical soils research and outreach to farmers early in the twentieth century, with the work of George Washington Carver at Tuskegee University. Teaching programs in soil science began as early as 1936 at North Carolina Agricultural and Technical (A&T) University, followed in the 1940s by Virginia State University. And despite the limited opportunities available, a few dedicated African-Americans persevered in following their career interests into professional employment in the SCS before 1964, including as soil scientists, and ultimately advanced as increasing professional, educational, and employment opportunities emerged.

EARLY AFRICAN-AMERICAN EMPLOYEES OF THE SOIL SURVEY

William (Bill) Shelton, a native of Buckingham County, Virginia, joined the soil survey in 1955. Having left his family's farm to serve in Italy during World War II, Shelton took advantage of the GI bill to earn a degree in agronomy at Virginia State University in 1949. Because of the segregation still in force in the South at that time, he could not go on for a graduate degree in soil science at Virginia Polytechnic Institute, the state's land-grant university where graduate programs in agricultural sciences were available. Instead, Shelton earned an M.S. degree in soil science from Michigan State University in 1953, partly paid for by the state of Virginia in lieu of his attendance at the state's own university (a routine practice in many southern states during this period).

Shelton joined the soil survey at Plattsburg, New York, where he surveyed Clinton County. Later he served as the soil survey party leader in Herkimer County. In 1965, the SCS sent Shelton to Nigeria as a soil scientist on a U.S. Agency for International Development team assembled to develop a conservation program. The four-person team provided training and on-site assistance to the Nigerians to carry out a soil and water conservation program. He returned to the United States in 1967 and worked another two decades in the SCS national headquarters on soil survey interpretations as a soil scientist on the Soil Survey Division staff (Shelton 1999).

Letember (Lee) McDowell, a native of Proctorville, North Carolina, and the son of a schoolteacher and farmer who had graduated from the institution now known as Fayetteville State University, earned an undergraduate degree in agriculture from North Carolina A&T in 1952. Having become interested in soil science, he went on to earn an M.S. degree in that subject from Michigan State University in 1954. Following a brief period as a state employee in Michigan, McDowell applied for the SCS jobs in Wisconsin, Michigan, and New York. He joined the SCS in New York in 1957 and worked as a soil scientist in that state until his retirement (McDowell 1999).

Plater T. Campbell grew up in the District of Columbia but developed an interest in an agricultural career during summer stays on his maternal grandparents' tobacco farm in Charles County, Maryland. From experiences on the farm, he knew about the dual, segregated system of county agents in the extension service and wanted to become a county agent in southern Maryland. At Pennsylvania State University, Campbell earned a degree in agronomy with an emphasis on crops and soils, including courses in tobacco diseases. Upon graduation, he served in the Marine Corps, after which he applied for soil conservationist positions with the SCS all along the East Coast. Campbell abandoned his original career plans because, according to his recollection, the extension service had begun requiring a master's degree.

Campbell received no job offers and found that the SCS did not hire African-Americans in positions that involved working directly with farmers. So he broadened his application to include the position of soil scientist in the midwestern states and was soon hired for a position in North Dakota. He mapped soils in North Dakota for 10 years before moving to Connecticut. After switching into the soil conservationist series, Campbell moved up the ranks, becoming state conservationist in New Jersey, then finishing his career as director of the Conservation Planning Division in SCS's national headquarters (Campbell 1999).

THE ROLE OF THE 1890 UNIVERSITIES

These few African-American soil scientists in the SCS prior to the Civil Rights Act of 1964 attended the northern land-grant schools for their graduate education. But two of the three developed their interest in soil science while undergraduates at 1890 land-grant universities. These universities, authorized by the 1890 Morrill Act (also called the 2nd Morrill Act), required and provided funds for states that refused admission to their land-grant university based on race, thus providing an alternative institution of higher education for excluded students (Mayberry 1991a).

The 1890 universities include Lincoln University in Missouri, Alcorn State University in Mississippi, South Carolina State College, University of Arkansas-Pine Bluff, Alabama A&M University, Prairie View A&M University in Texas, Southern University in Louisiana, Tuskegee University in Alabama, Virginia State University, Kentucky State University, University of Maryland-Eastern Shore, Florida A&M Univer-

Plater Campbell works on a soil survey in Connecticut, September 1967. (Courtesy of the Natural Resources Conservation Service, U.S. Department of Agriculture—CN-576-8)

sity, Delaware State College, North Carolina A&T State University, Fort Valley State College in Georgia, Langston University in Oklahoma, and Tennessee State University (see Figure 6.1).

Some research has suggested that historically Black colleges offer more opportunities and better preparation for African-American students to assume important leadership roles later in life than do other colleges (Fleming 1984). The leadership qualities apparent in 1890 graduates employed by the Natural Resources Conservation Service (NRCS) and other governmental agencies support that finding. A demand for trained scientists created opportunities for participation in the soil survey and also benefited the science programs at 1890 universities.

Graduates of 1890 universities have made major contributions in all areas of the soil survey program. They have served in almost every position involved with soil survey, including director of the Soil Survey Division and leader of the NCSS. Many of these graduates have published scientific papers on soil survey and related topics in the professional jour-

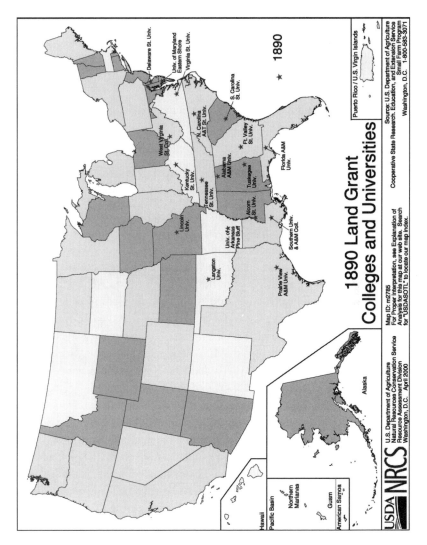

Figure 6.1 Map of 1890 land-grant universities. (Source: USDA, Cooperative State Research, Extension, and Education Service)

nals, and many have represented the soil survey on long-term and short-term international assignments on almost every continent on the globe.

The role 1890 institutions and their graduates have played and are playing in the soil survey, the Soil Science Society of America, and other worthwhile scientific endeavors can be easily documented, but it has received limited attention. It is hoped that emphasizing the contribution of these universities can raise awareness of the need for continued support of their programs.

Early Teaching Programs at the 1890 Universities

Although the 1890 universities contribute to the soil survey through all three of their missions—teaching, research, and extension/outreach—their undergraduate teaching program, which prepares minority students to work as soil scientists, administrators, computer scientists, and in other professional positions, is perhaps their greatest contribution.

Both Virginia State University and North Carolina A&T University, where Shelton and McDowell studied, introduced courses in soil science long before the 1964 Civil Rights Act provided the catalyst for developing soil science curricula throughout the 1890 university system.

As early as 1898, when James M. Colson of the Virginia Normal and Collegiate Institute at Petersburg (which became Virginia State University) learned about the work of the U.S. Department of Agriculture's Division of Soils, he wrote to the department, proposing to study in the division's laboratory to get instructions on developing the soil science component of the science and agriculture curriculum. Milton Whitney, chief of the U.S. Department of Agriculture's Division of Soils, advised him on readings and supplied some department publications, while Secretary of Agriculture James Wilson directed that Colson be received in the laboratory for a short time (Colson to Whitney 1898a, 1898b, 1899a, 1899b; Hill to Whitney 1898). The evidence is not clear about whether Colson actually visited in the lab with Whitney as was planned, and Colson left Petersburg before he had an opportunity to develop the soil science curriculum.

The long-standing teaching program at Virginia State University can be traced at least to 1948, however, with the graduation of Bill Shelton and Gregory Brockman. Other graduates from the Virginia State University program have included Sylvester Fletcher in the late 1950s, who was an early project leader in New Jersey and later entered into private business; David L. Jones (1962), who became state soil scientist for

As part of the District of Columbia's soil survey, Horace Smith (right) and Edward White take a soil sample in the National Park Service's Tulip Library, 1975. (Courtesy of Natural Resources Conservation Service, U.S. Department of Agriculture)

Mississippi; Garland Lipscomb (1963), who became the first African-American state soil scientist when he was appointed to that position for Pennsylvania in 1981; and Horace Smith (1964), who became director of the Soil Survey Division.

Horace Smith reached what many career soil scientists would consider the pinnacle of their profession when he became the seventh director of the Soil Survey Division in 1996. He was the first African-American to hold that position. Smith had already been, in 1977, the first African-American to serve as a soil correlator in a Regional National Technical Center and was one of the first two regional soil scientists to hold that position for the Southeast Region when he was appointed in 1995. A native of Clarkton, North Carolina, Smith joined the SCS in 1964 after completing his B.S. at Virginia State. He earned his M.S. in soil science at Ohio State University in 1972 while working for the SCS. Smith participated in one of the landmarks in soil surveying and interpretation of soils for urban uses when he served as party leader of the District of Columbia soil survey, undertaken with the intent of publishing the survey to coincide with the nation's bicentennial in 1976. Smith later served as state soil scientist for his native state of North Carolina.

Smith also represented the SCS/NRCS on many international assignments. In addition to presenting research papers at a number of international conferences in the 1980s and 1990s, he participated on several teams to provide expertise to nations attempting to develop soil survey and other conservation programs along the lines of those in the United States. He worked on a two-person team in Mexico in 1984 to develop a soil correlation document for the soils of the coastal region of Chiapis. In 1987, he provided assistance to the government of The Gambia, West Africa, in establishing a Soil Survey Division within their Ministry of Agriculture and Natural Resources, Soil and Water Management Unit. Smith traveled to Taiwan in 1994 to provide guidance to Taiwanese soil scientists on mapping mountain soils and on soil survey field review techniques. He visited and collaborated with soil scientists at the United Nations Food and Agriculture Organization and the European Soils Bureau in 1999.

Early soil science professors on the Virginia State faculty who influenced the lives of these men included Dr. John L. Lockett, Dr. Clarence C. Gray, III, Mr. Bennie A. Perry, Dr. Henry Ponder, and Mr. Millard T. Carter. Dr. T. Nelson Baker and Mr. Joseph H. Trotter, chemistry professors, also played an important part in their education and made a lasting impression on these men. The insight of these professors led to the development of a strong soils program at Virginia State University long before most other 1890 universities. For example, the 1961–1962 Virginia State University Annual Catalogue, describes the agronomy curriculum as follows:

> This department offers instruction in both crops and soils. Its objective is to provide a well-rounded practical as well as technical training for students in field crops, soils, fertilizers and other closely related subjects. Under crops are included such subjects as characteristics, production and improvement of farm crops. Courses in soils deal with chemical, physical and biological nature of soils and also with variation, modification and maintenance of productivity.
>
> The instruction is designed to train students for soil conservation, experiment station work, county agents and commercial farmers.

Soils courses listed in the catalogue included basic soils, soil classification, soil fertility, soil management, soil microbiology, soil analysis, soil conservation, soil chemistry, advanced soil fertility, and soil conservation, and land use. At the time of the passage of the Civil Rights Act, Virginia State University already had a full curriculum of soil science courses. As a result, a number of the first wave of African-American soil science trainees recruited by the SCS after 1964 were Virginia State graduates.

While North Carolina A&T University did not produce qualified soil scientists until 1968, their records show that they offered their first soils course in 1936. This was followed by their first course in soil erosion and control in 1937, and their first course designated as soil conservation in 1943. Among the early North Carolina A&T soils graduates were Robert Lee Powell, a 1968 graduate; James Dunn, who left federal service after serving as an SCS soil scientist in North Carolina and New Mexico; James H. Brown, who became state soil scientist in Maryland; and Dwight Holman, who became deputy chief for management of the NRCS. Willie Spruill graduated in 1969 and became an SCS resource soil scientist in North Carolina. Elliott Faison, Jr. worked with the SCS in Iowa and later in the cartographic unit in Lincoln, Nebraska, and Hyattsville, Maryland, before leaving the government for private employment. Tyrone Goddard became state soil scientist for New York. Goddard, like Horace Smith, represented the NRCS internationally, presenting a technical paper and participating in a 1998 symposium on sustainable agriculture and environmental protection in Herzegovina.

Alvin Parker and Robert Thomas both graduated in 1968; Thomas worked for the SCS in Iowa for a short time. Michael Mitchell graduated in 1969 and began his career with the SCS in Kentucky, where he continues to work. Carl Britt, a 1970 graduate, spent most of his career with the SCS/NRCS in Vermont, before leaving for a job in private industry. William Bowman, also a 1970 graduate, spent most of his career in Michigan. Bernard Benton graduated from the North Carolina A&T soils program in 1972. David Smith, another 1970 graduate, worked for the SCS in Illinois until 1974, when he joined the U.S. Forest Service. Smith became a regional soil scientist in Denver and later director of soil survey for the Forest Service in Washington, D.C. (Goddard 2001).

Some of the early professors in the soils program were Dr. W. E. Reed, the first professor of soils at North Carolina A&T, and Dr. Samuel Dunn, who came to the North Carolina A&T soils program in 1957.

Influence of the Civil Rights Act on Soil Science at the 1890 Universities

Despite the early programs in soil science offered by some of the 1890 universities, the most significant factor influencing the number of African-Americans being trained in soil science and entering the NCSS was the 1964 Civil Rights Act. The Civil Rights Act was a landmark in equal employment opportunity within the agency, and its passage provided the greatest boost to participation in the soil survey program by 1890 universities. Governmental agencies were directed to provide equal access to jobs and to programs that included financial support for education. Moreover, the implementation of student-trainee programs and the location of the USDA and the NRCS liaison officers on 1890 university campuses provided the agencies with better access to 1890 students and the opportunity to highlight many of the educational and research opportunities that exist on the campuses.

Student trainees, Temple, Texas, 1967. Sitting (from left) are Ronald E. Jarrett, Dewayne Mays, and Elmer J. Polk. Standing (from left) are Joseph J. Castille, Bobby J. Ward, Sylvester Bobo, George R. Hadnot, and Milton Lynn. (Courtesy of Dewayne Mays)

It was this event that prompted many of the 1890 universities to start offering programs with an emphasis in soil science, thus satisfying the minimum 15 semester hours needed to qualify as a soil scientist for federal service (Nettles 1988). The success of soil science teaching at 1890 universities can be attributed to the ingenuity and determination of the professors at these institutions, who developed new programs with limited resources. And their efforts have been greatly rewarded by the successes of their graduates.

After soil science curricula were established and soil survey student-trainee programs were implemented at many of the 1890 universities, there was a significant increase in the number of 1890 graduates who became soil scientists in the soil survey. By 1973, soil scientists from the 1890 universities numbered 79 career professionals and 14 student trainees out of a total of 1,300 soil scientists in the NRCS, although those years probably were the high point of employment of African-American soil scientists in the service (Smith 1999).

In the last four decades, that generation of employees, including soil scientists, has worked in a range of jobs throughout the agency. Three, Pearlie Reed, who graduated from the University of Arkansas-Pine Bluff in 1970, Dwight Holman, a North Carolina A&T graduate, and Charles Adams, who graduated from Southern University, worked in soil survey field parties early in their careers, and eventually rose to the top leadership ranks of the NRCS. Reed became chief of the NRCS, Holman became deputy chief for management, and Adams became regional conservationist for the South.

Southern University in Louisiana started to offer the necessary courses to qualify students for careers as soil scientists about 1959. Abe Stevenson and Lonnie Watson were the first two graduates in 1962. Stevenson began work for the SCS as a soil scientist in 1962 in New Mexico, later transferring to Wyoming. Other soil science graduates include Joseph Castille, who worked for the SCS in east Texas, Charles Adams, who worked on a soil survey party in Henderson County, Texas, and later became regional conservationist for the South, and Margaret "Gretta" Boley, who began her career with the SCS in New Mexico, before transferring to the U.S. Forest Service. Boley served on the Forest Service's National Soils Staff in Washington, D.C., then became deputy administrator at a national forest in Kentucky.

Rodney Peters, another product of the Southern University soils program, graduated in 1971 and joined the Forest Service. While there he mapped soils in New Hampshire, Missouri, and Arkansas, before becoming a forest watershed specialist in Texas. Bedford Cash, who graduated in 1974, mapped soils for the Forest Service in Arkansas's Ozark, St. Francis, and Kisathie National Forests and later was a ranger at the Tuskegee Ranger District. Earl Ford graduated in 1975 but began his career in 1972 with the SCS/NRCS in McCook, Nebraska, as a student trainee. He joined the Forest Service after graduation and worked in Texas, North Carolina, and Alaska, before becoming natural resources staff officer for California's Gifford Pinchot National Forest. Randy Moore graduated from the soils program at Southern University in 1976 and worked for the Forest Service as a soil scientist in Colorado, North Carolina, and at national headquarters. He left Washington, D.C., to become forest supervisor at Mark Twain National Forest. Art Bryant, another 1976 graduate, mapped soils for the Forest Service in Louisiana and North Carolina before becoming director of the Forest Service's Soil, Water, and Air Division in Washington, D.C. (Ford 2001; Vann 2001).

Early professors who had a major influence on graduates in soil science at Southern include Dr. Hezekiah Jackson, Dr. David Mayes, Dr. McKinley Mayes, and Dr. Mazo Price. More recent professors include Dr. Sughash Reddy, Dr. Thompson, Dr. Booker T. Watley, Dr. Kit Chin, and Dr. V. R. Bachireddy.

Ft. Valley State College in Georgia has produced a number of soil scientists who have worked in the NCSS program. It appears that Ft. Valley graduated its first soil scientist in 1965. In that year, the SCS employed Greenwood Hill as a soil scientist at Swainsboro, Georgia, and he was terminated 90 days later. He transferred to Ipswich, South Dakota, where he worked for four years. Other early graduates included Pete Davis (1965), who went to work as a soil conservationist, Thurman Sanders (1963), Hezakiah Benson, Donald Searcy, and William Bessent.

Cleo Stubbs started working for the SCS as a student trainee in 1966 before graduating from Ft. Valley and becoming a soil scientist with the SCS. He was the first African-American soil survey project leader in Alabama, but later left the SCS to become a private consultant. Joe Berry graduated from Ft. Valley in 1968 and began his career with the SCS in Jefferson County, Alabama. He later became a soil conservationist and rose to be the first GS-12 district conservationist in Alabama. Macarthur Harris, a 1970 graduate, joined the SCS and eventually became a soil survey project leader in Hail County, Alabama. Winford Andrews, a 1971 graduate, began with the SCS as a soil scientist, but soon converted to the soil conservationist series and became the first African-American District Conservationist in Alabama (McGhee 1994, 2001).

Hill and others credit their success and soils training to their soils teacher Mr. M. C. Blount. Dr. Mark Lattimore has led the soil science program since 1983.

The University of Arkansas-Pine Bluff (AM&N College) expanded its curriculum to qualify students as soil scientists in 1966. Dr. D. J. Albritton was the sole teacher of soils until the program grew in 1966, when Dr. M. S. Bhangoo joined the staff. More recent professors have included Dr. Mazo Price, Dr. Owen Porter, and Dr. Leslie Glover. The first graduate from the program was Dewayne Mays in 1968, who earned a Ph.D. in soil science from the University of Nebraska in 1982 and became head of the Soil Survey Laboratory in 1994. Other early graduates were Randal Buckner and Alex Winfrey in 1969, Bobby Ward, Pearlie Reed, and Andrew Johnson in 1970, and Emanuel Hudson, who began his studies at Tennessee State University, before finishing at the University of Arkansas-Pine Bluff and joining the U.S. Forest Service as a soil scientist. All of these men, except Hudson, were

Dewayne Mays at work, National Soil Survey Laboratory. (Courtesy Dewayne Mays)

employed in the soil survey program after graduation, with Reed rising through the ranks to become chief of the NRCS in 1997.

Bobby Ward became state soil scientist for Indiana and participated in a number of international projects for the SCS/NRCS. In 1993, Ward was part of a soil survey technology exchange in Zimbabwe and served as a technical specialist on a team that sampled benchmark soils that related to an agroforestry research project in Kenya and Malawi. In 1995, he served as team leader for a group of soil survey soil scientists involved in discussions with British soil scientists on issues of mutual concern. He also served, in 1999, as team leader for a group that assisted the South Africa's government in developing a field office technical guide. He, like Horace Smith and Dewayne Mays, also participated in international scientific conferences and field tours.

Florida A&M University elevated its soil science program in 1966 when they hired Dr. Charles Coultas to teach soils. Coultas trained with the SCS in Florida for five months, as part of an agreement between the SCS and Florida A&M to support the soils curriculum. Coultas taught from 1966 to 1985, during which time he spent a one-year sabbatical in Haiti (Coultas 2001). Florida A&M University students who graduated before 1966, including 1965 graduates Walter Douglas, Lloyd Law, and Robert Holmes, received additional soils training after graduation at 1862 land-grant universities. Douglas became state conservationist for South Carolina, while Holmes worked as an SCS soil scientist in New York and Virginia, and Law mapped for the SCS in Florida and Arizona before retiring (Vann 2001).

Coultas's first graduates included John Vann and Richard Ford, who graduated in 1968, William Taylor and Willie Pittman, who graduated in 1970, and Dennis Law, Ken Brady, and James Hart, who graduated in 1972. Vann and Bob Scott, another Florida A&M soils graduate, became soil scientists with the U.S. Forest Service, although Scott worked for the SCS in New Hampshire earlier in his career. Dennis Law began his career in Florida, moving later to Columbia, South Carolina. Ford, a 1968 graduate, played professional football with the Kansas City Chiefs for a short time before he joined the SCS as a soil scientist stationed in Florida (Vann 2001).

Dr. Kome Onokpise also taught soils at Florida A&M, and some of his students included Warren Henderson, who graduated in 1970 and became state soil scientist for Florida, and Al Roberts, who graduated in 1969 and mapped soils for the SCS in Connecticut, Rhode Island,

Texas, and Florida before becoming civil rights manager for the SCS/NRCS Southeast Regional Office. Edward Ealy, who became state soil scientist for Georgia, graduated from the soils program at Florida A&M in 1975.

Tennessee State University graduated its first qualified soil scientists in 1969. Soils professors Dr. Fred E. Westbrook and Dr. Desh Duseja had a strong influence on this program. Russell Barmore, who graduated in 1969 and became a soil scientist with the SCS in Nevada, then an SCS soil correlator in Arizona, is believed to be the first Tennessee State graduate with 15 semester hours of soils courses. Neil Williams graduated in 1970 and was employed by the SCS in Iowa. Travis Neely graduated in 1971 and eventually became state soil scientist for Indiana. Tommy Parham, who also graduated in 1971, rose to become state soil scientist for New Mexico, then director of the National Cartographic and Geospatial Center.

Other Tennessee State University graduates who went on to work with the SCS include Bobby Pirtle, who graduated in 1969 and worked as a soil scientist in Indiana; Thomas Jackson, a 1969 graduate; and Clarence Conners and George McElath, who both graduated in 1970 and became resource soil scientists, Conners in Tennessee and McElath in Indiana. Milton Allen, who also graduated in 1970, worked for the SCS in West Texas for two years before leaving to pursue a Ph.D. at the University of Tennessee. Leo Williams, another 1970 graduate, worked with the SCS in Iowa. Eddie Cummings, who graduated in 1971, worked with the SCS in Florida, and Charles Love, also a 1971 graduate, became a soil data quality specialist in Indiana. Emanuel Hudson, who finished his degree at the University of Arkansas-Pine Bluff and went to work as a soil scientist with the U.S. Forest Service, began his studies at Tennessee State (Barmore 2001).

Alcorn State University in Mississippi started to offer 15 hours of courses in soil science in about 1971, graduating their first soil scientists in 1975. Delmar Stamps, who advanced in the NRCS to the position of leader of the Regional Wetlands Team at Vicksburg, Mississippi, and a man named Whitorth, about whom little else is known, were the first graduates from the program.

Alabama A&M University's soil science curriculum started in 1972. Since 1972, the program has graduated a large number of soil scientists. Alabama A&M is the only 1890 university that offers a Ph.D. in soil science. Johnny Trayvic, believed to be the first graduate to concentrate in soils, joined the SCS in Selma, Alabama, after graduation and later became a soil survey project leader in Barbour County, Alabama. Milton Tuck and Delarie Palmer both graduated from the soils program in 1975. Tuck joined the SCS and eventually became a resource soil scientist in Alexander City, Alabama. Palmer worked for the SCS for about six years before leaving to become an environmental soil scientist at Ft. Rucker, Alabama. Lawrence McGhee joined the SCS after his graduation in 1982 and held the position of party leader in Alexander City, Alabama, in 2001 (McGhee 1994, 2001).

Prairie View A&M University in Texas has gone through phases during which 15 hours of soils courses were offered and phases when they were not. However, Nat Conner appears to have been the first qualified soil scientist to graduate. Conners graduated in 1955 and was the first African-American project leader and area soil scientist in Texas before he changed his career focus and eventually became state conservationist in Hawaii. Other graduates include Edward Griffin and Levi Steptoe, who both graduated in 1976 and became soil survey party leaders; Cleveland Watts, who graduated in 1977 and became an SCS soil scientist in Kansas and later state soil scientist in North Dakota; and Sam Brown, who graduated in 1979.

Langston University has one of the more recently implemented soils programs, which started only in 1993. Since that time, they have graduated three soil scientists who are working for the NRCS in Minnesota. Dr. Acquah is the principal coordinator for the program.

The only other 1890 university with a soils program is Tuskegee University, but few graduates have joined the SCS/NRCS or other federal services. One of their graduates, Felix Ponder, became a research soil scientist with the U.S. Forest Service at Lincoln University in Jefferson City, Missouri (Barmore 2001).

DEVELOPMENT OF SOIL SURVEY RESEARCH AND EXTENSION/ OUTREACH PROGRAMS AT THE 1890 UNIVERSITIES

While the important teaching role of the 1890 land-grant universities is well established, some of these institutions also offer advanced degrees, providing research in areas such as soil science and related earth sciences. Recently, the research role has become more important; the programs were small before 1967 mainly because of a lack of funding. Prior to this time, most of the research was relegated to 1862 institutions. Only a few 1890 universities had acquired financial resources to carry out limited research. Tuskegee University was an exception since it had been carrying out a variety of research since the turn of the twentieth century. The 1964 Civil Rights Act and public concern with the disparity between funding for 1862 universities and 1890 universities prompted a change in the administration of research and ultimately its application to the soil survey as it relates to 1890 universities (Neyland 1990).

Beyond the funds authorized through the original 1890 Morrill Act, additional financing has become available through other legislation. This legislation requires that the 1890 universities receive a share of federal funds that are designated for research at land-grant universities. The formation of the 1890 Association of Research Coordinators in 1972, which became the Association of Research Directors in 1977, provided a vehicle to coordinate and encourage research in the 1890 universities. With the passage of the Food and Agriculture Act of 1977 (PL 95-113), 1890 universities, including Tuskegee University, were provided funding to conduct research. Public Law 97-58 established the 1890 Facilities Program to provide $10 million a year to the 1890 universities with no requirement for matching funds. Federal grants to 1890 universities increased from $283,000 in fiscal year 1967 to $25,300,000 in fiscal year 1990 (Neyland 1990). These funds provided a great boost to the capacity of these universities to fulfill their research, teaching, and extension/outreach missions.

Special funding programs, like the 1890 Capacity-Building Grants, have helped these universities develop research programs that benefit students, faculty, and the larger community. Since 1990, the USDA's 1890 Capacity Building Grants program, has awarded 1890 universities more than $78 million, which supports more than 368 projects—$37.3 million for 217 teaching projects and $40.4 million for 169 research projects (Cooperative State Research, Education, and Extension Service 1998).

This additional funding has been particularly important to increasing capacity for training students in the science disciplines, including soil survey and its supporting research. Students have become better prepared to meet the employment and research demands of agencies like the SCS. The Geographic Information Systems (GIS) Centers of Excellence at Lincoln University and Alabama A&M University are just two examples of 1890 programs that are preparing students to meet the challenges of the modern soil survey program.

The first 1890 university to become a member of the NCSS, which brings the university into a direct research support association with the soil survey, was North Carolina A&T University. One year later, all 1890 universities became full members of the NCSS.

In 1998, the 1890 Institution Teaching and Research Capacity Building Grants program awarded funding to a number of projects that have benefited the soil survey. Teaching projects included: 1) Infusion of Precision Farming Concepts in Agricultural Concepts in Agricultural Curricula at Alabama A&M University; 2) Hands-on Experiences to Expand Undergraduate Scope in Environmental Hydrology at Florida A&M University, 3) Use of GIS to Enhance Urban Forestry Education at Southern University, and 4) GIS Training and Implementation at Tuskegee University. Research projects that have been of value to the soil survey include: 1) Effect of Best Management Practices on Primary Nutrient Delivery Pathways in the Manokin River Basin at the University of Maryland-Eastern Shore, and 2) Poultry Litter as a Soil Amendment: Possible Links to Foodborne Diseases at the University of Maryland-Eastern Shore (Cooperative State Research, Education, Extension Service 1998).

Other research at 1890 universities contributes to the program of the soil survey. At Florida A&M University, Dr. Y. P. Hsieh, Dr. Charles Coultas, and C. H. Yang have conducted research in the management and ecology of northern Florida wetlands. The results of their research were helpful in developing wetland criteria for government-administered programs, as well as for carbon sequestration studies and wetland restoration (Neyland 1990). Dr. Coultas attributes much of the information in his book *Ecology and Management of Tidal Marshes: a Model from the Gulf of Mexico* (Coultas 1997) to his former students (Coultas 1999).

Researchers at Alabama A&M University, studying remote sensing as it relates to agriculture and natural resources, have developed methods for quantifying soil differences acquired from satellite imagery. Research in GIS methodology in soil survey and inventory monitoring, led by Dr. Tommy Coleman also at Alabama A&M, will be very important in developing newer and more efficient ways of making soil surveys.

North Carolina A&T University's research with sewage sludge promises to provide helpful information for the development of soil interpretation criteria for animal waste disposal in soils and for the effects of resulting heavy metals. Dr. M. R. Reddy has studied contaminant migration in groundwater, nitrogen fixation effects on eutrophication in ponds, and the use of sewage sludges (Neyland 1990).

The practice of examining soil behavior and transferring the knowledge and experiences between and among areas is important to the success of soil survey programs in this country. This practice has a long history at the 1890 universities, dating back to the movable schools created in 1906 by George Washington Carver at Tuskegee. Through practical demonstrations at working farms, Dr. Carver used his teaching and research skills, including an understanding of soil behavior differences, to encourage crop rotation and soil conservation practices. These practices became nationally recognized and are credited with having a positive impact on the economic success of the South (Mayberry 1991b). More recent examples of outreach efforts that provide assistance to improve soil management include the small farmer programs at Florida A&M University, Ft. Valley State College, and the University of Arkansas-Pine Bluff.

THE FUTURE FOR THE NRCS AND THE 1890 UNIVERSITIES

In 1998, the NRCS made an analysis of its workforce and determined that a large number of its scientists, especially soil scientists, were reaching retirement age. The analysis also revealed that the number of minorities in soil scientist positions was dwindling fast, and the prospect of maintaining a diverse cadre of qualified soil scientists to support the soil survey in the future did not look promising. As a result, the NRCS began providing resources to support a Soil Science Scholars Program at five of the 1890 universities—Virginia State University, Tennessee State University, Alabama A&M University, Prairie View A&M University, and University of Arkansas-Pine Bluff. At the same time, the NRCS began providing support for a Soil Science Scholars Program at an Hispanic-serving university and a Native American-serving university, in hopes that these programs will help attract highly qualified students into the soil science discipline and eventually lead them to become scientists and leaders within the soil survey.

In addition, the Soil Science Institute, a four-week highly intensive training course sponsored by the NRCS, was held during March 2000 at Alabama A&M University. This course, conducted by Dr. Andrew Manu, was designed to update midcareer soil scientists in all areas of soil science—soil chemistry, soil physics, soil fertility, and soil classification—and to expose them to cutting-edge topics and technologies relevant to soil survey such as GIS, global change, animal agriculture, and Fuzzy Logic, among other concepts. Although the course was not new, this was the first time in the history of the course that it was held at an 1890 university.

The soil survey program has grown and profited as a result of its increased association with 1890 universities. African-Americans' documented leadership, research, and other scientific contributions to the development of this nation can also be seen in the soil survey's efforts to provide nationwide soil survey information of the highest quality to its customers. Graduates from 1890 universities have successfully served at all levels within the soil survey program.

The soil survey experience serves as an example of the success that can be achieved when the opportunity and nourishment necessary to succeed are provided. The soil survey's success should be credited in equal parts to 1) the foresight of those pioneers at 1890 universities who were determined to see their students succeed; 2) the graduates who dared to take a chance on a future in soil science; and 3) the governmental agencies and individuals who were willing to put forth the extra effort needed to achieve success. Future cooperation and partnerships between the soil survey and 1890 universities look extremely promising, and constraints appear to be few.

REFERENCES

Barmore, Russell, telephone interview by Dewayne Mays, 25 May 2001.

Campbell, Plater, telephone interview by Douglas Helms, 5 February 1999.

Colson, James M. to Milton Whitney. 26 August 1898a. Letters Received, Division of Soils; Record Group 54, Records of the Bureau of Plant Industry, Soils, and Agricultural Engineering. National Archives at College Park, Maryland.

Colson, James M. to Milton Whitney. 5 September 1898b. Letters Received, Division of Soils; Record Group 54, Records of the Bureau of Plant Industry, Soils, and Agricultural Engineering. National Archives at College Park, Maryland. National Archives at College Park, Maryland.

Colson, James M. to Milton Whitney. 15 May 1899a. Letters Received, Division of Soils; Record Group 54, Records of the Bureau of Plant Industry, Soils, and Agricultural Engineering. National Archives at College Park, Maryland. National Archives at College Park, Maryland.

Colson, James M. to Milton Whitney. 4 July 1899b. Letters Received, Division of Soils; Record Group 54, Records of the Bureau of Plant Industry, Soils, and Agricultural Engineering. National Archives at College Park, Maryland. National Archives at College Park, Maryland.

Cooperative State Research, Education, and Extension Service (CSREES). 1998. *1890 Institution Teaching and Research Capacity Building Grants*

Program Annual Summary FY 1998. CSREES Higher Education Programs, U.S. Department of Agriculture, Washington, D.C.

Coultas, Charles, telephone interview by Dewayne Mays, September 1999.

Coultas, Charles, telephone interview by Dewayne Mays, 3 July 2001.

Coultas, Charles L., and Yuch-Ping Hsieh. 1997. *Ecology and Management of Tidal Marshes: A Model from the Gulf of Mexico*. Delray Beach, Fla.: St. Lucie Press.

Fleming, Jacqueline. 1984. *Blacks in College*. San Francisco: Jossey-Bass.

Ford, Earl, telephone interview by Dewayne Mays, 25 May 2001.

Goddard, Tyrone, telephone interview by Dewayne Mays, 23 May 2001.

Helms, Douglas. 1991. Eroding the color line: The Soil Conservation Service and the Civil Rights Act of 1964. *Agricultural History* 65(2):35–53.

Hill, William to Milton Whitney. 8 September 1898. Letters Received, Division of Soils; Record Group 54, Records of the Bureau of Plant Industry, Soils, and Agricultural Engineering. National Archives at College Park, Maryland. National Archives at College Park, Maryland.

Mayberry, B. D. 1991a. *A Century of Agriculture in the 1890 Land-Grant Institutions and Tuskegee University, 1890–1990*. New York: Vantage Press.

Mayberry, B. D. 1991b. The Tuskegee movable school: A unique contribution to national and international agriculture and rural development. *Agricultural History* 65:85–104.

McDowell, Lee, telephone interview by Douglas Helms, 5 February 1999.

McGhee, Lawrence. 1994. History of Black SCS employees in Alabama. February. Unpublished manuscript, Soil Conservation Service, Alexander City, Ala.

McGhee, Lawrence, telephone interview by Dewayne Mays, 15 June 2001.

Nettles, M. T., ed. 1988. *Toward Black Undergraduate Student Equity in American Higher Education*. New York: Greenwood Press.

Neyland, L. W. 1990. *Historically Black Land Grant Institutions and the Development of Agriculture and Home Economics 1890–1990*. Florida A&M University Foundation, Inc., Tallahassee, Fla.

Shelton, William, telephone interview by Douglas Helms, 9 March 1999.

Smith, Horace, information supplied to Douglas Helms, 23 May 1999.

Vann, John, telephone interview by Dewayne Mays, 24 May 2001.

7

SOIL SURVEY AND THE U.S. FOREST SERVICE

Dennis Roth

INTRODUCTION

Innovation accompanied by control and verification are necessary features of any modern science. Institutionally, these functions may be performed separately, as has historically been the case for forest soil surveys. When the U.S. Forest Service became involved in soil investigations, it discovered that methods devised for agricultural land were not always appropriate for upland mountain soils. It needed to innovate and it did. The Soil Conservation Service (SCS, renamed the Natural Resources Conservation Service, NRCS, in 1994), on the other hand, has functioned as the control to this innovation. Sometimes perceived as a "rigid" barrier to change, it has also played the necessary role of providing methodological checks and balances.

CREATION OF THE U.S. FOREST SERVICE

In 1905, the Congress established a new land management agency in the U.S. Department of Agriculture (USDA) by transferring 18 million acres of public forestland and accompanying personnel from the Department of the Interior's General Land Office to Gifford Pinchot's Bureau of Forestry in the USDA, renamed the U.S. Forest Service after the transfer. During the next two years, President Theodore Roosevelt, following the advice of his friend Pinchot, greatly enlarged the size of the national forest system by setting aside (reserving) forestland from the public domain. In 1907, however, Congress asserted its authority over the creation of national forests. It required that further additions to the system be subject to its approval, bringing expansion to a virtual standstill. When Roosevelt left office in 1909, the national forest system had essentially assumed its current shape, except in the eastern United

States where national forests were created later through purchase from private owners.

From its beginning, the U.S. Forest Service has been charged with managing national forestlands for multiple uses. Over the years, however, the relative priorities given to these uses have changed. Once, timber and grazing were dominant; today, recreation and wildlife are in the ascendancy.

EARLY SOILS WORK ON NATIONAL FORESTLAND

During its first decade, the Forest Service had to contend with various legal and political challenges to its existence from ranchers, loggers, and others. In all of these cases, the courts upheld the right of the federal government, through the Forest Service, to establish grazing fees, regulate timber harvesting, and control the construction of private dwellings and resorts. In only one case was the agency forced to retreat. In their haste to get land into the system, Roosevelt and Pinchot had included some potentially arable land within their acquisitions; Congress directed that these lands be opened to homesteading.

Foresters had done a good job of selecting forestland but were not always qualified to identify potential cropland, usually found in and around mountain valleys. For help, they called on another young USDA agency, the Bureau of Soils. During the summers from 1913 to 1917, scientists from the Bureau of Soils conducted what today would be called large-scale reconnaissance surveys of many of the national forests. Their principal task was to locate agricultural land, but in the process they also produced maps of soils for many of the forests. How the Forest Service used these maps is not recorded. However, since interpretations did not accompany maps of forest soils, as was routine for maps of agricultural soils, they likely were of limited value to the Forest Service, except in their capacity to show the location of potential agricultural land. Today, these maps reside in the National Archives in College Park, Maryland (USDA 1914).

For the next 30 years, the Forest Service had little contact with the Bureau of Soils or the SCS, which was created in 1935. During these years of "custodial" management, when relatively little commercial or recreational use was being made of the national forests, the Forest Service felt no need to call on other disciplines and agencies. In addition, foresters often took introductory courses in disciplines such as soil science and prided themselves on being able to do virtually any job in the forests. ("Foresters can do anything" was a slogan occasionally heard as late as the 1970s.) This lack of demand for knowledge of forest soils was mirrored in soil scientists' indifference to its study. Soil survey, until the early 1960s, was almost totally preoccupied with the classification of agricultural land, consigning forests in their surveys to the realm of undifferentiated "rough and stony ground." World War II, however, brought an end to the relatively static "custodial" era.

BEGINNINGS OF HOME-GROWN WORK

The war and post-war construction boom placed a tremendous demand on commercial forests. Timber companies began to look increasingly to the national forest system for their supplies, which they had previously ignored because of their remoteness and high costs of access. Opening up the national forests required the building of more roads, but the building of logging roads on slopes could lead to landslides, which began to happen with increasing frequency in California, Oregon, and Washington. It was quickly realized that the proper engineering of logging roads depended on some knowledge of the soils on which they were constructed.

Bill Wertz, a "first-generation" Forest Service soil scientist, has opined that "everything good and bad in America seems to begin in California and that was also the case with Forest Service soil science" (Wertz 1997a). In addition to failing logging roads, California had several other characteristics that predisposed its Forest Service personnel to become pioneers in forest soils, including: 1) a strong Forest Service research program; 2) the University of California's active extension program; 3) the fact that there was only one state SCS office to deal with, thus simplifying interagency interactions (other Forest Service regions encompassed several states); and 4) the existence of fairly rapid landscape transitions, which brought agricultural and forest soils into closer association than in other regions (Bradshaw 1997; Goudey 1997; Sherrell 1997; Leven 1997; Anderson 1997).

Jack Fisher, who became the first soil scientist in Region 6 (R-6; Oregon and Washington) in 1959, received a degree in soil science from the University of California at Berkeley, where he first developed an interest in the engineering aspects of soils. After World War II, he worked for the state of California in research, and then, in 1949, he joined the Forest Service's Arcata Laboratory, where he became involved in road-building tests. Throughout the rest of his career, Fisher maintained a strong interest in soil engineering. In 1955, he transferred to the Forest Service's Pacific Southwest Region, where he worked for Ken Bradshaw, who had just become the California regional soil scientist in the Division of Watershed Management in the San Francisco office (Fisher 1997). (Fisher had an office at the Pacific Southwest Experiment Station in Berkeley, but actually worked for the regional office, which administered California's national forests.)

Bradshaw, who had also graduated from Berkeley (with a degree in forestry), was a temporary worker for the SCS in the late 1930s. After the war, he joined the Pacific Southwest Station, where he worked for A. E. Wieslander, who had formed a partnership with Professor Earl Storey of the Soil Science Department at Berkeley. Both were interested in the association between soils and vegetation, but found that the SCS's findings on upland areas were too general. Something more specific was needed. Thus began the California Soil and Vegetation Survey, which combined the efforts of the station, Berkeley, the SCS, and the USDA Extension Service. Most of these early surveys took place on private forestland, although some national forestland was also covered. In 1949, Professor Storey conducted the first soils training school for foresters, initiating "a few green foresters" into the relationships between upland soils and vegetation (Bradshaw 1997).

Also in 1949, Herbert A. Lunt, another pioneer in forest soils, and C. L. W. Swanson, both of the Connecticut State Agricultural Experiment Station, published "Mappable Characteristics of Forest Soils" in the *Journal of Soil and Water Conservation*. The authors reasoned that the surveying of forest soils had been neglected because forest soils were of lesser economic value and the "need for information on farm soils was great" (Lunt and Swanson 1949). They referred to a recent conference on forest soils in Portland, Oregon, where it had been agreed that the disappearance of virgin timber meant forest soils could no longer be taken for granted, especially soils in the Douglas-fir region. They went on to say that traditional cropland survey methods needed to be modified for a forest survey program so that, among other things, soil types having similar characteristics could be grouped together for mapping purposes (Lunt and Swanson 1949).

BECOMING ESTABLISHED

It took six more years for momentum to build. In 1955, the same year Bradshaw became the first regional soil scientist, John Retzer, a Ph.D. in soil science and a soil scientist since 1946 at the Rocky Mountain Research Station in Fort Collins, Colorado, was transferred to headquarters in Washington, D.C., to become the service's first national soils coordinator. Bradshaw has speculated that Wieslander's work on the California survey, which he promoted effectively, had convinced Forest Service officials in Washington of the need to begin a national program in soils. Charles Kellogg, director of USDA's Soil Survey in the SCS, was also undoubtedly an important influence in establishing the program. Several first-generation soil scientists believe that Kellogg recommended Retzer for the job, and that thereafter Retzer was anxious to remain on good terms with him. But Retzer's close relationship with Kellogg may not have been just a matter of gratitude because, as several of these same men remember, Kellogg was a strong, intellectually dominant personality, with a professional reputation unequaled among federal bureaucrats (Wertz 1997a; Richlen 1997; Olson 1997; Meurisse 1997).

Within a year of his transfer to Washington, Retzer began to recruit regional soil scientists for each Forest Service region. Known as a scholarly individual who encouraged publication but was not particularly good at public relations, Retzer looked for capable men who could write well. He also wanted his program to comply with the standards of the SCS's National Cooperative Soil Survey (NCSS). Because of his work on the relationship between landforms and hydrology, he could sympathize with the desire of some of his colleagues in the regional offices to go beyond the SCS soil survey and classification standards, but he did not encourage it in practice.

Between 1955 and 1959 he selected 15 men who, in addition to Bradshaw already in place in California, would lead the new soil science program. They were Mac Maconnell, Region 1 (Northern Region); Charlie Fox and John Nishimura, Region 2 (Rocky Mountain Region); Jack Williams, Region 3 (Southwest Region); Olaf Olson and Dick Alvis, Region 4 (Intermountain Region); Ken Bradshaw and Art Sherrell, Region 5 (California Region); Jack Fisher, John Arnold, and Rich Richlen, Region 6 (Pacific Northwest Region); Bob Reiskie, Region 7 (North Central Region, which was broken up and merged with Region 8 and Region 9 in 1967); T. W. Green, Region 8 (Southern

Region); Bill Wertz and Ed Newman, Region 9 (Great Lakes Region); and Freeman Stevens, Region 10 (Alaska Region).

Bill Wertz has given a portrait of some of his colleagues in the first generation:

> Olaf Olsen (Ole) was with Forest Service Research in California working on erosion problems and was a good pick for Region 4. Ole was an excellent field man and liked field work. He was also an excellent mixer, well-liked with friends all over.
>
> The oldest I think was Mac Maconnell in Region 1. Mac was an old-time soil mapper from the Dakotas, very thorough and competent, but not a pushing program leader. He retired in the early 60s and was followed by Rich Richlen, who transferred from Portland. Richlen was himself a good technician with lots of energy.
>
> Region 2 was always low-key, a reflection perhaps of Charlie Fox, quite competent with soils but again not a program driver. Charlie had been all over and had lots of stories. His best is of talking about the old-timers in the original Soil Survey and their foibles, and then he would sadly say, "but there aren't any real characters left around anymore." He should have looked in the mirror.
>
> In Region 3 was John Williams. He came out of a SCS soils lab in New Mexico and was pure gold. Some years older than myself and most of the rest, he was respected by all, although some of the younger guys thought him a little overbearing. Possibly he was the best of the lot in his job. He had a number of other nicknames, including "Straight Arrow" and "Zuni Jack," always a mark that you are liked or respected or at least that people know you are there.
>
> In Region 6 were Jack Fisher and John Arnold. Fisher was a kind of outlaw of the group, had all the attributes of a true Californian and probably invented the movie and television concept of Happy Days. A good guy to know and like. Arnold went down to Ogden and then to Boise in Idaho. His early work in Oregon was brilliant and the groundwork for much to come. John was a forester, went to work in soils and became a self-taught geomorphologist. He was always dedicated to what he saw as right and vehement to anyone who couldn't see it.
>
> Region 8 was T. W. Green, a true son of the South, a colonel in the Marine Air Corps who came up through the old soil survey. T. W. fit in like a glove. He was at home with the good ol' boys and did well. I called him "Daddy Rabbit." (Wertz 1997b)

As can be gathered from Wertz' description, this was a friendly and congenial group of men. All, except John Arnold, came from working-class or rural backgrounds, and most began their careers in the SCS. (Forest Service soil science was an all-male domain until 1975, when Penny Luehring became a temporary employee at the Santa Fe National Forest.) They came together at the annual meeting, which rotated among regional offices until 1966 when Retzer left the Forest Service and the practice stopped for a number of years. They learned from each other's experiences, especially the younger men who profited from their association with soil science veterans such as Jack Williams and Olaf Olson. Retzer promoted the idea of uniform standards, but he was aware that the regional soil scientists, like other agency professionals, had to adapt to local conditions and thus did not attempt to impose any particular standards on them (Wertz 1997a). After Retzer's departure and the temporary cessation of annual meetings, regional practices began to diverge even more.

For its first six years, Retzer's program remained a "pilot" program and was housed in the Division of Watershed Management Research. As a result of this status, the 11 regional soil surveys under his direction carried the designation "pilot surveys." In 1961, soil survey shed its "pilot" status when the agency's leadership, convinced of the program's usefulness, transferred it to the Division of Watershed Management. (Several regional soil scientists were already a part of Watershed Management.) Also in that year, the Forest Service signed its first formal national agreement with the SCS, stating that Forest Service would follow the procedures of the NCSS (Russell 1961).

Over the next 30 years, the relationship between the Forest Service and the SCS was marked by varying degrees of cooperation, indifference, and hostility. Prior to its 1961 cooperative agreement with the Forest Service, the SCS had entered into cooperative agreements with academic departments, state experiment stations, and other state agencies. A few states, including California and Illinois, chose not to participate in the federal program, but most acceded to federal standards and the leadership of the Bureau of Soils. In the Forest Service, however, the SCS encountered an agency nearly 10 times its size. The service had little patience for work not directly supportive of its multiple-use management objectives. Moreover, the Forest Service was divided into 10 (nine after 1967) semi-autonomous regions, each of which had its own way of surveying and interpreting soils. Therefore, this history in its

A soil scientist and forester obtain data for soil site correlations, Eldorado County, California, November 1961. (Courtesy of the National Archives at College Park, Maryland; Print 114-H-Cal-7,304; Record Group 114, Records of the Natural Resources Conservation Service)

particulars is best told from the regional perspectives. First, however, there are some general ideas, applicable to some extent throughout the Forest Service and the SCS/NRCS that need to be mentioned.

WORKING WITHIN TWO DIFFERENT CONTEXTS

The soil survey within the NRCS is primarily a scientific endeavor and "correlation," the placement of newly surveyed soils into the established taxonomy, is its most important component. Correlation is the scientific method of control in soils work because without it, investigations cannot be accurately compared or replicated, and individual reports become one-of-a-kind and relatively useless. As Dick Arnold has pointed out, the correlation of soils throughout the United States is a function that only the federal government is equipped to perform. Labor intensive and demanding, it is beyond the means or interests of other organizations (D. Arnold, 1997).

Therefore, as the official correlator of all U.S. soils, the NRCS has always been concerned that this work be done to proper standards. These standards include collecting enough soil samples or profiles within a given unit of land, the proper completion of Form 5s and 6s, and, as much as possible, that soils are classified taxonomically down to the "series" level (roughly analogous to the "species" level in biology). If these steps are not followed and NRCS cannot correlate the results, they are not published as part of the NCSS.

The Forest Service soils program has not been a part of a research division since 1961. The agency has always taken a very practical approach to soils work, asking how it can "help us manage that chunk of land out there" or how it can be used in a national forestland management plan. If enough information can be gathered short of a standard Order 2 detailed survey, then Forest Service managers have often favored lower intensity Order 3 or Order 4 surveys to save time and money. The fact that inaccessible forestland is more expensive to survey than cropland undoubtedly influences this approach. When agency soil scientists became too zealous for their managers' liking, they could be brought up short with the sarcastic "who are you working for, the Forest Service or the SCS?" Or more specifically, a forest supervisor could tell them, "If I can't meet my [timber] cuts without soil surveys, then they're important. Otherwise not" (Corliss 1997).

Soil science has had to defend itself within the Forest Service. On at least one occasion in the late 1960s, it faced possible elimination or radical downscaling. But when Forest Service soil scientists failed to follow SCS standards, they were not always just responding to management pressure. They often felt that SCS correlators were "book-driven," or overly rigid, wanting to make soil distinctions in areas where they were not appropriate, such as steep, gravity-mixing slopes, which were composed of intermingled "composite" soils (Cline 1997).

The SCS's insistence on making such distinctions must have been ironic to some in the Forest Service, since Forest Service soil scientists were now being asked to make fine differentiations on land that had been previously relegated in the SCS Soil Survey to the amorphous catch-all category of "rough and stony ground." To the SCS, however, it just meant that a previously neglected type of land now had to be surveyed to the same standards as agricultural land.

Related to the above is another consideration, which Dick Arnold (1997) pointed out: the difference between studying soils for themselves (NRCS) and studying them as indicators of other things (Forest Service). For instance, on mountain slopes the Forest Service was just as interested in the "substratum" material as it was in the soil layers above because the tendency for slippage is dependent on both factors. Of course, the SCS and its predecessors had been interested in the associations between soils and commercial crops. But over the years, these became well known and were perhaps taken somewhat for granted. Moreover, the increase in the use of irrigation and chemicals had made these relationships less deterministic. The Forest Service wanted to know what soils implied about tree growth, reforestation, browse yields, water flows, and wildlife habitat. If this could not be done (or done cheaply enough) within the confines of a standard soil survey, then other ways had to be found. These different approaches were also reflected in the "languages" the agencies used. Motivated by its need to integrate multiple uses, the Forest Service by the late 1960s was increasingly speaking the jargon of systems theory, which as Dick Arnold recalls, the SCS soil scientists found unintelligible (D. Arnold 1997).

Interestingly, the combined effect of geology, personality factors, and history appeared to create a rough north-south gradient for the degree that the Forest Service accepted the SCS/NRCS standard soil survey methods. For instance, California (R-5) has generally been much more compliant than Oregon or Washington (R-6). The same has been true for the South (R-8), as well as for the Southwest (R-3) (at least until the late 1970s) and Colorado (R-2), compared with the Great Lakes (R-9) and the northern Rockies (R-1).

PACIFIC NORTHWEST REGION (R-6)

The main stream in the history of the Forest Service soil program begins in the Pacific Northwest Region (R-6), runs through the Northern (R-1) and Intermountain (R-4) Regions and then terminates in the Great Lakes Region (R-9). From there it has turned back and washed over most of the National Forest System, except the Southwest (R-3), which has gone its own way.

In 1957, John Arnold initiated one of the first pilot surveys on the South Umpqua section of the Umpqua National Forest in Oregon. He established preliminary map units using photographs through a process called "premapping" and a map legend for "mass wasting" because he believed slope was a "composite" matter, transcending individual soil types and needing its own indicator. When the SCS later "correlated out" that legend, Arnold felt that "war had been declared" because the Forest Service, the principal user of the map, should "have had the right to determine its own legends" (J. Arnold 1997).

Arnold's next assignment took him to work on forests in Washington, where he was strongly influenced by the landforms approach of Warren Starr, professor of soil science at Washington State University. Retzer, afraid that Arnold's rambunctious frankness (e.g., "I've never seen a soil survey that meant anything to us.") would damage relations with the SCS, had him transferred in 1959 to the Wasatch National Forest in Region 4. There he conducted a standard soil survey until his transfer to the Boise National Forest, also in Region 4 (J. Arnold 1997).

Rich Richlen completed the Umpqua survey, which became an important part of the Umpqua Multiple Use Plan because of the information it provided on potential location of roads, tree nurseries, and campgrounds. The survey was issued as an in-house Forest Service report, and the SCS only published it later because of its backlog of unpublished manuscripts. The Forest Service often used the problem of publication backlog in later years as a rationale for bypassing the NCSS.

The Umpqua survey pioneered several techniques for Forest Service soil mapping that soon became standard procedure, at least in the West. These included premapping of landforms and the stereoscopic viewing of color and infrared photographs. In Richlen's words, "We didn't just dig holes, because we were not making interpretations for corn. We were interested in compaction and timber stands, and therefore we had contrastingly different interpretive needs." Although Richlen did not always follow standard soil survey methods, he, unlike the more confrontational John Arnold, remained on cordial terms with his SCS colleagues throughout his career (Richlen 1997).

The Umpqua survey also set the stage for the Mantle Stability Survey (1960–1964), also called the Cut-Bank Survey, which was the brainchild of Region 6's Jack Fisher. In 1960, Dick Alvis was transferred to

Region 6 from Region 4, where he had worked with Arnold on the Wasatch. During the next four years, Alvis, assisted by Loren Herman, conducted the first large-scale reconnaissance surveys, living out of his Forest Service car while covering five Oregon national forests west of the Cascades, missing only the Mt. Hood National Forest. These surveys were also the first nonstandard Forest Service surveys, and they focused primarily on road stability and erosion. The SCS never published them, but they were well received by Forest Service engineers, for whom Alvis conducted many training sessions. By the time he had finished in 1964, Alvis concluded that "it was no question to me that reconnaissance was the way to go in the northern forests." Following completion of the Cut-Bank Survey, he was transferred to Region 1 (Northern Region) in Missoula, Montana, where he continued to work on reconnaissance surveys with Rich Richlen (Alvis 1997).

By 1965, when John Corliss took over from Jack Fisher in Region 6, the split with the SCS had reached the point where both agencies had agreed to go separate ways. Relations remained cordial but distant until the late 1970s, when new regional soil scientist, Bob Meurisse, who arrived in 1975, oversaw efforts to cooperate more closely with the SCS. According to Herman, "We tried to correlate some of our soils, but we didn't have enough information, and we didn't know enough about soil series in the uplands to do that because the SCS had never done uplands" (Herman 1997).

Corliss, who held a Ph.D. in soil science and was a well-respected scientist, recognized that his superiors preferred surveys that covered whole forests in a relatively short period of time to standard soil surveys. He, along with Tom Glazebrook, the head of the Division of Soil and Water Management in Region 4, who joined him in 1969, chose to describe these surveys with the term Soil Resources Inventory (SRI). This technique became the standard Forest Service procedure. As Corliss remembered years later, "They were not interested in fine detail, rather something rough and quick. If we had followed standard surveys, we would have had postage stamps on the Empire State Building" (Corliss 1997). The pressure to move to such surveys intensified in the 1970s as Congress passed more planning and environmental legislation that affected the national forests.

The SRIs extended Mantle Stability Surveys by providing for many more interpretations ("capability" measures for various resources), in effect becoming something in between standard and reconnaissance surveys. According to Corliss, they were done differently and with more consistency in Region 6 than in other regions that used the term SRI (Corliss 1997). All the national forests were completed by 1981, six years after Bob Meurisse took over from Corliss, making Region 6 the first region to survey its entire land base. The SRIs proved to be very useful in regional planning, especially in delineating areas not suitable for timber harvesting, either from the standpoint of stand regeneration or slope stability. In one case, the Region 6 staff were able to rebut a regional forester's claim that the Willamette National Forest had no "unsuitable" land by showing with an SRI that at least 100,000 acres should be removed from timber plans (Meurisse 1997).

By the late 1970s, SCS personnel had come to recognize the important role that SRIs were playing in the Forest Service, while the Forest Service began to experience the need for more detailed and correlated soil surveys. Increasingly intense forest management practices and growing environmental scrutiny, reflected in both the National Environmental Policy Act (1969) and the National Forest Management Act (1976), changed the way planning and management was conducted on the national forests. Both laws contained specific language about sustaining productivity and protecting the soil resource, which required more detailed soil information to implement (Meurisse 2000).

At the same time, the SCS began to implement changes that resolved some of the issues of map unit design, level of sampling, and kinds of interpretations that divided the SCS and Forest Service. In the early 1970s, the NCSS adopted the five soil survey orders that allowed surveys at varying levels of intensity and established an approved correlation process for these orders. The SCS also established "minimum survey standards" in 1978, replacing the requirement for many detailed tables, as well as the necessity of interpreting for roads, septic tanks, etc., "things that didn't make sense in a forest" (Meurisse 1997, 2000).

As a result, the channels of communication began to open again between the two agencies. Under the leadership of Bob Meurisse, who, like his predecessor Corliss, held a Ph.D. in soil science and a reputation as a strong scientist, several national forests in Region 6 made agreements with the SCS for Order 2 or 3 correlated soil surveys. During this period, Region 6 Forest Service staff were regular participants in the annual work planning conferences for the NCSS in the region. Their contributions at these conferences and in field reviews led to increased sharing of ideas between soil scientists in the Forest Service and those in other agencies engaged in the NCSS, which in turn helped shape the NCSS into a more flexible program that could better meet specific user needs (Meurisse 2000).

The appointment in 1979 of Dick Arnold, formerly a professor at Cornell University, to be director of Soil Survey in the SCS encouraged this gradual thaw in relations. Arnold promoted dialogue and flexibility, resulting in attempts to reach compromise throughout both agencies, including Region 6, where he chaired a memorable interagency meeting in 1979. By his own account, his bridge-building efforts landed him in some difficulty, due in part to his naivete in the ways of the federal bureaucracy. In one such incident, on returning in 1980 from a fruitful discussion with Region 1 staff soil scientist Dick Cline (an early "second generation" Forest Service soil scientist), he found a note on his desk severely chastising him for not alerting the Montana State Soil Conservationist about his visit (D. Arnold 1997; Collins 1997).

Useful as the SRIs were in overall regional planning, some problems began to emerge as specific national forest plans were written. Tim Sullivan arrived on the Malheur National Forest in eastern Oregon in 1980 to find the nonstandard, uncorrelated SRI had already been published. For his first few years at the forest, the information in the SRI proved adequate for his work, but when he and others began drafting the Malheur forest plan, he discovered that the SRI did not correlate with those in the three neighboring forests in the Blue Mountains.

Among other things, this was "driving loggers crazy" because the mitigations they were required to apply to prevent such effects as erosion, runoff, and stream damage, changed from forest to forest depending on the soil analysis of each plan. According to Sullivan, "That was my re-education in following the NCSS process." As a result, in recent years, Region 6 has been resurveying its national forests to NRCS standards. In most cases, contractors have been doing this work because the declining number of soil scientists are stretched thin doing special project work, such as the development of sampling procedures for measuring soil compaction caused by logging equipment, which Sullivan was the first to implement (Sullivan 1997).

INTERMOUNTAIN REGION (R-4)

In 1956, Olaf Olson became the first regional soil scientist in Region 4. Olson was only the second person in the Forest Service, after Retzer, to receive the title soil scientist. He had previously participated in the

California Soil and Vegetation Survey and then moved to Region 4's Engineering Division, where he worked on the soils aspects of road construction. As was the case in other regions, most of his time was spent on special project work, which, especially in the early years, justified soil science's existence in the agency. Olson, however, was at first eager to cooperate with the SCS in survey work, but gradually began to look for other ways of doing it. Olson remembered:

> We went along fairly well doing their standards, and it took a long time doing it, so we began looking for some latitude. In some places standard surveys worked, others we needed more or sometimes less information. For road jobs, standard surveys met the first level of investigation, but then we needed more foot-by-foot information. The same was true with range studies. The physical aspects of soils were more important to us than classification, and we could calculate yields from them. We were way out ahead of standard surveys. We never had any disputes with SCS. They didn't even know what we were doing. SCS probably thought we should be going faster on standard surveys, but I never got the idea they were over us, which, of course, they were not. (Olson 1997)

Olson replaced Retzer as the Washington Office coordinator in 1967, and Bill Wertz transferred from Region 9 (Great Lakes Region) to become the new Region 4 regional soil scientist. Although he knew little about the West, he was glad to be there. In those days, he thought soils work would be more appreciated in Region 4 "because a mistake lives forever in the West but not in the East." Wertz continued the experimentation begun under Olson, which culminated eventually in the idea of the Land Systems Inventory, the product of a collaboration among Wertz, John Arnold, Dick Alvis, and Rich Richlen. Published in 1972 under the authorship of Wertz and Arnold as *Land Systems Inventory,* the report was one of the few times Arnold committed his ideas to paper (Wertz and Arnold 1972). As Wertz remembers, Arnold had an almost instinctive feel for landscapes but he was reluctant to write. More than 25 years after it was issued, it is considered one of the most important documents in the history of Forest Service soil science, not so much for what it immediately accomplished as for what it later stimulated (Wertz 1997a).

Only 12 pages long, the report drew together concepts of landscape and geomorphology from sources throughout the world and established a hierarchy of categories working down from a "physiographic province" to an individual "site." Following Alvis's formulation, the land system was defined as "a conceptual device which achieves an integrated overview of the relationships between geologic history, soils and plant ecology, as an aid in understanding land resources" (Wertz and Arnold 1972). Since the beginning of soil science in the Forest Service, its practitioners had been "proto-ecologists" concerned with interrelationships rather than individual resources. The publication of this report solidified that orientation.

According to Bill Wertz, the Land Systems Inventory was needed because "you could never build a car from just its parts. You need the concept of a car first" (Wertz 1997a). In 1977, Canadian geomorphologist Stan Rowe, who may have been reacting to the debate in the United States, explained some of the conceptual differences between soil survey and land systems taxonomies and why it was often necessary to develop categories not found within soil classifications:

> Most formal taxonomies stress the use of inherent properties in classifying from below . . . just as in soil taxonomy the soil profile individuals "created" by the confines of a pit are grouped in series, families, and subgroups. In such strict taxonomies, the higher classes (more generalized) are logically dependent on the individuals with which the classing starts. . . .
>
> The situation is different with surface spatial units such as land areas. At different scales and for different purposes we create a variety of "individuals" from the geographic continua that constitute the geomorphologic surface. . . .
>
> Suppose, for example, that a geomorphologist were to attempt a formal classification of landforms based only on inherent properties such as surface shape (elemental flats, convexities, and concavities), attempting to build the system up from below. He would never arrive at the unit "floodplain," for it is an illogical pattern of spatially associated flats, convexities, and concavities. *The unit known as a floodplain only comes into existence through the understanding of a significant process* (emphasis in original). (Rowe 1977)

Rowe also pointed to another significant difference—the absence of an agreed taxonomic system in the Land Systems Inventory because every "practitioner has his own classification that usually appears as an appendage to a map." The danger of the Land Systems Inventory, which did not include an agreed taxonomic system for soils, was that it could easily degenerate into arbitrariness. More seriously, it could become an excuse for less rigorous work, as Jerry Ragus, regional office correlator (1981–1986), discovered when he suggested at a 1981 training meeting the need for some Order 2 surveys. Several staff members attending the meeting responded, "That's why we left SCS, so we wouldn't have to do all that stuff." Later at that same meeting, the 35 attendees were given a map of a section of the Boise National Forest and asked to identify map units. The resulting Tower of Babel in which 115 unit names were given, with even the personnel from the Boise forest disagreeing on the nomenclature, "told us that there was no consistency and that the training program was inadequate. There was no quality control, and that is what has made soils work a success. Fifteen forests doing 15 different things and you get a hodge-podge" (Ragus 1997).

A contributing reason for consistency problems with the Land Systems Inventory was the increasing number of new hires in the 1970s who were placed in ranger districts to work for months or years in relative isolation from their colleagues. Another was that the system was initially weak on providing vegetative associations for helping to define map units, which also increased the amount of variability, as Dick Alvis recently acknowledged (Alvis 1997). Owen Carlton, Southwest regional soil scientist, was especially critical of it for that very reason (Carlton 1997).

The 1981 meeting was a surprise and a revelation both to Ragus and to his boss, Tom Collins, the regional soil scientist. As a result, they increased training and leavened the Land Systems Inventory with more traditional soil survey work (Ragus 1997).

GREAT LAKES AND EASTERN REGION (R-9)

During Bill Wertz' tenure in the Great Lakes Region in the 1960s, the region's small group of soil scientists worked as a team, moving from forest to forest. The team completed Order 3 medium-intensity surveys that conformed to SCS soil survey standards, beginning in the Eagle River District on the Nicolet National Forest, then in the Hiawatha National Forest. Close cooperation with the SCS came naturally in the Midwest and East, where the national forests were built gradually through purchases from private owners. Under these conditions of intermingling national forest and private land, less common in the West, it made good sense for the two agencies to cooperate, or to at least keep each other informed about their respective activities (Wertz 1997a; Green 1997).

Following Wertz' departure in 1967 from what was now the expanded Eastern Region, the function of regional soils coordinator was divided between Ed Newman, who handled survey work, and Merv Stevens, who was in charge of special projects. The following year, at the insistence of supervisors of individual forests who wanted their own permanent soil scientists on staff, the regional team was broken up and its members divided among forests. By the end of the decade, quiet rumblings had surfaced, suggesting that traditional survey methods could not provide enough of the "right" kind of information. In 1970, Ed Newman introduced a modified Soil Resource Inventory into the region but it did not catch on. According to Jim Jordan, soil scientist attached to the Ottawa National Forest at that time, it was viewed as being imposed from above (Jordan 1997). Jordan and Don Prettyman on the Superior National Forest were two of the most respected soil scientists in the region, primarily because of their careful, almost perfectionist approach to their work, and they needed to be persuaded by something more than an amendment to the regional handbook to give up what they had been doing.

Devon Nelson transferred to the region in 1972 from the Boise National Forest in Region 4 to become the Region 9 soil survey coordinator. A year later, Dick Alvis, dissatisfied with a reorganization in Region 1, took a "downgrade" and moved to the White Mountain National Forest in New Hampshire. Together they became allies in moving Region 9 toward a Land Systems Inventory approach.

Nelson had worked with John Arnold on the Boise National Forest and shared his ideas about the importance of geomorphology and the need to clearly identify map units that land managers could understand. Nelson and Arnold had adopted some ideas about vegetation from Washington State University ecologist Rexford Daubenmire. According to Nelson, physiography and vegetation were what "we used to get a handle on things because soils themselves were so variable." Nelson was especially iconoclastic when it came to the practical value of correlating soils:

Management land units are visible to land managers while a soil survey is not. The three requirements are: 1) relevance to the issue; 2) visibility to the manager; and 3) consistent mappability. SCS talked about reproducibility but when it comes to mapping, their areas are just as arbitrary as the rest of us. SCS can reproduce taxonomically (theoretically) but not practically on a map. One really gets into the subjective when drawing lines on a map. In any case, lack of reproducibility doesn't necessarily imply lack of functionality. (Nelson 1997)

From his new position in Region 9, Nelson began to evangelize. In the early 1970s, he held a meeting on the While Mountain National Forest that was attended by John Arnold, Bill Wertz, Bob Bailey (another developer of Land Systems Inventory), among others. A similar meeting took place later in the Great Lakes area. Nelson's first impression led him to believe that most Region 9 soil scientists "thought they were working for their forest supervisors. They just wanted to do their own things." By the time Nelson left the region in 1977, "some had picked up Land Systems enthusiastically, others lagged, and a few resisted" (Nelson 1997).

Outside of the White Mountains, several forests in the Great Lakes area, as well as the Mark Twain National Forest in Missouri, began to embrace the Land Systems Inventory idea. Its acceptance in this area surprised Nelson somewhat, since he believed that the Great Lakes would be a difficult place to use it because of the "subtlety of the geomorphology there." He thought places that had some relief, such as the Allegheny forests, would pick it up first, but, "ironically, it was the Great Lakes" (Nelson 1997). It may have been ironic, but it was also fortunate because once Land Systems had been introduced into the Great Lakes' forests, the subtlety of their relief compelled soil scientists, such as Jim Jordan, to make careful investigations about the relationships among geology, soils, and vegetation—something that had not yet been done in the more dramatic landscapes of the Rocky Mountain West.

In remembering this period of his career, Nelson recalled the rather convoluted, semicircular path by which the Land Systems approach reached the Forest Service soil science program in the Great Lakes Region. In the 1930s, soil scientist J. O. Veatch had formulated some of the basic ideas of the Land Systems approach in his work on the cut-over lands in the Great Lakes states. By integrating geographic, geologic and soil information, Veatch hoped to construct a natural division of land that would be "more directly useful to the agriculturist, the ecologic geographer, the economist and land planner than is the separate soil type units of the Soil Survey. . ." (Veatch 1937). Charles Kellogg, director of the soil survey, followed the progress of this work and acknowledged its potential usefulness.

After World War II, Australian and Canadian soil scientists adopted these ideas and used them in the reconnaissance surveys of the vast interiors of their countries. Bill Wertz and John Arnold adapted the Australian/Canadian work to Forest Service soil inventory efforts in the American West. Nelson himself, their protégé, then brought it back to the Great Lakes where colleagues and successors, including Jim Jordan, Dave Cleland of the Huron-Manistee National Forest, and Steve Fay of the White Mountain National Forest, developed it further (Nelson 1997).

When Walt Russell succeeded Nelson in 1978, he found Region 9 divided between the Land Systems Inventory and standard survey factions. Russell had come from the Southern Region (R-8), where he had generally followed the standard model. He soon discovered, however, that the way in which the Region 9 staff were developing the Land Systems approach was not antithetical to the careful methods of the standard survey, but rather refined and extended them (Russell 1997). According to his contemporaries, Russell was largely responsible for mediating between the factions and for gradually bringing all of the soil scientists of Region 9 into the new era of the Ecological Classification System, which is differentiated from its predecessors by the fact that land, soils, and vegetation are studied and mapped in an integrated manner (Nelson 1997).

In the 1980s, Jim Keys, quality control specialist, and Jerry Ragus, regional coordinator, considerably refined and enhanced Region 8's soil survey methodology, making it one of the most complete in the agency (Keys 2000).

SOUTHWEST REGION (R-3)

Under Jack Williams' leadership (1957–1975), the Southwest Region closely followed standard soil survey procedures. According to Dick Alvis, this was possible because the older southwestern geologic formations produced a better fit between soils and landscape than was the case in the younger formations of the northern Rockies. Williams' assistant, Owen Carlton, also a strong SCS adherent, planned to continue with the standard survey when he took over in 1975. It soon became clear to Carlton, however, that despite these favorable geological conditions, some of the identified SCS soil series were crossing life and temperature zones, which meant that soils alone were not good indicators of important ecological relationships. Since he believed climate, rather than soil series, was the most important factor to consider, he began to investigate intensively the relationships between soil moisture/temperature and vegetation.

Having established these relationships to his satisfaction, Carlton began to think of surveying in ecological units that interrelated climate, vegetation, and soils rather than soil series alone (Carlton 1997). In 1980, he christened his approach as the Terrestrial Ecosystem Survey (TES), which was later defined as follows: "the systematic description, classification, and mapping of terrestrial ecosystems. A terrestrial ecosystem is an integrational representation of the ecological relationships between climate, soil, and vegetation. Phases of terrestrial ecosystems form the mapping units of the TES. Categories of soil climate (moisture/temperature) regimes are aligned along a gradient and used in the initial separation of continuums into segments. Indicator plants are correlated to these separations through field measurements. . ." (Carlton 1986).

The Region 3 soil survey program's emphasis on vegetation as a "surrogate" indicator for climate regimes and its unwillingness to classify down to the series level of soils displeased local SCS personnel and led to a break in cooperative relations, lasting from 1978 to 1993. Memos disputing one another's scientific findings went back and forth during this period, but positions had hardened and did not begin to loosen until after the departure of the major principals in the dispute. Since then, the Region 3 soil science program, under coregional directors Wayne Robbie, formerly Carlton's assistant, and Penny Luerhing, coregional coordinator and the first female soil scientist to hold a regional coordinator position, has redeveloped a cooperative working relationship with the NRCS. Says Robbie: "We have redefined what it is to be a cooperator. We no longer dwell on our differences. We don't want to jeopardize our relationships. Now cooperation just doesn't involve arguing over correlation. Our cooperation is based on technology transfer. We share data and it does not matter if it exactly fits their criteria. When we do joint studies, we try to be objective and not fight over who is right" (Robbie 1997).

The TES has been employed successfully in the Southwest for a wide range of Forest Service resources and activities. Most recently, it has been helpful in locating potential archaeological sites, a very important concern in the region. It has also been "reproducible," in the sense that the soil/climate/vegetative relations that were discovered by Carlton and his colleagues obtain throughout the region (Robbie 1997). But TES largely developed autonomously within Region 3. Carlton made some limited efforts to spread it beyond the Southwest but without much success, in large part because of the difficult and time-consuming nature of vegetation/soil moisture/temperature measurements. Current Forest Service regulations, however, strongly reflect the soil science work of Region 9 and Region 3. It is the product of careful and exacting research spurred by both their positive and negative interactions with the SCS/NRCS (Russell 1997; Avers 1997).

CURRENT TRENDS

In recent years, the soil survey programs of the Forest Service and the NRCS have gradually converged. On the one hand, the period of experimentation in the Forest Service program is now over, making open discussion between the two agencies easier. According to Jim Keys, current director of the Forest Service Soils Program, "the Forest Service strives for consistency in ecological classification across the United States and requires all soil inventories to meet the standards of the National Cooperative Soil Survey" (Keys 2000). On the other hand, the NRCS is more receptive to different survey approaches, having learned lessons from the Forest Service and from its work for and with the Bureau of Land Management. The Forest Service is still more landscape oriented than the NRCS, but both agencies have been working to reach common definitions of eco-regions that will coordinate their work, which may lead to the coordination of soil classification work among all federal agencies (Avers 1997; Russell 1997; Calhoun 1997; Smith 1997).

REFERENCES

This history is based entirely on oral interviews. I would have preferred to have had more documents to cite, but did not have the opportunity to access records in the Federal Records Center. However, I found that all my interviewees told very similar stories and did not disagree about facts or interpretations. In addition, three veteran Forest Service soil scientists (two retired and one still active) have reviewed this manuscript. Therefore, I am confident of its general accuracy.

Alvis, Dick, interview by the author, 10 July 1997.

Anderson, Jerry, interview by the author, 21 August 1997.

Arnold, John, interview by the author, 10 July 1997.

Arnold, Dick, interview by the author, 24 June 1997.

Avers, Pete, interview by the author, 15 August 1997.

Bradshaw, Ken, interview by the author, 22 August 1997.

Calhoun, Tom, interview by the author, 25 June 1997.

Carlton, Owen. 1986. *Agronomy Abstracts.* American Society of Agronomy Annual Meeting, New Orleans, Louisiana. November 30–December 5.

Carlton, Owen, interview by the author, 4 August 1997.

Cline, Dick, interview by the author, 26 June 1997.

Collins, Tom, interview by the author, 24 July 1997.

Corliss, John, interview by the author, 9 July 1997.

Fisher, Jack, interview by the author, 26 August 1997.

Goudey, Chuck, interview by the author, 18 August 1997.

Green, T. W., interview by the author, 8 August 1997.

Herman, Loren, interview by the author, 17 July 1997.

Jordan, Jim, interview by the author, 15 September 1997.

Keys, Jim, personal communication with the author, 26 April 2000.

Leven, Andy, interview by the author, 18 August 1997.

Lunt, Herbert A., and C. L. W. Swanson. 1949. Mappable characteristics of forest soils. *Journal of Soil and Water Conservation.* 4(1, September):5–12.

Meurisse, Bob, interview by the author, 30 July 1997.

Meurisse, Bob, personal communication with the author, 7 July 2000.

Nelson, Devon, interview by the author, 28 August 1997.

Olson, Olaf, interview by the author, 14 July 1997.

Ragus, Jerry, interview by the author, 5 August 1997.

Richlen, Rich, interview by the author, 9 July 1997.

Robbie, Wayne, interview by the author, 28 July 1997.

Rowe, Stan. 1977. Canada committee on ecological (biophysical) land classification working group on methodology/philosophy. Working paper. June.

Russell, Walt. 1961. Minutes of the fourth annual technical soils conference for Forest Service soil scientists. Atlanta, Georgia, October 16–27. (From personal collection of W. Russell.)

Russell, Walt, interview by the author, 27 August 1997.

Sherrell, Art, interview by the author, 25 August 1997.

Smith, Horace, interview by the author, 18 July 1997.

Sullivan, Tim, interview by the author, 25 July 1997.

U.S. Department of Agriculture (USDA). 1914. Reports on Forest Land Classification; Entry 219; Record Group 54. National Archives at College Park,
Maryland.

Veatch, J. O. 1937. The idea of the natural land type. *The Proceedings of the
Society of American Soil Science.* 2:499–503.

Wertz, William, interview by the author, 14 July 1997a.

Wertz, William, interview by the author, 20 July 1997b.

Wertz, W. A., and J. F. Arnold. 1972. *Land Systems Inventory.* Forest Service
Intermountain Region, Ogden, Utah.

8

A History of Soil Surveys and Soil Science in the Bureau of Land Management

James Muhn

Soil science is a critical aspect of the Bureau of Land Management's (BLM) administration of the public lands and resources for which it is responsible. As Paul J. Culhane points out in *Public Land Politics* (1981), "Protection of soil systems to protect watersheds may be the most important feature of agency planning for all users." It is "the heart of multiple use management," for the quality and stability of soils determines the uses to which public lands can be put and the intensity of those uses (Culhane 1981, 128). Healthy soil conditions are essential to the BLM's fulfillment of the Federal Land Policy and Management Act mandate, which requires that BLM-administered lands be managed in a manner that not only provides for their multiple use and the sustained yield of their resources but also provides for their protection and conservation.

An effort has been made to understand public land soils since the earliest days of the nation. The Land Ordinance of 1785 directed public land surveyors to assess and provide information on "the quality of the lands" over which they ran township and section lines (GLO/USDOI 1838, I:12,50). This soils information was intended to provide prospective purchasers and settlers reliable information on the character of public lands that they might obtain. Eventually, deputy surveyors were instructed to grade soils according to their agricultural capability. Soils were rated in survey notes as either first, second, third, or fourth rate in character, although because the instructions to deputy surveyors provided no guidelines on what constituted good or unfavorable soils, such ratings were not always reliable. The determination of soil quality rested solely upon the shoulders of the deputy surveyors and depended on their individual knowledge of the agricultural capability of various soils (White n.d.).

After 1900, the U.S. Geological Survey (USGS) joined in the work of identifying soils on the public lands. The agency had responsibility for classifying public lands subject to entry under the terms of the Enlarged and Stock Raising Homestead Acts. These land classifications took into account the "arability of the lands," but the soils information gathered consisted of little more than a description of their texture and ability to retain moisture (USGS 1912, 101; USGS 1917, 155).

The passage of the Taylor Grazing Act in 1934 marked a new stage of federal government interest in public land soils. Among the new law's primary purposes was "to stop injury to the public grazing lands by preventing overgrazing and soil deterioration, to provide for their orderly use, improvement, and development to stabilize the livestock industry dependent upon the public range, and for other purposes." To prevent further soil deterioration, Congress authorized the secretary of the interior to study soil erosion problems within grazing districts that were established under the terms of the law and to use funds for the rehabilitation of eroded public lands.

The Grazing Service, which had responsibility for the administration of public lands within grazing districts, understood the relationship between range forage conditions and the quality and stability of soils. Thus they conducted range surveys to determine forage and soil conditions. The surveys emphasized collection of data on forage type and density, but also collected information on generalized soil types and erosion conditions (Rutledge 1941). Early results of this range survey work led the Grazing Service to conclude that there was "much to be learned about the productivity of the various soils on Federal ranges" (USDOI 1940, 344).

By stopping overgrazing, range survey work was intended to improve the condition of public lands. The Grazing Service also constructed erosion control projects to improve its lands. These projects were among the earliest range improvements constructed by the Civilian Conservation Corps camps assigned to grazing districts. The work involved construction of diversion ditches and check dams, range reseeding, and other projects to retard water and wind erosion (Division of Grazing/USDOI 1937). The Soil Conservation Service (SCS) conducted this erosion control work on public lands until April 1940, when a reorganization transferred the work to the agencies administering those lands. The work continued in close cooperation with the SCS, but at an accelerated pace (USDOI 1941).

The work implemented under the Taylor Grazing Act also affected public lands outside the grazing districts, which were administered by the General Land Office (GLO). Like the Grazing Service, the GLO conducted range surveys and was involved in constructing soil and moisture conservation projects (Muhn and Stuart 1988; GLO/USDOI 1945). In 1946, the Grazing Service and GLO were merged to create the BLM (Muhn and Stuart 1988).

The new agency continued the emphasis on soil erosion projects. In 1951, the BLM characterized about half of the rangeland it administered as being either severely or critically eroded and advocated more aggressive range management techniques to halt overuse, as well as more erosion control projects (BLM 1951). The BLM also continued the land classification and range management work along the same lines as its predecessor agencies; consequently, while soils data were collected, the information gathered continued to be largely superficial (BLM 1957).

In 1957, the BLM considered revising the range survey methods developed in 1937. The Forest Service and the SCS had abandoned use of the 1937 standards, and the BLM saw a need for reestablishing "some uniformity in methods and standards in order to avoid confusion among users of public owned grazing land" (BLM 1957). The agency also wanted more accurate forage surveys because the strength of the agency's range adjudication decisions rested upon the conclusions reached in those surveys (BLM 1957). The BLM also inaugurated a program of range condition and trend studies in conjunction with the effort to revise range survey methods (BLM 1960a; Deming 1960).

These changes in the BLM's range management program resulted in the agency taking a closer look at the role of soils in range management. By 1960, range studies were determining the best methods to meet not just the "natural requirements" of vegetation, but also the types and conditions of soils (BLM 1960b, 37). And under the leadership of Secretary of the Interior Stewart Udall of the Kennedy Administration, the BLM established its own credible soils program. Udall significantly changed the emphasis of public land policy. The BLM's approach shifted from seeing resources primarily as commodities to an awareness of the need to balance the demands of mankind and nature. Udall called for a more "thoughtful approach" to the utilization of resources through comprehensive land-use plans that achieved the "wisest and highest use" of public land resources (Udall 1962; Muhn and Stuart 1988, 104–106).

Udall and his new conservation philosophy had far-reaching consequences for the establishment of a formal soils program in the BLM. In

1961 and 1962, the BLM entered into cooperative soil survey projects with the SCS in Montana, Nevada, and New Mexico. For the cooperative soil surveys in Nevada and Montana, the BLM assigned range conservationists to accompany the SCS field crews, but in New Mexico, the BLM's first soil scientist participated in the fieldwork. The BLM also had two projects in California where it paid the SCS to inventory the soils in two soil conservation districts that had mixed public and private lands (SCS et al. 1968; Pomerening 1980).

The BLM viewed these cooperative soil surveys as a means "to determine the feasibility of using standard soil survey procedures or some modification of them in the BLM resource inventories" (USDOI 1963, 60). A 1964 BLM assessment of the projects found that soil surveys did provide "basic information for resource management plans," but concluded that intensive soils surveys should not be conducted for all BLM projects or land-use plans (BLM 1964). One reason for such a conclusion was the expense of soil surveys, but more importantly, the authors of the report did not believe detailed soil surveys were necessary. "Such information," the report's authors argued, "[was] not essential as a general rule," and the BLM should limit soil surveys to locales where serious resource problems required detailed soils information. For such cases, the report recommended that the BLM develop its own soil survey expertise (BLM 1964).

The BLM Director Charles Stoddard received the 1964 report favorably, adopting the staff report recommendations in an instruction memorandum. The BLM, he noted, recognized "the value of soils data in areas with critical edaphic problems and where successful management of the public lands is *dependent* on knowledge of *all* the important factors affecting the resources" (emphasis in original) (BLM 1964). However, since most land-use plans did not need to be "predicated on the need for soil surveys prior to the initiation of intensive management and rehabilitation," the BLM would not fund any cooperative soil surveys with the SCS without the demonstration of a clear need and justification for the data (BLM 1964).

Furthermore, recommended soil surveys were to be "primarily limited to areas proposed for long-term multiple resource management and where it has been demonstrated that soil surveys [would] furnish additional information needed for the solving of critical [resource] problems" (BLM 1964). More importantly, Stoddard believed that soils data interpretations for BLM-administered lands were solely the agency's responsibility, and asserted that the agency's understanding

A soil scientist samples soil on a marsh range unit, 1964. (Courtesy of the National Archives at College Park, Maryland; Print 114-H-LA-62,640; Record Group 114, Records of the Natural Resources Conservation Service)

of the information would guide its land management plans (BLM 1964).

The latter point posed a problem for the BLM since the agency had little soil science expertise. The BLM had established a Soil and Watershed staff in 1964, but its responsibility at the time dealt primarily with soil erosion control, and it had no soil scientist on staff. In fact, while many BLM employees had taken basic soil science courses in college when studying to be range conservationists, foresters, and other resource specialists, the agency still employed only one soil scientist, the one it had hired for New Mexico in 1961 (Pomerening 1980; Turcott 1988).

The Soil and Watershed Staff, however, was determined to implement the new soil survey policy. The staff recognized that soils data could be used for most BLM activities and recommended that the BLM adopt two levels of soil surveys: broad-brush reconnaissance for most areas, and detailed surveys for areas in need of significant rehabilitation measures.

James Pomerening has criticized Stoddard's instruction memorandum, calling it a "severe set-back" for the soils program in the BLM (Pomerening 1980). Dr. Pomerening, however, ignored the fact that the staff report and instruction memorandum did recognize a need for soil surveys and that the limitations imposed upon the conduct and use of soil surveys were a consequence of personnel and fiscal constraints. Rather than being a "set-back," the 1964 instruction memorandum was a step forward in establishing a soils program in the BLM. Pomerening has also characterized the memorandum from the Soil and Watershed Staff as a disagreement with the soil survey policy promulgated by Stoddard. The memorandum, however, was clearly nothing more than that staff's proposal to director Stoddard on how to implement the soil survey policy.

After issuing its 1964 soil survey policy, the BLM began to hire more soil scientists. In 1965, the agency hired soil scientists for its service centers in Denver, Colorado, and Portland, Oregon, to act as soil specialists for the BLM throughout the West and Alaska. Adding to the soil scientists already on staff in New Mexico, the BLM offices in Colorado, Arizona, Idaho, and Oregon also hired soils specialists during the 1960s, but these individuals could not possibly meet all of the BLM's soil survey needs (Pomerening 1980). Thus, the task these first soil scientists faced was daunting. In 1966, the chief of the Soil and Watershed Staff estimated that 50 million acres of BLM-administered lands could be classified as "frail," and in the "advanced stages of erosion" (Turcott 1966). Their condition demanded corrective action and required the BLM to "evaluate the ability of [those lands] to respond to rehabilitation" (Turcott 1966).

The National Environmental Policy Act of 1969 (NEPA) further increased the work of soil scientists by requiring the BLM and other federal agencies to assess the potential environmental impacts of their actions. Through the preparation of environmental impact statements (EISs), agencies examined the resources likely to be affected by a particular action and considered and chose management alternatives for mitigating adverse impacts (Muhn and Stuart 1988; Hofman 1972).

One of the BLM's challenges in preparing the necessary EISs was their lack of information on "ecological and environmental relationships" (Hofman 1972, 121). Soils were among the subjects on which the BLM had little information. The BLM entered into cooperative agreements with the SCS for some of the necessary soil survey work but also significantly increased the number of its own soil scientists (Pomerening 1980).

The BLM also needed more soil science professionals for its expanding coal program. In 1974, the Interior Department gave the BLM responsibility for the Energy Minerals Rehabilitation Inventory and Analysis (EMRIA) program. Under EMRIA, the BLM collected soil and other data to determine the impact of coal development on established ecosystems and to determine the reclamation potential of public lands that might be disturbed by coal leasing activity. The EMRIA investigations focused on areas in the western states where coal leasing might occur in the next 20 years (BLM 1978a).

The soils program got a further boost with the passage of the Federal Land Policy and Management Act (FLPMA) in 1976, a law that required the BLM to administer the lands under its jurisdiction on a "basis of multiple use and sustained yield . . . in a manner that will protect the quality of scientific, scenic, historical, ecological, environmental, air and atmospheric, water resources, and archeological values" (FLPMA of 1976, Public Law 94-579; 90 Stat. 2744–2745). To accomplish that objective, the FLPMA directed the BLM to develop more comprehensive land-use plans, which in turn required the agency to conduct more resource inventories.

In order to provide better and more comprehensive resource inventories, the BLM reevaluated its range of inventory procedures. The BLM's Denver Service Center began in 1977 to develop a new soil and vegetation inventory technique that came to be called the Site Inventory Method (SIM). "It was," in the words of one of those involved in its development and implementation, "meant to be a once [and] for all inventory for the rangelands, including soils, vegetation, wildlife habitat, and forestry—the whole business, was [to be] included in one inventory" (Baker 1991). For the soils component of the SIM, the BLM adopted SCS soil mapping techniques that were used for determining potential plant communities and required that soils be mapped as phases of a soil series. From that data, those analyzing the inventories could determine the potential vegetation for each soil phase (Baker 1991).

There were many problems with the SIM. It was labor intensive, field crew training was difficult, and the per acre cost of the inventory methods was high. The BLM modified the inventory method in 1978. The new Soil-Vegetation Inventory Method (SVIM) did not attempt a comprehensive resource inventory, concentrating instead on the resources in its title—soils and vegetation. The SVIM also looked at sample areas for each rangeland type in an area, then used a computer to analyze the collected data before determining resource allocations for a locale. Information from the sample sites could then be extrapolated to similar rangelands in a locality. In the opinion of one BLM employee, "This was [a] really good inventory method" (Baker 1991). The agency agreed with that assessment and adopted the SVIM as its basic soil and vegetation inventory method in 1978 (Pomerening 1980; BLM 1978c).

At the same time that the SVIM was being developed, the BLM entered into a "memorandum of understanding" with the SCS in which the BLM recognized that proper management of the public lands under its jurisdiction required it to obtain "information on the nature and distribution of soils in such public lands and their potential suitabilities and limitations for different uses and activities" (BLM/SCS 1978). To accomplish that, the BLM agreed in 1978 to conduct public land soil surveys in accordance with the technical standards of the National Cooperative Soil Survey (NCSS) and to provide its soil scientists with appropriate training. The BLM also consented to cooperate with the SCS and other government entities in its soil surveys (BLM/SCS 1978).

After signing the memorandum with the SCS, the BLM took a serious look at the implications of the agreement. The BLM immediately recognized that the soil survey needs of the agency were beyond both its own capabilities and those of the SCS, given currently available data. Therefore, the BLM began to develop a long-term plan for the coordination of soil survey work with the SCS that would provide the SCS with data sufficient to meet all of the BLM's soil information needs. The agency set the ambitious goal of completing the basic soil surveys needed for all the lands it administered by 1989.

To achieve this goal, the BLM and the SCS developed a "Master Plan Framework for Soil Surveys on Public Land," a document intended to let the BLM develop a Soil Survey Plan of Operation for all its land outside of Alaska (BLM 1978b). Soil scientists in the BLM offices were to determine the acreage to be surveyed each year to meet the agency's 1989 soil survey goal and to coordinate that soil survey work with the soil inventory needs of other BLM resource programs.

The BLM's Washington office recognized that its soil survey goal would place increased demands upon soil scientists and their already strained budgetary resources. They also fully expected that other problems would arise over time. However, by determining the rate at which

Table 8.1   Bureau of Land Management (BLM): progress of
soil surveys to 1989

State	BLM-Administered Lands by Acres	BLM Lands Mapped by Acres	Percentage of Area Mapped
Arizona	12,428,584	9,569,421	77.0
California	17,204,689	4,262,500	24.8
Colorado	8,276,890	8,046,988	97.2
Idaho	11,867,773	11,179,123	94.2
Montana	8,070,658	6,052,933	75.0
Nevada	47,962,636	42,734,000	89.1
New Mexico	12,869,913	12,869,913	100.0
North Dakota	67,030	61,740	92.1
Oregon	15,691,674	8,379,212	53.4
South Dakota	279,595	270,946	96.9
Utah	22,141,908	17,077,184	77.1
Washington	312,582	304,098	97.3
Subtotal	**175,577,966**	**134,799,558**	**76.8**
Alaska	92,740,505	3,989,502	4.3
Total	**268,318,471**	**138,789,060**	**51.7**

Data compiled by the Bureau of Land Management, Soil Survey Progress Review Team.

soil surveys needed to be completed and by remaining aware of potential difficulties, the BLM prepared itself to meet its goal (BLM 1978b).

All of these developments in the late 1970s indicated that the BLM's soils program was becoming better established. As Pomerening (1980, 192) confidently proclaimed, the soils program's future was "bright." Since 1964, the number of soil scientist and technician positions in the BLM had risen to 122. By 1980, the SVIM project had inventoried over 110 million acres of the 170 million acres of BLM-administered rangeland, and soil survey work on BLM-administered lands was also progressing rapidly (see Table 8.1) (Pomerening 1980; Pomerening 1992; Barton 1987).

"The soil survey program," as a 1980 study noted, "[was] just the beginning of the BLM's soil management program" (Pomerening 1980). With the completion of the soil surveys, the collected data would need interpretation, and management applications would need implementation. Each BLM field office would, therefore, need a soil scientist able to provide managers with information on the capabilities and limitations of soils in particular locales. Soil science professionals educated in soil physics, soil fertility, soil interpretations, and other aspects of soil science and less specialized in soil classification would now be needed (Barton 1987; Pomerening 1980).

But rather than expanding, the number of BLM soil science professionals declined in the early 1980s, although not all BLM field organizations experienced this reduction. The BLM's offices in western Oregon hired additional soil scientists to help control erosion caused by timber harvesting and logging road construction. Agency-wide, however, only 74 soil science professionals remained in the BLM by 1986. The decline came as a consequence of several changes in the BLM soils program: the end of the EMRIA program, the abandoning of the SVIM as a technique to assess resource conditions, and the near completion of soil survey work in many areas. This situation challenged the soils program to devise methods to accomplish more with fewer personnel (Barton 1987; Baker 1991; Pomerening 1992).

One such technique for doing more with less involved modifications in soil mapping methods. The BLM had found that standard soil mapping techniques were not well suited to agency-administered lands. The BLM managed large areas of rough terrain that were often difficult to access and were put to a wide variety of uses. The agency needed a better means of mapping and classifying public land soils that would be not only efficient and cost-effective but would provide accurate and useful soils information (Amen and Foster 1987; Amen 1998).

As a result, the BLM's soil scientists, in cooperation with the USGS and the SCS, initiated the Soil Landscape Analysis Project (SLAP). SLAP used Geographic Information Systems (GIS) technology to integrate elevation data, Landsat thematic mapper data, and photographic, climatic, geological, and vegetative information. The soils maps created by the computer model conformed to NCSS standards, and testing in the early 1980s showed that the SLAP methodology could be utilized to supplement and update existing soil survey work (Amen 1998).

The success of SLAP gave the BLM's soil scientists a means to move their program beyond conducting soil surveys. They began to stress more forcefully to BLM managers how a proper understanding of soils affected the entire spectrum of resource and land use on BLM-administered lands. Soils information, they pointed out, had uses beyond assessing range and forestry resource conditions or protecting watersheds and assuring water quality. Wildlife habitat depended on soils and in turn influenced animal populations. Mining rehabilitation efforts needed soil data to assess the reclamation potentials of disturbed lands. Soils information could also assist in preventing recreational use of public lands from causing erosion damage. Predictive models using soil survey data could be developed to help archeologists determine the probability of prehistoric sites in certain areas. But to do all these things effectively required soil scientists working in cooperation with other resource specialists (Amen 1989).

At the same time, the BLM's soil scientists did not abandon basic soil services during these years. Soil surveys were still important as a means of classifying public land soils, despite the fact that the agency had failed to meet the 1978 goal of having completed basic soil surveys for all its lands in the lower 48 states (BLM 1978b); by 1989, the year that goal was to be accomplished, only 76 percent of BLM-administered lands, exclusive of Alaska, had some level of soil survey inventory. Completion rates ranged from 100 percent for the agency's lands in New Mexico, to only a quarter of the agency's lands in California (see Table 8.1).

Watershed management and rangeland inventory work also continued to be an important part of the soils program agenda. On projects such as the Sagers Wash Watershed near Moab, Utah, BLM soil scientists predicted soil erosion, sediment, and salt yields under a variety of erosion control and grazing management practices. At Muddy Creek near Rawlins, Wyoming, among other places, soil scientists conducted studies to see how soil moisture and vegetative data could be better correlated for more effective application to range management (Amen 1992).

In 1994, the agency's ecosystems management initiative further enhanced the value of soil science to the BLM. Developed under the leadership of Acting Director Mike Dombeck, ecosystems management integrated ecological, economic, and social principles to sustain natural systems and processes "in order to meet the social and economic needs of future generations" (BLM 1994a). Accomplishing this required that the BLM have a better "understanding of the relationship among land management activities, site capability, social and economic demands, and ecological health and sustainability" (BLM 1994a). This in turn demanded that the BLM "gather and use the best available scientific information as the cornerstone for resource allocations and other management decisions" (BLM 1994a).

With adoption of ecosystems management, the BLM undertook major shifts in its land management practices. Rather than looking to technical and engineering solutions for short-term means of protecting and restoring public lands, the agency sought to identify "the primary causal factors that degrade land health in the first place" (Van Haveren et al. 1997). The BLM worked to refine its watershed analysis techniques so it could determine the causal factors for different landscapes. After discovering the underlying reasons for the deterioration of watersheds and their resources, the agency could adopt land-use practices that would prevent further deterioration and lead to more comprehensive and long-term restoration practices (Van Haveren et al. 1997).

The BLM realized that soils would be a key element of any watershed analysis method. Therefore, soon after the ecosystems management initiative was introduced, the soils program released its *Strategic Action Plan for Soil Resource Management in the Bureau of Land Management* (BLM 1994b). Soils, commented Dombeck at the launch of the plan, were the foundation upon which ecosystems were built, and that foundation was affected by how the BLM managed its lands. It was therefore essential that the BLM learn how much change public land soils could tolerate before endangering a particular kind of ecosystem.

The *Strategic Action Plan* was intended to guide the BLM in "fully assimilating soil ecology into ecosystems management" by 2000 (BLM 1994b). To accomplish this, the BLM would focus on protecting, maintaining, and improving the soil quality of public lands through a soils program that worked toward better integrated ecological inventories. A new Soil Information System (SIS) would make soils data more easily accessible and usable, and the Water Erosion Prediction Project (WEPP) would allow land managers to better determine how various land uses influence soil erosion.

Perhaps more important was that the *Strategic Action Plan* recognized the need to better integrate soil scientists into the BLM. Soil scientists needed to become part of interdisciplinary teams, "which would allow for exchange of ideas among the various resource professionals and help soils to be seen as a part of "the holistic science of ecology" (BLM 1994b). Such developments would in turn lead to better land and resource decisions (BLM 1994b).

The BLM has moved forward on implementing the 1994 *Strategic Action Plan*. The need for soil surveys continues, since land exchanges and acquisitions are constantly bringing new lands under BLM juris-

Table 8.2   Bureau of Land Management (BLM): progress of soil surveys to 1993

State	BLM-Administered Lands by Acres	BLM Lands Mapped by Acres	Percentage of Area Mapped
Arizona	14,257,940	9,699,836	68.0
California	17,301,768	4,410,500	25.5
Colorado	8,303,010	7,822,900	94.2
Idaho	11,855,480	11,469,537	96.8
Montana	8,075,850	6,052,933	75.0
Nevada	47,959,301	45,977,000	95.9
New Mexico	12,890,539	12,890,539	100.0
North Dakota	61,377	60,102	97.9
Oregon	15,726,434	9,212,919	58.6
South Dakota	279,085	270,852	97.1
Utah	22,167,464	17,090,785	77.1
Washington	352,332	312,064	88.6
Wyoming	18,392,533	14,298,762	77.7
Subtotal	177,623,113	127,984,825	72.1
Alaska	88,296,012	115,540	0.1
Total	265,919,125	128,100,365	48.2

Data compiled by the Bureau of Land Management, Soil Survey Progress Review Team. Data differs from Table 8.1 because of land exchanges, disposals, acquisitions, and other actions.

diction (see Table 8.2). Soil science professionals in the BLM continue to provide basic soil services, such as soil interpretations and consultations, to help other resource disciplines within the agency. The BLM's soil scientists, however, continue to look for better methods to gauge the condition of public land soil and associated resource conditions.

One resource assessment method they are exploring is the National Resource Inventory (NRI) developed by the Natural Resources Conservation Service (NRCS, formerly the SCS). The NRI provides for intensive inventorying of selected sample sites of various ecological types. The data collected includes soil verification and quality, ground cover, vegetation production, and other ecological information, which is then digitized and analyzed in a manner that will tell users the condition of soil, water, and vegetation of the various ecological site types (Davis 1998).

The BLM field-tested the NRI in Colorado during 1997 to determine its effectiveness in helping the agency's resource professionals identify whether land management methods are enhancing or decreasing the productivity of public land resources (BLM 1997). While the BLM is still evaluating the results of its NRI field test, for some involved with the field test, the NRI has demonstrated that it can provide the BLM with broad brush information on the ecological status and condition of its lands over time (Davis 1998).

The BLM soil scientists and other resource professionals are also considering an enhanced SLAP to help achieve the goals set forth in the 1994 *Strategic Plan*. This new approach, called the Integrated Landscape Analysis (ILA), is intended to improve the BLM's management and monitoring of rangelands and wildlands. The ILA incorporates recent advances in GIS and digital technologies to better integrate landscape, soil, vegetative, and other resource data. Pilot tests of the ILA in several western states have been favorable. In addition to enhancing the BLM's soil inventory information, the ILA can help users better determine the condition of specific environments and identify those uses that increase erosion and other stresses on those landscapes (Amen and Kuka 1995).

The BLM's commitment to initiatives like the NRI and the ILA underscores the agency's recognition of the role soil surveys and science play in making land-use decisions. As Acting Director Dombeck noted in 1994: "Soil processes determine, to a large extent, the structure and function of ecosystems. Soil quality, health, productivity, and sustainability are strongly influenced by our land management practices. It is critically important to know how much change in soil processes can be tolerated without jeopardizing the functioning of the entire system" (BLM 1994b, 2). Without good soils information and interpretation, the BLM fully understands that it cannot effectively manage the lands under its jurisdiction or meet its congressional mandate to administer those lands, using the principles of multiple use and sustained yield, while at the same time protecting their natural values.

ACKNOWLEDGMENTS

The author would like to recognize the assistance of Alan Amen in preparing this study. A soil scientist with the BLM, Mr. Amen provided the author with numerous documents and information from his personal files, and answered many of the author's questions.

REFERENCES

Amen, Alan. 1989. National Cooperative Soil Survey presentation. Draft. Unpublished paper. Historian's Office, Soils Program File, Bureau of Land Management, National Applied Resource Sciences Center, Denver, Colo. (hereafter cited as BLM-NARSC).

Amen, Alan. 1992. Bureau of Land Management (BLM)—Soil survey and related soil correlation activities. Paper presented at a National Soil Correlation Workshop, Lincoln, Neb. Historian's Office, Soils Program File, BLM-NARSC.

Amen, Alan. 1998. Technology & NCSS. Unpublished paper. Historian's Office, Soils Program File, BLM-NARSC.

Amen, Alan E., and John W. Foster. 1987. *Soil Landscape Analysis Project (SLAP) Methods in Soil Surveys.* Bureau of Land Management Technical Note Number 379. Bureau of Land Management, U.S. Department of the Interior, Washington, D.C.

Amen, Alan, and Joe Kuka. 1995. An integrated landscape analysis approach for land management and land health assessment. Unpublished report. Historian's Office, Soils Program File, BLM-NARSC.

Baker, John. 1991. Interview by James Muhn, Land Law Historian, BLM Service Center, Denver, Colo. BLM Library, National Science and Technology Center, Denver, Colo. Transcript.

Barton, Katherine Barton. 1987. Bureau of Land Management. In *Audubon Wildlife Report, 1987*, edited by Roger L. Di Silestro, pp. 3–59. Orlando, Fla.: Academic Press.

Bureau of Land Management (BLM). 1951. *Rebuilding the Federal Range: A Resource Conservation and Development Program.* Washington, D.C.: Government Printing Office.

Bureau of Land Management (BLM). 1957. Improvement of range survey methods under way. *B.L.M. Management Notes* VI(July):5–6.

Bureau of Land Management (BLM). 1960a. Bureau Range Management program show's significant growth. *B.L.M. Management Notes* VIII(October):4.

Bureau of Land Management (BLM). 1960b. *Project Twenty-Twelve: A Long Range Program for Our Public Lands.* Washington, D.C.: Government Printing Office.

Bureau of Land Management (BLM). 1964. *Soil Surveys on BLM Administered Lands.* [with Staff Report] Instruction Memorandum Number 64-555, Washington Office. Library, BLM-NARSC.

Bureau of Land Management (BLM). 1978a. *Public Lands Digest: Montana, North Dakota, South Dakota—1977–78.* Billings, Mont.: Bureau of Land Management, Montana State Office.

Bureau of Land Management (BLM). 1978b. *Soils—Long Term Planning Com-*

pletion by June 30. [Includes Attachment: Master plan framework for soil surveys on public land] Instruction Memorandum Number 78-315, Washington Office. Library, BLM-NARSC.

Bureau of Land Management (BLM). 1978c. *Soil Surveys on BLM Administered Lands,* 31 July 1978. Instruction Memorandum Number 78-406, Washington Office. Library, BLM-NARSC.

Bureau of Land Management (BLM). 1994a. *Ecosystem Management in the BLM: From Concept to Commitment.* Bureau of Land Management, Department of the Interior, Washington, D.C.

Bureau of Land Management (BLM). 1994b. *Strategic Action Plan for Soil Resource Management in the Bureau of Land Management.* Bureau of Land Management, Department of the Interior, Washington, D.C.

Bureau of Land Management (BLM). 1997. National Resource Inventory: Colorado test pilot handbook. Unpublished manuscript. BLM-NARSC.

Bureau of Land Management and Soil Conservation Service (BLM/SCS). 1978. Memorandum of understanding between the Bureau of Land Management and Soil Conservation Service relative to the making of soil surveys on lands administered by the Bureau of Land Management. Library, BLM-NARSC.

Culhane, Paul J. 1981. *Public Lands Politics: Interest Group Influence on Forest Service and the Bureau of Land Management.* Baltimore, Md.: Johns Hopkins University Press.

Davis, Scott. 1998. National Resource Inventory (NRI) for Colorado BLM. Unpublished manuscript. Historian's Office, Soils Program File, BLM-NARSC.

Deming, Milo H. 1960. Condition and trend surveys: Scientific data for sustained yield range management. *Our Public Lands* 10(October):3–5; 12–13.

Division of Grazing, U.S. Department of Interior (Division of Grazing/USDOI). 1937. Emergency conservation work under the Division of Grazing. *The Grazing Bulletin* 1(January):14–17.

General Land Office, U.S. Department of Interior (GLO/USDOI). 1945. The General Land Office in World War II. Unpublished report. Historian's Office, General Land Office File, World War II Folder, BLM-NARSC.

General Land Office, U.S. Department of Treasury (GLO/USDOI). 1838. *General Public Acts of Congress, Respecting the Sale and Disposition of the Public Lands, with Instructions Issued, from Time to Time, by the Secretary of the Treasury and Commissioner of the General Land Office, and Official Opinions of the Attorney General on Questions Arising under the Land Laws,* 2 vols. Washington, D.C.: Gales and Seaton.

Hofman, Ronald D. 1972. Implementation of the National Environmental Policy Act. *Publius: The Journal of Federalism* 2(fall):119–128.

Muhn, James, and Hanson Stuart. 1988. *Opportunity and Challenge: The Story of BLM.* Washington, D.C.: Government Printing Office.

Pomerening, James A. 1980. The Soil Survey Program of the Bureau of Land Management: Past, present, and future. Unpublished manuscript. Historian's Office, Soils Program File, BLM-NARSC.

Pomerening, James A. 1992. Special report: Changes in the BLM's soil scientist positions and personnel between the spring of 1980 and the fall of 1991. Unpublished manuscript. Historian's Office, Soils Program File, BLM-NARSC.

Rutledge, R. H. 1941. Classification of grazing land for its future use and management. *The Grazing Bulletin* 4(January):26–32.

Soil Conservation Service (SCS), Bureau of Land Management, and New Mexico Agricultural Experiment Station. 1968. *Soil Survey: Cabezon Area, New Mexico*. Washington, D.C.: Government Printing Office.

Turcott, George L. 1966. Cover for frail lands. *Our Public Lands* 15(winter):20–21.

Turcott, George L. 1988. Interview by Hans Stuart, Writer/Editor, BLM Service Center, and James Muhn, Land Law Historian, BLM Service Center. BLM Library, National Science and Technology Center, Denver, Colo. Transcript.

Udall, Stewart. 1962. The national land reserve: Today—Tomorrow. *Our Public Lands* 11(January):3–4; 16.

U.S. Department of the Interior (USDOI). 1940. *Annual Report of the Secretary of the Interior, Fiscal Year Ended June 30, 1940*. Washington, D.C.: Government Printing Office.

U.S. Department of the Interior (USDOI). 1941. *Annual Report the Secretary of the Interior, Fiscal Year Ended June 30, 1941*. Washington, D.C.: Government Printing Office.

U.S. Department of the Interior (USDOI). 1963. *Annual Report of the Secretary of the Interior, Fiscal Year Ended June 30, 1963*. Washington, D.C.: Government Printing Office.

U.S. Geological Survey (USGS). 1912. *Annual Report of the Director of the United States Geological Survey for the Fiscal Year Ended June 30, 1912*. Washington, D.C.: Government Printing Office.

U.S. Geological Survey (USGS). 1917. *Annual Report of the Director of the United States Geological Survey for the Fiscal Year Ended June 30, 1917*. Washington, D.C.: Government Printing Office.

Van Haveren, Bruce P., Jack E. Williams, Malka L. Pattison, and John R. Haugh. 1997. Restoring the ecological integrity of public lands. *Journal of Soil and Water Conservation* 51(4):226–231.

White, C. Albert. n.d. *History of the Rectangular Survey System*. Washington, D.C.: Government Printing Office.

9

SOIL SURVEY AND SOIL-GEOMORPHOLOGY

Vance T. Holliday, Leslie D. McFadden,
E. Arthur Bettis, and Peter W. Birkeland

Soil-geomorphology emerged as a distinct subdiscipline at the interface of pedology, geology, and geography only in recent decades and remains an active and evolving field of study (e.g., Birkeland 1974; Birkeland 1984; Birkeland 1999; Gerrard 1992; Daniels and Hammer 1992). Technically, soil-geomorphology is "the application of geologic field techniques and ideas to soil investigations" (Daniels and Hammer 1992, 1) or, more simply, "an assessment of the genetic relationships of soils and landforms" (Gerrard 1992, 2). More broadly defined, soil-geomorphology involves study of the coevolution of soils and landscapes, and can include the use of soils for reconstructing paleoenvironments and paleolandscapes and for dating or estimating age (Boardman 1985; Catt 1986; Knuepfer and McFadden 1990; Birkeland 1999). Judging from the number of books by a diverse group of earth scientists using the terms "soils and geomorphology" in their titles, it has already attracted a wide following (e.g., Birkeland 1984; Birkeland 1999; Daniels and Hammer 1992; Gerrard 1992).

Soil-geomorphic research is diverse and carried out by individuals with differing disciplinary backgrounds and viewpoints, but a common link among all is the concept of the soil and the relationship of soils to the factors of soil formation and landscape evolution (Arnold 1983; McFadden and Knuepfer 1990, 197; Daniels and Hammer 1992, 1; Gerrard 1992, 2–5; Johnson 1993; Birkeland 1999). Soil-geomorphology has many links to pedology, particularly its field focus on soil-forming processes and on the spatial relationships of soils. As an academic field, however, it tends to be institutionally located in geology or geography departments rather than with pedology, which is generally found among the soil science subdisciplines in agronomy departments, despite its intellectual root connections to geology and geography (Tandarich and Sprecher 1994; Tandarich 1998a; Tandarich 1998b). Academically, pedology represents a component of the agricultural sciences and soil-geomorphology a subdiscipline of the earth sciences.

Research in both areas can include soil genesis and its relationship to geomorphic setting, but the research questions in soil-geomorphology tend to focus on regional landscape relationships among soils, on the age and evolution of landscapes, and on past environments. While pedology tends to focus on surface soils, soil-geomorphology includes the study of buried soils. Moreover, soil-geomorphologists usually work with soils as either formal or informal soil-stratigraphic units; they use soils, either buried or at the surface, for subdividing deposits in terms of their age and for correlating deposits.

Despite the academic bifurcation of pedology from soil-geomorphology, soil-geomorphic research has been conducted by specialists in pedology, geology, and geography since at least the 1930s. Of particular focus in this chapter is the research of the Soil Survey Division of the U.S. Department of Agriculture's (USDA) Soil Conservation Service (SCS; later changed to the Natural Resources Conservation Service, NRCS) and its influence on soil-geomorphology in the United States, especially in the geosciences outside of pedology. The three most significant of those influences examined in this chapter are the development of *Soil Taxonomy* (Soil Survey Staff 1975), the production of county soil surveys, and the soil-geomorphic research projects from the 1950s through the 1970s. The purpose of the chapter is to illustrate and discuss both the positive impacts of SCS/NRCS activities in soil-geomorphology as well as the lack of impact or the difficulties of implementation of SCS/NRCS results. A wide gulf exists between soil-geomorphology and much SCS/NRCS work, and the ultimate intent of the chapter is to explain why this is so in the hope that some areas of the gulf can be bridged.

SOIL TAXONOMY, SOIL HORIZON NOMENCLATURE, AND SOIL-GEOMORPHOLOGY

The development of *Soil Taxonomy* (Soil Survey Staff 1975) undoubtedly has been the SCS/NRCS's most influential contribution to pedology, both nationally and internationally. It is probably one of the most significant advancements in the field in the second half of the twentieth century. Because it is the official classification system of the SCS/NRCS and the National Cooperative Soil Survey, and because of the links between these entities and academic soil science and agronomy programs, *Soil Taxonomy* was almost immediately adopted by the bulk of soil investigators in the United States (Birkeland 1984, 42; Birkeland 1999, 49; Hallberg 1984, 55). It became the *lingua franca* for all U.S. pedologists.

Some geoscientists initially rejected the new terminology of the original *Seventh Approximation* (Soil Survey Staff 1960) (e.g., Hunt 1972, 180–182; Morrison 1978, 100–101), but most investigators quickly recognized the utility of a widely used, well-organized, comprehensive set of well-defined, mutually exclusive, largely nongenetic terms for describing and classifying soils. The advent of this soil classification system has had many advantages for soil-geomorphology. The newer system eliminated some of the ambiguity in classification and horizon nomenclature in the older, 1938 system (Buol et al. 1997, 32–33; Bartelli 1984), and the concept of the diagnostic horizon, probably the most useful aspect of the new system, provided an efficient means of communicating a wealth of information about a soil or components of a soil-stratigraphic unit.

Because diagnostic horizons are quantified, soil development, and by inference landscape development, can be quantified to at least a general level. For example, identification of an argillic horizon in a soil indicates that a specified, minimal amount of clay was illuviated. Further, because of the quantitative nature of *Soil Taxonomy*, labs have generated reams of data on soils throughout the country (e.g., the USDA National Soil Survey Center/Soil Survey Laboratory database). Researchers can return to the data and use it for additional purposes.

Adoption of the *Soil Survey Manual* (Soil Survey Staff 1951; Soil Survey Division Staff 1993) to meet the requirements of a nationwide soil survey for standardized and detailed soil profile descriptions also assured a vast amount of nationally comparable information on field properties of soils. The well-known, well-defined standardized set of genetic horizons (the master horizons and subhorizons such as A and Bt) for field descriptions in soil survey has proven to be one of the most widely used and convenient systems of shorthand field nomenclature in the geosciences (e.g., AGI 1982). This information allowed soil geomorphologists to create a variety of soil development "indices" as a way of semiquantitatively expressing soil development for purposes of correlation or to assess soil development as a function of the soil-forming factors (Bilzi and Ciolkosz 1977; Harden 1982; Harden and Taylor 1983; Schaetzl and Mokma 1988).

Soil Taxonomy has created some disadvantages for soil-geomorphology as well. Most generally, the considerable efforts during the 1960s and 1970s that went into developing and refining the new classification system that became *Soil Taxonomy* seem to have stifled other pedologic research (Swanson 1993; Paton et al. 1995, 1). The SCS supported several extensive soil-geomorphic research projects from the 1950s to the 1970s, but otherwise most governmental and academic pursuits in pedology focused on classification as an end in itself (Runge and McCracken 1984; Daniels 1988; Daniels and Hammer 1992, xvi; Swanson 1993). This focus, and the absence of geologists and physical geographers in soil science departments, may help explain why soil-geomorphology evolved largely in geoscience departments throughout those years.

Applying the USDA-approved diagnostic and genetic horizons in soil-geomorphic research has had a few disadvantages as well. Most problems stem from the need for arbitrary rules or decisions inherent in any attempt to categorize and classify parts of a continuum. For example, the amount of illuvial clay in a soil can be an excellent indicator of landscape stability and landscape age (Birkeland 1999). For a horizon to qualify as argillic, however, only a small amount of illuvial clay must be present. Therefore, identification of a subhorizon as argillic conveys some information about a minimal degree of pedogenesis, but within the designation of argillic there can be a significant range of clay content, potentially masking important soil-geomorphic distinctions.

As another example, soil-geomorphologists working in dry environments have found the concept of the K horizon as a master soil horizon very useful (Birkeland 1999; Machette 1985; Harden et al. 1985; Birkeland et al. 1991; Reheis and Kihl 1995). It serves as a convenient field designation that provides a qualitative assessment of a zone of massive carbonate accumulation, which is typically found on older geomorphic surfaces and represents the culmination of the time-dependent evolutionary process of calcic soil development. The SCS, however, never adopted the K horizon as part of the official nomenclature for genetic horizons (Soil Survey Division Staff 1993). One of the reasons the K horizon designation was not adopted was to avoid setting a precedent that might lead ultimately to renaming other horizons; Bt horizons would become T horizons, for example (personal communication, K. Flach to P. Birkeland 1989).

The soil classification system itself has proven of less utility in soil-geomorphology than have the diagnostic and genetic horizons. The system was designed to facilitate classification for soil survey and land-use purposes, and is geographically biased toward the agriculturally productive soils of the midlatitudes. It was not designed to be a tool in soil-geomorphic or other geoscientific research. The taxonomy serves as a very useful nomenclature for referring to or describing surface soils, conveying a tremendous amount of relatively specific information.

Some studies have shown that soil taxonomy can be applied to soil-geomorphic research (Birkeland 1999, 31, 39). Taxonomic classification can be used in soil geomorphology to describe facies or lateral variations in a soil-stratigraphic unit. For example, L. Gile and coworkers used taxonomic classification to identify pedogenic variability (soil facies) within soil-geomorphic units (which could also be soil-stratigraphic units) in a variety of alluvial and eolian deposits in south-central New Mexico (e.g., Gile et al. 1981) (see Figure 9.1) and in northwestern Texas (Gile 1979; Gile 1985). Sometimes the classification differentiates soils of different ages in a chronosequence. For example, the sequences Hapludoll-Argiudoll-Paleudoll or Cambid-Haplargid-Paleargid can denote development in the thickness, clay content, and color of the argillic horizon with time. Long-term pedogenic pathways can even express themselves at the order level of soil classification, e.g., Entisol-Inceptisol-Alfisol-Ultisol-Oxisol.

On the other hand, the degree of specificity and the rules and requirements in *Soil Taxonomy* have serious drawbacks for soil-geomorphology uses. As Hallberg (1984, 53) notes, "classification involves . . . the Tyranny of the Pigeonhole." He further points out (Hallberg 1984, 57) that ". . . the institutional, or bureaucratic implementation of the U.S. system of soil taxonomy . . . has often had the effect of making [it] inflexible; its implementation often rigid and legalistic." In other earth sciences, in contrast, there are a number of "scientific codes or guidelines put forth by professional societies, which are freely debated in the scientific literature [and at] professional meetings" such as the *Code of Stratigraphic Nomenclature* (e.g., NACOSN 1983)

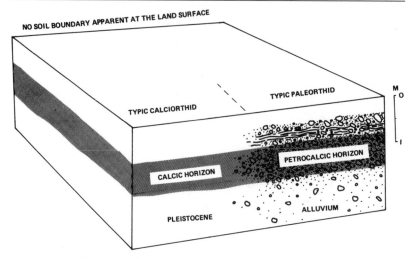

FIGURE 9.1 An illustration of pedologic variability (soil facies) in a soil-stratigraphic unit developed on late Pleistocene alluvium along the Rio Grande of southern New Mexico. Changes in lithology within the alluvium result in soils with two different morphologies of carbonate accumulation (a calcic horizon on the nongravelly alluvium versus a petrocalcic horizon in the very gravelly alluvium), therefore, producing two soil mapping units (a Calciorthid and Paleorthid, respectively) across a landscape otherwise similar in topography, ages, and climatic history. (Source: Gile 1975, Fig. 5; Courtesy of the *Soil Science Society of America Journal*)

in geology (Hallberg 1984, 57). Because many soil-geomorphologists are trained in geology or physical geography, we often alter terms from *Soil Taxonomy* to suit our needs. For example, we might provide adjectives such as "weak argillic horizon" (Btj or juvenile Bt in field nomenclature) for horizons that barely meet argillic criteria.

Roger Morrison, one of the pioneering soil-geomorphologists in the western United States, and one of the few workers who attempted to establish principles of Quaternary soil stratigraphy (Morrison 1967; Morrison 1978), further notes that "*Soil Taxonomy* suppresses or distorts certain pedologic distinctions which can be important to geologists by various legalistic and narrow restrictions; for example, the thickness limitations for most epipedons and diagnostic subsurface horizons . . . and the minimum allowable concentration of CaCO3 for calcic horizons" (Morrison 1978, 100).

A specific example of this problem can be found among surface soils across the high plains of Texas and New Mexico, where a group of well-developed soils formed in Pleistocene eolian sediments. Two of the more common surface soils are the Amarillo and Acuff series, both with thick, well-expressed argillic and calcic horizons. The only significant difference between these soils is the thickness of the A horizon: the Amarillo has a thinner A, the Acuff a thicker A (Mathers 1963; Allen et al. 1972; Stoner and Dixon 1974; Blackstock 1979; Holliday 1990). Otherwise, they are the same soil geomorphically and stratigraphically. These differences in A horizon thickness probably are due to wind erosion (Holliday 1990), but this single difference between otherwise identical soils results in their classification in two orders: Paleustolls (Acuff) and Paleustalfs (Amarillo).

Some of the terminology of *Soil Taxonomy* may cause confusion for soil-geomorphologists. Occasionally, terms given nongenetic definitions in *Soil Taxonomy* may hold genetic implications in soil-geomorphology research. For example, the "Pale" prefix at the Great Group level should indicate old soils in a soil-geomorphologist's view, but the definitions of Paleboroll and Paleboralf refer to the depth to the top of the argillic horizon. This probably has more to do with geomorphic position than soil age. In any case, defining how old is "old" would further impose arbitrary distinctions that would serve no useful purpose. In another example, the Fluvent suborder was designed to include floodplain soils with multiple buried A-C profile, but the definition omits landscape setting and thus Fluvents can occur in eolian deposits such as sand dunes.

The study of buried soils, especially for stratigraphic correlation and for environmental reconstruction, is probably the oldest and perhaps the best-known component of soil-geomorphology and serves well to illustrate some of the difficulties of applying *Soil Taxonomy* in soil-geomorphic research. The study of buried soils requires application of essentially the same level of care and attention to detail as the study of surface soils, including use of much of the same terminology. Applying both the diagnostic horizon nomenclature and taxonomic classification to buried soils is significantly more problematic, however, because of erosion of near-surface horizons or of postburial diagenesis of the soils, which are both greater the longer the soil or sediment has been buried, and because *Soil Taxonomy* is explicitly designed for surface soils only.

Components of buried soils can be described in terms of diagnostic horizons, but the characteristics of the horizon may be different from its preburial state. Erosion or compaction changes horizon thickness, for example, which is a significant component of the requirement for mollic and calcic horizons. Color of a mollic epipedon, also a classificatory requirement, usually changes after burial due to oxidation of organic matter. Furthermore, pedogenesis in the deposits that bury a soil may modify the buried soil in a process known as "soil welding" (Ruhe and Olson 1980); a calcic or argillic horizon can be superimposed over a buried mollic or argillic horizon, for example.

Because the genetic and diagnostic horizons are well defined and widely used, all investigators must take care to apply the terminology appropriately. Zones in buried soils that are reddish and high in clay have been identified as argillic horizons (Retallack 1986; Retallack 1988; Retallack 1990; Lehman 1989) but with no clear documentation for clay translocation (see discussion in Dahms and Holliday 1998; Dahms et al. 1998). Patterson (1991) documented the diagenetic origin of some of these zones in Eocene sediments in Wyoming. Furthermore, mollic epipedons have been designated in the absence of adequate color, thickness, base saturation, or organic carbon requirements in soils affected by millions of years of diagenesis (Retallack 1990; Retallack 1993; Retallack 1997); it is therefore questionable whether the term "mollic," with all that it implies, should be used for these soils (Dahms and Holliday 1998; Dahms et al., 1998).

Taxonomic classification often is possible if a complete buried profile is preserved and the burial was recent, but the classification will differ from the preburial classification over the long term. Burial changes characteristics of the diagnostic horizons and almost always changes soil moisture and soil temperature, environmental characteristics necessary for much classification. Significant changes in soil and water chemistry can also accompany or follow burial and can significantly affect classificatory characteristics such as base saturation. Further problems can be encountered when classifying buried soils in terms of horizons and taxonomy for paleoenvironmental interpretations because few types of horizons or taxonomic categories are associated with unique environmental conditions.

Because of the difficulty of applying *Soil Taxonomy* to the classification of buried soils, alternative classifications have been proposed (Mack et al. 1993a; Nettleton et al. 1998). Interestingly, they use terms, concepts, and a structure from or similar to *Soil Taxonomy*. Which system, if any, becomes the *lingua franca* of paleosol studies awaits extensive field testing.

SOIL SURVEYS AND SOIL-GEOMORPHOLOGY

The soil survey work of the SCS/NRCS is probably the agency's best-known activity. This work has produced tremendous amounts of data on soils from throughout the United States, exemplified by the USDA National Soil Survey Center/Soil Survey Laboratory database. The most widely known and widely used products of the soil survey research, however, are the many published soil surveys themselves. They contain a wealth of information regarding soils and land-use capability, and many include descriptions of soil properties of interest to engineers. In some cases, they also contain data of relevance to soil-geomorphic studies. For example, soil surveys in the midwestern United States often characterize loess deposits, and these data have been used in soil-geomorphic studies of loess thickness and origin, soil erosion, and textural controls on soil-geography (Fehrenbacher et al. 1986b; Fehrenbacher et al. 1986a; Mason 1992; Mason and Nater 1994; Mason et al. 1994). The surveys also are a remarkable source of maps and aerial photographs.

For a variety of reasons, however, soil surveys have not been used to any significant degree in soil-geomorphic research, probably because the purpose of the soil surveys is for assessing land capability and management, especially in agricultural contexts. The soil surveys were never intended as geoscientific resources. As Swanson (1990a, 17) notes, however, "In the course of mapping, soil scientists learn much about the nonsoil properties of the landscape. Some of these properties are described in a general way in the soil survey report; however, much of the information does not find its way into the manuscript and is subsequently lost." As a result, data on regional geomorphology and geology in the surveys often are secondary and skimpy at best and, in most cases, are a very minor component of the survey. Discussion of all five factors of soil formation typically occupies no more than two to three pages out of perhaps 50 to 100 pages of text in a soil survey, with only a few paragraphs on the most general characteristics of the geology and geomorphology (e.g., Huckle et al. 1974; Blackstock 1979; Rector 1981).

Admittedly, however, there are many regions in the U.S. where only minimal data on surficial geology and geomorphology were available prior to the soil surveys. Soil surveys sometime contain more substantive discussions of geomorphic and geologic characteristics of the soil landscape in regions with a long history of surficial geological studies, such as the glaciated Midwest (e.g., Hole 1976) or where intensive soil-geomorphic studies were carried out (discussed below) prior to the soil survey work (e.g., Gerig 1985; Branham 1989).

Further inhibiting soil-geomorphic applications of soil surveys is the emphasis on the soil series and the lack of emphasis on the soil-forming factors. The soil series, though the lowest level of soil classification, is the basic mapping unit of soils surveys and typically is the highest level of grouping "soil individuals" (Swanson 1990b). As the soil surveys developed through the twentieth century, genetic and factorial relationships among series as well as the aforementioned geologic and "physiographic" aspects, were expressly de-emphasized (Simonson 1997), despite Jenny's (1946) work that showed a good relationship.

With publication of *Soil Taxonomy*, soil series were defined strictly as subdivisions within the classificatory system and were intended to

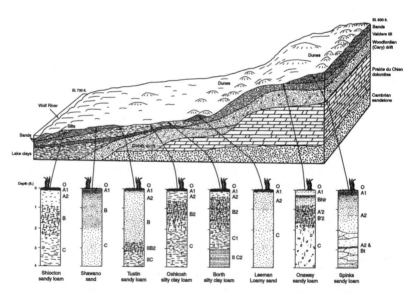

FIGURE 9.2 An illustration of soil-geomorphic relations mapped in east-central Wisconsin by F. D. Hole. The soil series are linked to the soil forming factors of slope position, parent material, and age of the landscape. (Key: Shioctin = Aquollic Eutrochrept; Shawano = Typic Udipsamment; Tustin = Arenic Hapludalf; Oshkosh = Typic Hapludalf; Borth = Typic Hapludalf; Leeman = Typic Udipsamment; Onaway = Alfic Haplorthod; Spinks = Psammentic Hapludalf) (Source: Hole 1976, Fig. 11-5; Courtesy of the University of Wisconsin Press)

"record pragmatic distinctions, i.e., to be keyed to soil usefulness" (Simonson 1997, 80). The result is that mapping units (soil series) are rarely related to one another except taxonomically. "To illustrate, the soil classification relates an Aquept in a certain map unit to other Inceptisols throughout the world. But it does not address the relationship between the map unit containing the Aquept and other map units that occur adjacent to it in the survey area" (Swanson 1990b, 52). There are a few notable exceptions to this approach in soil mapping, such as F. D. Hole's work in Wisconsin (e.g., Hole 1976) (see Figure 9.2).

Moreover, some series may include a variety of soils that are genetically and geomorphically distinct, a notable problem in valley landscapes because of the inherent variability over short distances of valley fills and geomorphic surfaces, compounded by the lack of adequate landscape models in use by soil mappers. Examples include some soils on the floors of dry valleys on the southern high plains that are mapped as the Berda Loam, a series which can include a group of geomorphically and stratigraphically distinct soils (Holliday 1985), and the Napier series applied in the thick loess region of western Iowa (Bettis 1995). On the other hand, a group of series may comprise what is geomorphically one soil (i.e., a single soil-stratigraphic unit), as seen on the southern high plains surface (Holliday 1990).

There are several other reasons why soil-geomorphologists have not readily used soil surveys in their research to any great extent. An important one is the different spatial perspectives of geoscientists versus pedologists. Geomorphologists and stratigraphers, on the one hand, tend to view the world horizontally, emphasizing features such as geomorphic surfaces and landforms, as well as three-dimensionally, studying stratigraphic units. Few of these specialists include the soil in their studies. Many pedologists, on the other hand, traditionally tend to view soils as "independent entities occurring at specific points" (Daniels and Nelson 1987, 289) and focus on the vertical dimension of soils.

Pedology field training and field experience often deals with soil pits and soil profiles rather than soil landscapes (Daniels and Hammer 1992, xv–xvi), probably because 1) so much training and research in pedology involves digging soil pits or taking soil cores for mapping and studying profiles for taxonomic classification (Swanson 1990a; Paton et al. 1995, 1), 2) *Soil Taxonomy* defines soils as single points (Daniels and Hammer 1992, 77), and 3) many soil-forming processes promote downward movement. For example, in histories and historical perspectives on soil

survey and soil classification, some of the key players in the postwar
decades (Arnold 1984; Bartelli 1984; Simonson 1987) emphasized the
soil profile, the pedon and polypedon, and the "soil individuals."

Though landscape position is an important component of field investigations of soils, soil surveys in recent decades have de-emphasized
viewing or investigating soils as components of landscapes, i.e., dealing
with soils as three-dimensional, contiguous bodies (Daniels and Nelson
1987; Daniels and Hammer 1992, 77; Paton et al. 1995, 5–8). Soil
surveys place even less emphasis on the soil parent material or on soil
evolution through time (i.e., as four-dimensional bodies formed in sediment or rock) (Daniels and Hammer 1992, 10, 77; Simonson 1997).
The underemphasis on the landscape and soil parent materials probably is because taxonomic classification emphasizes "soils as units unto
themselves" (Daniels and Hammer 1992, 77) or "soil individuals," a
view which developed along with the arbitrary subdivision of soils into
pedons and the resulting separation of soils and pedons from natural
landscapes (Knox 1965; Daniels and Hammer 1992, 77). Exceptions to
this view of soils include some of the early soil survey work, where
different soil series were identified if the parent material changed
(Simonson 1997) (predating the concepts of the pedon and soil individuals), soil surveys in areas of substantial soil-geomorphic research (e.g.,
Gerig 1985; Branham 1989), and soil surveys by geologically trained
investigators such as Francis D. Hole (e.g., Hole 1976) (see Figure 9.2),
one of the country's preeminent pedologists whose formal education
was in geology rather than soil science (Tandarich et al. 1988).

There are a few examples of soil surveys being used to make Quaternary geologic maps. One is a map of the San Joaquin Valley, California
(Birkeland 1999, Fig. 2.4). It was made by combining information from
the soil maps, basically using soil series to denote general geologic age
and geomorphic relationships. Another example is where Maat (1992)
mapped various ages of dune sand, combining information from soil
maps, geomorphic relationships, and radiocarbon ages (Birkeland
1999, Fig. 2.5).

IMPACT OF THE USDA SOIL-GEOMORPHOLOGY PROJECTS

Soil Taxonomy and soil mapping are clearly the most widely known
contributions of the SCS/NRCS to pedology, but the U.S. Soil Survey

also supported some of the most intensive, systematic investigations of soil-geomorphology in North America. Effland and Effland (1992, n.d.) have described these investigations and their origins and support within the soil survey. The SCS/NRCS's initial involvement in soil-geomorphic research began in the 1930s as a component of erosion studies. Significantly, scientists in disciplines other than pedology initiated the work. This early research arose from the recommendations of the eminent geographer Carl O. Sauer, a member of Franklin Roosevelt's Presidential Science Advisory Board, that pedology, geology, and climatology should be integrated in land research. The SCS implemented Sauer's proposal by establishing a Division of Climatic and Physiographic Research and appointing the noted climatologist C. Warren Thornthwaite as head. The projects administered by Thornthwaite produced valuable data, but were limited by declining funds, lack of baseline data on soils and geomorphology, and the onset of World War II.

Soil-geomorphic research in the SCS resumed in 1953 in a program created under the impetus of Charles D. Kellogg and Guy D. Smith. They ". . . advocated a fully developed multi-site series of soil-geomorphology studies as the basis of a research program in support of the soil survey" (Effland and Effland 1992, 204). Seven soil-geomorphology studies were authorized over the next 25 years. In 1953, Robert V. Ruhe was hired to begin the first of these studies and to direct the entire soil-geomorphology research program, which he did until leaving the SCS in 1970.

Ruhe's unique contribution to the study of landscape evolution was to emphasize the characteristics and geomorphic distribution of soils in studies of landscapes and thus lay the foundation for the development of modern soil geomorphology. From a pedologic standpoint, ". . . Ruhe was instrumental in pioneering the process and quantification of landscape studies and integrating them into modern soil science studies" (Olson 1997, 415; see also Olson 1989). Ruhe's view of landscape evolution began to gel during geomorphological research in the Belgian Congo from 1951 to 1952 (Ruhe 1954a; Ruhe 1956a). He was heavily influenced by Lester King's pedimentation concepts (slope back wearing) (King 1949, 1950, 1953) and especially by Milne's catena concept (Milne 1935a, 1935b).

An unusual aspect of Ruhe's approach was to ". . . completely separate descriptions of the soil, and the geomorphology and geology in a study area . . ." (Effland and Effland 1992, 204), then later integrate them to interpret the history of the landscape and to predict the landscape occurrence of geologic materials, weathering profiles, and soils. "[This approach] reduced the tendency of soil surveyors to describe the geology based on characteristics of the overlying soil and the soil characteristics based on the underlying geology" (Effland and Effland 1992, 204).

The study areas were selected to represent a wide range of climatic and topographic features. Four of the planned seven projects were completed: the humid glaciated midcontinent of southwestern Iowa, the arid and semiarid basin-and-range country of south-central New Mexico, the humid Coastal Plain Province of North Carolina, and the humid maritime west coast environment of western Oregon. For the humid-tropics component of these studies, Ruhe investigated soil-geomorphic relations on Hawaii (e.g., Ruhe 1964a; Ruhe 1965; Ruhe et al. 1965a; Ruhe et al. 1965b), but this work was not as extensive or intensive as the other four projects. In 1971, following completion of the New Mexico "Desert Project," principal investigators Leland Gile and John Hawley initiated the "Southern Plains Project," for the study of soil-geomorphology on the flat, semiarid high plains landscape of northwestern Texas and eastern New Mexico. SCS support for this program was terminated in 1975, however, bringing to a close the large-scale SCS soil-geomorphology studies. The only published work related to the Southern Plains Project was Gile's "sandhills" study (Gile 1979, 1985).

Most of these soil-geomorphology investigations were in cooperation with local universities, agricultural experiment stations, and state geological surveys. These were landmark investigations of the relationship of soils, geology, and landscapes. They demonstrated the benefits of detailed soil-geomorphology studies for interpreting landscape history and facilitating soil survey. A unique aspect of this work was the emphasis on quantitative laboratory methods and mathematical expression of physical and chemical relationships across landscapes. These studies established linkages among deposits, soils, and geomorphic processes that set the stage for the blossoming of process geomorphology. Such studies became the standard for North America and the world.

The following sections summarize the interpretations and impact of the four soil-geomorphology projects. The emphasis is on the "Iowa Project" and the New Mexico Desert Project because these two had the most far-reaching impact of the four and were two of the most influential soil-geomorphic studies in the history of the subdiscipline. The success of those two studies is arguably because both geologists and geomorphologists were involved.

THE IOWA PROJECT

The soil-geomorphology studies initiated in Iowa by geologist Robert V. Ruhe were the prototype for modern landscape evolution and soil geomorphology research. The Iowa Project involved soil-geomorphic studies in three stratigraphically, geomorphically, and pedologically distinct parts of the state, with sites in southwestern, north-central, and eastern Iowa.

The southwest Iowa study area, where the project was initiated in 1953, was the site for the study of the relationships among Quaternary stratigraphy, geomorphology, and soils in 58 railroad cuts along a 72-kilometer transect between the towns of Bentley and Adair (Ruhe et al. 1967; Ruhe 1954b; Ruhe 1969). An expansion in 1955 added the Greenfield quadrangle area of southwestern Iowa to the study and brought R. B. Daniels to the project team as a soil scientist. The study area extended from the Loess Hills region eastward into the southern Iowa drift plain (Prior 1991), within the region of the classic pre-Illinoian (Kansan and Nebraskan) glacial sequence (Ruhe 1969; Hallberg 1986). Across the area, late Wisconsin loess thinned from about 15 meters in the west to 9 meters in the east, burying older landscapes and soils developed in Illinoian loess and pre-Illinoian glacial deposits. Soil development also varied from west to east, with deeper decalcification and more intense oxidation to the east. A significant result of this research was demonstrating that the soil variability was not just related to the west-east climate gradient, but was also due to the eastward fining of the loess and due to the lower sedimentation rates to the east.

The Iowa Project's recognition of the soil landscape and stepped erosion surfaces (Ruhe 1960) laid the conceptual groundwork for modern soil geomorphology. These concepts hold that because weathering takes place from a land surface, soils and the landscape elements on which they occur are intimately linked. Professional papers and field trips resulting from the southwest Iowa study demonstrated the critical role of soil-morphologic and soil-geomorphic studies in elucidating the geomorphic history of landscapes. It also began the melding of geomorphology and pedology, which led to the development of the subdiscipline of soil-geomorphology.

The concept of "stepped erosion surfaces," which increase in age as they ascend toward divide areas (Ruhe 1956b; Ruhe et al. 1967), was one of the most far-reaching and influential geomorphic concepts developed during the southwestern Iowa study. The relative ages of these geomorphic surfaces were demonstrated on the basis of lithostratigraphic, soil stratigraphic, and pedologic arguments rather than solely on the basis of topographic relationships. Principles of ascendancy and descendancy were applied to hill slopes to establish relative age relationships among the stepped surfaces (Ruhe 1975). This premise, that a slope element is younger than the higher surface to which it ascends but the same age as the sediments to which it descends, is a fundamental relative dating concept in modern geomorphology.

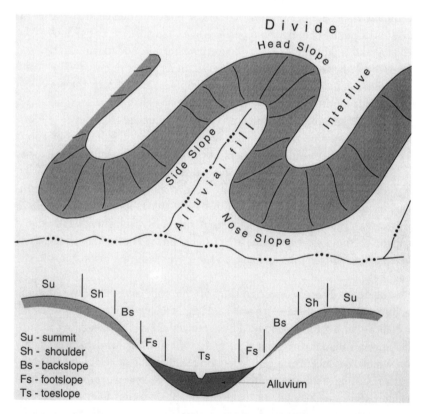

FIGURE 9.3 The classic model of the "hillslope profile" developed by R. V. Ruhe as part of the Iowa soil-geomorphology project. (Source: Ruhe 1969, Fig. 4.5; Courtesy of Iowa State University Press)

The term "pedisediment," referring to the surficial sediment above the stone line on a pediment, was first introduced to American geoscience as a result of this study (Ruhe 1956a; Ruhe et al. 1967). Ruhe demonstrated the morphostratigraphic linkage of the pedisediment, stone line, and pediment, bringing together slope evolution concepts developed in arid, humid, and tropical landscapes (Bryan 1940; King 1953; Ruhe et al. 1967). Another important concept introduced during the southwest Iowa Project was the "hillslope profile," a commonly used system describing the components of a slope profile, which include summit, shoulder, backslope, footslope, and toeslope (see Figure 9.3) (Ruhe 1960).

The north-central and eastern Iowa soil-landscape study began in 1960 (Ruhe 1970), examining landscape history and the relationships among geologic deposits, geomorphology, and soils on a young, glaciated till plain in the north-central part of the state (late Wisconsin Des Moines Lobe) and on what was then interpreted as an earlier Wisconsin ("Iowan") till plain in eastern Iowa (Ruhe 1969). The study involved G. F. Hall, R. C. Schuman, and E. Robello (who relieved Schuman) from the SCS Washington staff as active participants and also trained many students who later became prominent pedologists and soil geomorphologists, including G. H. Simonson, T. E. Fenton, P. H. Walker, and W. J. Vreeken.

Using extensive transects of borings at five "bogs" (fens) associated with different moraines of the Des Moines Lobe, the team studied stratigraphic, sedimentologic, pedologic, chronologic, and palynologic relationships (Walker 1966). Investigators related the Holocene history of slope evolution and basin infilling to vegetation and inferred climatic changes and established linkages among vegetation cover, soil erosion, and sedimentation that have strongly influenced Holocene landscape evolution concepts in recently glaciated humid to subhumid continental regions (Gerrard 1992). The study also demonstrated that the effects of regional climate and vegetation change on the soil-geomorphic system produced a predictable sequence of slope sediments and basin fills that strongly influence the geography of surface soils (see Figure 9.4). Using the principle of ascendancy and descendancy, in combination with radiocarbon ages on basin fills, the work showed that the region's slopes, and the soils formed on them, were less than 3,000 years old (Walker 1966; Walker and Ruhe 1968).

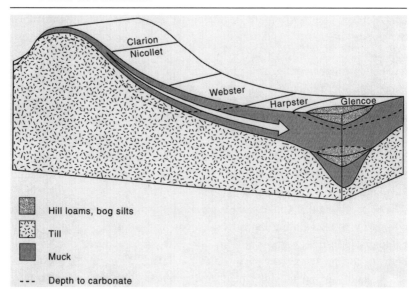

FIGURE 9.4 Generalized model of the soil-geomorphic system around a bog in north-
central Iowa based on slope evolution and basin infilling in the late Holocene. Key:
Clarion = Typic Hapludoll; Nicollet = Aquic Hapludoll; Webster = Typic Haplaquoll;
Harps (not Harpster) = Typic Calciaquoll; Glencoe = not used anymore and usually
included in a general muck mapping unit. (Source: Ruhe 1969, Fig. 4.15 [based on
Walker 1966]; Courtesy of Iowa State University Press)

Investigations in the "Iowan Drift" region of eastern Iowa built on
the landscape evolution concepts developed in southwestern Iowa,
addressed long-debated Quaternary stratigraphic issues, and brought
"stone lines" to the attention of North American geologists and pedologists. A long history of controversy surrounded the relatively low-
relief, loess-free or thin-loess-mantled landscape of the Iowan Drift
region and the relationship of the ubiquitous stone line to the underlying glacial drift (McGee 1891; Calvin 1899; Kay and Apfel 1929;
Leighton 1933; Leverett 1942). By applying soil stratigraphic principles
and radiocarbon dating to transects of deep borings extending from
elongate, loess-capped prominences (paha) onto adjacent parts of the
Iowan Drift plain, where the loess was thinner or absent, the study
established that the Iowan was a late Wisconsin erosion surface developed on much older glacial deposits (Ruhe et al. 1968).

The research team also concluded that the stone line formed as an
erosional lag during development of the erosion surface and was there-

fore much younger than the underlying drift and closely related in time to the overlying loess. The Iowan Drift plain was shown to consist of a series of discrete multilevel erosion surfaces that step down from divides to an integrated drainage net (Vreeken 1975; Hallberg et al. 1978). The concept of stepped erosion surfaces developed in southwest Iowa also provided the key to understanding and predicting the distribution of geologic materials and soils in the Iowan region.

In addition, the Iowa Project confirmed the effectiveness of radiocarbon dating and of core-drilling transects in landscape evolution studies, which had significant impacts on the direction of subsequent soil-geomorphology research, especially in the midcontinent of North America. Radiocarbon dating has become a standard method for correlation and for assessing rates of soil development in many soil geomorphology studies, while core-drilling has permitted studies in areas where natural exposures are rare or excavations are impractical (e.g., Holliday 1995; Mandel 1995; Bettis and Autin 1997).

Soil-landscape models produced during this project greatly benefited the soil survey by significantly increasing the predictive capabilities of soil mapping. Indeed, Ruhe's soil-landscape model was still applied in relatively recent soil surveys in Iowa (e.g., Branham 1989). Reciprocally, geomorphology also benefited from soil surveys that incorporated soil-landscape models and thereby provided continuous spatial data on soil-geomorphology.

THE NEW MEXICO DESERT PROJECT

The New Mexico soil-geomorphic studies, universally known as the Desert Project, ran from 1957 to 1972. The Desert Project was the longest sustained study of desert soils to date, and it lives on through recent publications of the New Mexico Bureau of Mines and Mineral Resources (e.g., Gile et al. 1995). It had and continues to have a significant influence on geomorphological research in deserts. Robert Ruhe served as the initial leader of the Desert Project, developing the overall research strategy and publishing some of the earlier results (Ruhe 1962; Ruhe 1964b; Ruhe 1967). Beginning in the early 1960s, however, pedologists Leland H. Gile, recruited by Ruhe out of the Ph.D. program at Cornell University, and Robert B. Grossman and geomorphologist John W. Hawley became the key investigators. Hawley, after working as a geologist for the state of Nevada and replacing Fred Peterson on the Desert Project in 1962, was the only principal researcher, besides Ruhe, in the soil survey's soil-geomorphology projects who had formal, graduate training and experience in geology and geomorphology. This teaming of formally trained pedologists and geomorphologists may in part account for the success of the project.

Although SCS scientists initially proposed the desert region near Tucson, Arizona, as the desert soils and geomorphology research study area, complications arose that led to the selection of the area in and around Las Cruces, New Mexico, where investigators could call on support from an office of the Agricultural Research Service, the Jornada Experimental Research Station, and New Mexico State University's College of Agriculture. Ultimately, the area provided an ideal setting for desert soil-geomorphic research. Extensive geomorphic surfaces on terraces of the Rio Grande and alluvial fan remnants of nearby piedmonts, all mostly associated with lithologically similar parent materials, enabled development of soil chronosequences. Some surfaces spanned a wide enough range of elevations to provide pronounced climatic and vegetation gradients, allowing evaluation of these factors in pedogenesis. In addition, variation in bedrock types allowed for studies of soil development on alluvium derived from local lithologies such as limestone, in comparison with soils formed on the more common gravelly arkosic sediments.

Most of the data from the Desert Project was published in the massive, 984-page *Desert Project Soil Monograph* (Gile and Grossman 1979), and the key soil-geomorphic relations (including maps), concepts, and conclusions were included in the *Guidebook to the Desert Project* (Gile et al. 1981). These publications vividly illustrate the role that soil-geomorphic research can play in soil survey. Klaus Flach, the assistant administrator for the soil survey at the time, noted that such soil maps are unlike those that typify a standard soil survey precisely because the emphasis of the project was soil-geomorphic research (Gile and Grossman 1979, iv).

The finely drafted, high-resolution soil maps from selected areas superbly illustrate how soils are related, however subtly, to diverse desert landforms. The maps are superimposed on vertical aerial photographs and often include diagrammatic cross sections of the most important study sites and the locations of pedons. In some cases, block diagrams were included to better display the relations of surface age, geomorphic position, and soil type. These diagrams are particularly effective in revealing the underlying cause of variation in soil properties, as well as classification at various taxonomic levels (e.g., Gile 1975) (see Figure 9.1).

Although the monograph and guidebook are important comprehensive publications of the Desert Project research, the one publication most frequently cited in the desert soils and geomorphology literature, and the one that arguably had the greatest impact on desert geomorphology, is "Morphologic and Genetic Sequences of Carbonate Accumulation in Desert Soils," by Gile et al. (1966). This paper first presented the now classic "stage" concept for carbonate accumulation in desert soils, illustrating the morphological evolution of calcic and

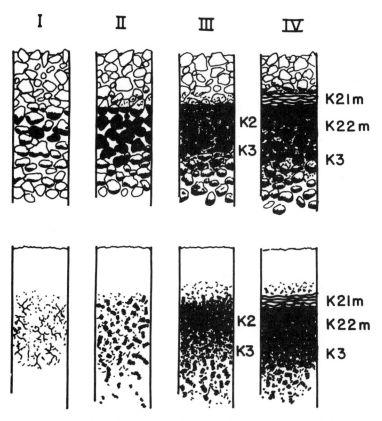

FIGURE 9.5 The classic illustration of the stages of carbonate accumulation in high-gravel and low-gravel parent material for south-central New Mexico. (Source: Fig. 5 in Gile, Leland H., Fred F. Peterson, and Robert B. Grossman. 1966. Morphological and genetic sequences of carbonate accumulation in desert soils. *Soil Science* 101:347–360)

petrocalcic horizons as a function of time and parent material, proceeding from stage I to stage IV (see Figure 9.5; see also Figure 9.1). Subsequent research also demonstrated the importance of landscape position on carbonate morphology (Gile et al. 1981). In their discussion of the "stage concept," Gile et al. (1966) demonstrated that these stages reflected a time-dependent evolutionary process of calcic soil development. Recognition of this time dependency provided a powerful means to estimate the age of alluvial fan and terrace deposits of desert regions because they seldom contain materials suitable for the numerical dating techniques available in the 1960s to 1980s. Development of this new technique was particularly significant because dating is one of most critical types of information required in the solution of many geoscientific problems in geomorphology and related earth sciences.

In the year preceding publication of this important paper, the Desert Project team had introduced the K horizon as a new master horizon (Gile et al. 1965). They recognized that a zone of massive carbonate accumulation (a K horizon) was primarily observed on the older geomorphic surfaces in the region (usually late Pleistocene and older) and represented the culmination of calcic soil development (stages III and IV). They also showed that the K horizon, once formed, had a profound influence on subsequent soil development and landscape evolution.

Many earth scientists working in deserts, especially North Americans, quickly adopted the concept of morphologic stages of carbonate accumulation and the K horizon. Both the stage designations and the K horizon were very handy shorthand field designations that conveyed a lot of qualitative information regarding the morphology and amount of carbonate accumulation and also provided clues to relative soil and landscape age. As mentioned above, the SCS/NRCS did not accept the K horizon (Soil Survey Division Staff 1993), designating such a soil horizon as a Bk or Bkm horizon. According to the *Soil Taxonomy*, soils with stage II and III horizons are calcic horizons, whereas a stage IV horizon is usually a petrocalcic horizon.

Another very significant result of the Desert Project research was demonstrating the importance in arid region soil genesis of dust and other external additions to soils. The entrapment and progressive accumulation of calcium derived from dust (and to a smaller extent, rainwater) was the key to explaining the remarkably systematic development of horizons of carbonate accumulation. Prior to the Desert

Project studies, scientists attributed many key aspects of desert soils to slow chemical weathering of calcium-containing silicate minerals that, given the limited depth of leaching of desert soils, favored calcification at a relatively shallow depth ("pedocals").

The Desert Project research showed that this explanation in most circumstances could not be correct. Ruhe, Gile, and their colleagues observed an enormous mass of pedogenic carbonate in sequences of soils that were formed in gravelly arkosic deposits dominated by silica-rich but calcium-poor minerals. Mass balance studies showed that chemical weathering of tens of meters of parent material would be required to produce this carbonate, yet observations could not provide evidence for this magnitude of weathering. Moreover, the Desert Project team observed that relatively large amounts of carbonate had accumulated in soils (stage I) that were formed in parent materials no older than late Holocene. Studies of modern dust samples collected using dust traps showed that modern dust contained both the kinds and amounts of materials required to form pedogenic carbonate, as well as to form the silicate clay in the overlying Bt horizons in middle Holocene soils. Although other researchers, notably Israeli scientists (e.g., Yaalon and Ganor 1973), had recognized the potentially large influence of dust on the genesis of calcareous and/or clay-rich horizons of desert soils, the Desert Project, by linking the atmospheric with the soil data, provided the data to critically test and ultimately accept the hypothesis. Finally, investigations showed that both the Bt and Bk horizons formed rapidly, and, in places, the Bt horizons formed more quickly in desert than in humid environments (Birkeland 1999).

The popular Desert Project field trips conducted annually well into the 1980s helped convey key concepts of the research to a wide array of scientists, including many from outside the United States (Gile et al. 1981, vii; personal communication, J. Hawley to L. D. McFadden 1996). These interactions not only helped provide an international audience for the Desert Project, but they also provided an opportunity for feedback and critique from accomplished desert soil scientists and geomorphologists from, for example, Israel and Australia.

The Desert Project findings significantly increased understanding of soil-forming processes in deserts and provided the basis for development of conceptual models of Quaternary soil-landscape evolution in desert regions. These models reflect the influences of Quaternary climate changes on soil development and geomorphic processes (e.g., changes in regional base level) and other important geologic factors. Figures by Peterson (1981), one of the early researchers in the Desert Project, and Wells et al. (1985) provide visual summaries of many key aspects of these models. All of the above acted as catalysts in the development of subsequent, innovative research in geomorphology.

The studies of Michael Machette and his colleagues exemplify some of this research. Bachman and Machette (1977) and Machette (1985), for example, applied the morphological-stage approach in mapping, identification, and correlations of Cenozoic deposits in the southwestern United States. During this research, they encountered soils that exhibited carbonate morphology much more advanced than stage IV. Consequently they defined two additional stages of development (stages V and VI), showing that at least in some deserts the soil-landscape connection could be sustained for several million years.

Machette (1978; 1985) also developed a means to procure numerical age estimates from calcic soils based on the determination of the total accumulated mass of pedogenic carbonate and some assumed rate of calcareous dust input. These were used to determine the ages of alluvial deposits, including seismogenic deposits such as fault scarp colluvium, and to estimate fault recurrence intervals. The 1978 study is one of the first to demonstrate the critical importance of soils in paleoseismologic research (see also Machette 1988). Machette (1985) also developed a more advanced conceptual model for calcic soil formation that combined dust input and climate to predict how climate changes might effect the evolution of calcic soils in the Quaternary.

Other advances that strongly reflect the influences of the Desert Project research include the development of a soil profile index that could be used for noncalcic and calcic soils (Harden and Taylor 1983; McFadden et al. 1986), numerical models for the evolution of calcic soils (McFadden 1982; McFadden and Tinsley 1985; Marion et al. 1985; Mayer et al. 1988; McDonald et al. 1996; McFadden et al. 1998), a new model for the formation of desert pavements (Wells et al. 1985; Wells et al. 1995; McFadden et al. 1986; McFadden et al. 1987; McFadden et al. 1998), new applications of soil-stratigraphic studies in geoarchaeological research (e.g., Holliday 1994), derivation of climate proxy and paleoatmospheric data from stable isotopes in calcic soils (e.g., Cerling et al. 1989; Quade et al. 1989; Amundson et al. 1988; Wang et al. 1996), studies of modern rates, compositions, and origin of desert dust (Reheis and Kihl 1995; Reheis et al. 1995), and numerical dating of soil carbonates (Ku et al. 1979; Amundson et al. 1994; Wang et al. 1996).

Ironically, the success of the Desert Project has also indirectly caused a few problems. Many earth scientists, for example, have had an unfortunate tendency to rely too heavily on observed stages of carbonate accumulation for age estimates of Quaternary alluvial deposits. Many of these researchers apparently do not recognize that there are other means by which nonpedogenic features that resemble petrocalcic horizons can form in surficial deposits or in the shallow subsurface environments (Birkeland 1999). Nor do they seem to recognize that rates of pedogenic carbonate accumulation can vary widely in many landscapes and that the observed progression of morphological stages in southern New Mexico (see Figure 9.5) does not even occur in some regions (Machette 1985; McFadden 1981; McFadden 1988; McDonald and McFadden 1994; Holliday 1995, 94). They may also overlook the fact that, whereas knowledge of age control for Holocene deposits, and therefore rates of Holocene carbonate accumulation, is generally good, age data for Pleistocene deposits and calcic soils is sparse, as well as the fact that many important questions remain concerning the actual genetic processes (e.g., biogenic versus inorganic) of pedogenic carbonate accumulation (Monger et al. 1991; Monger and Adams 1996; Verrecchia et al. 1995). These problems are partly due to the typically sparse treatment of soil science in the curricula of most earth science departments (McFadden 1993) and to the fact that earth scientists using this strategy are understandably motivated by the need to answer geologic questions and not by a desire to understand the soils or soil-like features themselves.

Another problem associated with the carbonate stage concept is that some researchers have attempted to base models of the timing and nature of alluviation and their linkage to Quaternary climate changes in deserts on temporal constraints provided by carbonate stage data. However, in the Desert Project study area most of the older deposits have little or no numerical age control. Subsequent research forced major changes in the original age estimates for upper Pliocene to middle Pleistocene deposits in southern New Mexico. This is because soils and sediments of more than one age are within the general age range of some geomorphic surfaces (e.g., Gile 1987) and because of subsequent studies in tephrachronology and magnetostratigraphy (e.g., Mack et al. 1993b). For example, Gile (Gile 1987; Gile et al. 1995) reported much wider age ranges and older ages for the Jornada II surface (25,000 to

150,000 years) and the Lower La Mesa surface (500,000 to 900,000 years) versus the earlier estimates of 25,000 to 75,000 years and 500,000 years (Gile et al. 1981). These very broad age ranges preclude attempts to associate deposition of the alluviums associated with these surfaces with particular climate changes in the Quaternary.

The ongoing value of the Desert Project is marked in several other ways. The New Mexico Bureau of Mines and Mineral Resources recognized the significance of the research by sponsoring all Desert Project publications beginning in 1967. The study area became a part of their environmental geology program in 1977, when they also assumed sponsorship of all subsequent field trips (Gile et al. 1981, vii). More broadly, the Desert Project research is prominent in some of the most widely read volumes on soils, geomorphology, and Quaternary geology (Birkeland 1984; Bull 1991; Gerrard 1992) and in the major geomorphology textbooks (e.g., Ritter et al. 1995; Easterbrook 1993).

This exposure ensures that new generations of students in geomorphology and other areas of the earth sciences will be introduced to the ideas of the Desert Project and made aware of its legacy. Some of these students will be inspired to seek out the original publications of the Desert Project and will see how detailed soil survey investigations provided the essential foundation for ideas that have so strongly influenced scientists who work in deserts.

COASTAL PLAIN PROJECT

Begun in 1960 with R. B. Daniels as project leader and the assistance of pedologists E. E. Gamble and W. D. Nettleton from the SCS Washington office, the "Coastal Plain Project" investigated deposits, landscapes, soils, and paleoenvironments of geomorphic surfaces, ranging in age from Pliocene to Holocene, along the Atlantic Coastal Plain Province of North Carolina. Among other contributions, this investigation produced one of the first comprehensive studies of Ultisol genesis. Systematic increases in mineral weathering, solum thickness, and gibbsite and plinthite content were demonstrated to be associated with increasing age of the geomorphic surface on which soils occurred (Daniels et al. 1970; Daniels et al. 1978; Daniels and Gamble 1978), although these relationships were not linear because of water-table regime complications. For example, they concluded that the thickness of the E horizon and the depth to the B horizon was related to the position of the water table. This study also showed that soil variability decreased from young to old surfaces as a result of parent material uniformity and the convergence of soil profile characteristics over extended periods of weathering and soil development (Gamble et al. 1970). The "edge effect" concept was developed during investigation of soil variability within geomorphic surfaces on this project. This concept explains variations in soil color, E horizon thickness, and gibbsite and free iron oxide content with respect to water-table flux in different landscape positions (Daniels et al. 1967). These studies also found that the relationship between soil color, saturation, and oxidation state were very complex and not evident solely through interpretation of soil morphology (Daniels et al. 1973).

The Coastal Plain Project led to subsequent, more extensive studies of soil-geomorphology on the Atlantic Coastal Plain and neighboring areas. From 1979 to 1984, the SCS and the U.S. Geological Survey (USGS) conducted cooperative regional studies of the relations between soils and geology on the Piedmont and Coastal Plain Provinces of the middle Atlantic and southeastern states (Markewich et al. 1986; Markewich et al. 1987; Markewich et al. 1988; Markewich et al. 1989; see also Markewich and Pavich 1991; Markewich et al. 1990). The work showed that soils and other aspects of physical and chemical weathering could be used to estimate the ages of landscapes with a variety of siliceous parent materials spanning the past several million years. In particular, they provided substantial data on the evolution of Ultisols and Spodosols, on the impact of parent material and drainage on pedogenic processes, and on the soil characteristics best suited to estimating the ages of land surfaces in that environment. Closer to the setting of the Coastal Plain Project, Phillips et al. (1994) tested the "edge-effect model" by looking at local-scale variability in soils. The model successfully predicted some aspects of soil variability, but not all patterns could be accounted for by water-table position, drainage, or topography. Phillips et al. proposed that local soil variability may be linked to biological factors such as tree throw and faunalturbation.

THE OREGON PROJECT

Located in the Willamette River valley of northwestern Oregon, the "Oregon Project," led by pedologists C. A. Balster and Roger B. Parsons, took place largely in the 1960s. Most of the work focused on soils associated with late Pleistocene and Holocene geomorphic surfaces in the valley (Balster and Parsons 1968; Parsons et al. 1970), although they did conduct some studies on such nearby areas as the adjacent mountains (Parsons 1978). These landscapes evolved through a complex series of events related to base-level changes controlled by the Columbia River, catastrophic floodwaters and sediment input from the Columbia River system, and sediment input from glaciers within the Willamette watershed (summarized by McDowell and Roberts 1987).

Of particular interest in the studies were investigations of soil variability as a function of lithologic and stratigraphic variability in parent materials. The buried soil-stratigraphic record exerted considerable influence on the geography and interpretation of surface soils. For example, relatively flat smooth surfaces were shown to have complex origins that included both erosion and deposition; below such surfaces, the lithologies vary significantly. Parsons and Balster (1967) established that discontinuous lenses of primary clay were often mistaken for argillic horizons, and they demonstrated that the complex evolution of some surfaces included pedogenesis and subsequent soil burial, with surface soils now welded to buried soils (Parsons et al. 1968).

The model of fluvial landscape evolution that resulted from the Oregon Project had a significant impact on soil survey in Oregon. In the decades since the end of the SCS project, the model has been used for mapping soils and describing their distribution along rivers (e.g., Gerig 1985) and on the Pacific Coast (e.g., Shipman 1997). Indeed, Parsons prepared the discussions of geomorphic surfaces and soil development for some soil surveys (e.g., Gerig 1985, 182–187), illustrating well the significant contributions of basic soil-geomorphic research in the preparation and publication of soil surveys.

CONCLUSIONS

The SCS/NRCS's soil survey was probably the single most important program influencing the direction of pedological research in the United States in the twentieth century. As such, the soil survey influenced soil-geomorphic research, though to a lesser degree. The soil survey provided most of the basic language of pedologic research used in soil geomorphology, in the form of horizon nomenclature, guidelines for soil profile descriptions, and a taxonomy for soil classification. Soil survey activities have also provided a wealth of data on soils, in the form of field and laboratory data, county soil surveys and maps, and statewide soil maps.

In terms of understanding fundamental aspects of soil-geomorphic relations, soil survey work has been less successful, however. Much soil survey work focused on mapping, describing, and classifying soils, and surveyors were discouraged from pursuing important geomorphic or other geologic and environmental relationships, despite the fact that soil factorial functions and an understanding of soil-forming processes had been shown to successfully predict soil patterns (Jenny 1941; Jenny 1946; Arnold 1994). Input by geomorphologists familiar with the region or a means of communicating the geomorphic and geologic information that the soil mappers acquire would significantly enhance the usefulness of soil surveys.

The biggest impact of soil survey activity on soil-geomorphology was through the SCS soil-geomorphology studies from the 1950s to the 1970s, an "extraordinary collaboration between academic pedologists and soil scientists of the National Cooperative Soil Survey" (Jacob and Nordt 1991, 1194). The Iowa Project and the New Mexico Desert Project were the most notable achievements, helping to formalize and provide direction for the new field of soil geomorphology and influencing the broader disciplines of geomorphology and Quaternary geology. Much of the success of these projects resulted from the teaming of pedologists and geomorphologists.

The significance of these studies has been formally recognized by the Quaternary Geology and Geomorphology Division of the Geological Society of America. Every year since 1958, the division has given the prestigious Kirk Bryan Award for an outstanding publication. Two publications that won the award, Ruhe's 1969 book *Quaternary Landscapes in Iowa* (1974 award) and Gile, Hawley, and Grossman's 1981 *Guidebook to the Desert Project* (1983 award), resulted directly from the SCS soil-geomorphology projects. Additionally, the soil-geomorphology projects heavily influenced other award-winning books by Birkeland (1984) (1988 award) and Holliday (1995) (1998 award). The success of the SCS soil-geomorphology projects and the subsequent joint SCS-USGS surveys on the Coastal Plain and Piedmont Provinces indicate that pedologists, geomorphologists, and Quaternary geologists would substantially benefit from future soil-geomorphology projects, perhaps as cooperative ventures among the NRCS, the USGS, state geological surveys, and academic departments of soil science, geology, and geography.

The story of the influence of the Soil Survey on soil geomorphology reflects a strong dichotomy between soil survey and soil geomorphology, one representing the fundamental descriptive level of soil mapping and classification, the other the large-scale, long-term, interdisciplinary research projects. We believe that for soil-geomorphology to continue to benefit from the Soil Survey, and vice versa, the integration of the two needs to fall more often in the middle ground, particularly at the level of teaching and training fundamental principles of pedology and geoscience in both soil science and geoscience departments, at both the undergraduate and graduate levels.

Moreover, soil survey and soil mapping should more actively incorporate geomorphology and geomorphologists in its work, as was the case years ago. Few earth scientists get to know a landscape as intimately as soil mappers; they should be encouraged and trained to make geomorphic observations and to link soils and soil-forming processes to the landscape and geomorphic processes. Closer ties between agronomy-trained pedologists and geoscience-trained soil-geomorphologists will further both disciplines. As our friend John Hawley once put it: "You can't understand geomorphology without understanding soils, and you can't understand soils without understanding geomorphology."

ACKNOWLEDGMENTS

We thank John Hawley, Bill and Anne Effland, Carolyn Olson, and Scott Burns for information and insights on the history and impact of the SCS-sponsored soil-geomorphologic projects. Jim Bockheim, John Jacob, Dan Muhs, and Dave Swanson reviewed the manuscript and shared their unique perspectives on soil-geomorphology in the Soil Survey. We also thank Douglas Helms for inviting us to prepare this chapter and Anne Effland for her substantial editorial help.

REFERENCES

Allen, B. L., B. L. Harris, K. R. Davis, and G. B. Miller. 1972. *The Mineralogy and Chemistry of High Plains Playa Lake Soils and Sediments*. WRC-72-4. Water Resources Center, Texas Tech University, Lubbock, Tex.

American Geological Institute (AGI). 1982. *AGI Data Sheets*. Falls Church, Va.: American Geological Institute.

Amundson, R. G., O. A. Chadwick, J. M. Sowers, and H. E. Doner. 1988. The relationship between modern climate and vegetation and the stable isotope chemistry of Mojave Desert soils. *Quaternary Research* 29:245–254.

Amundson, R., Y. Wang, O. A. Chadwick, S. Trumbore, L. McFadden, E. McDonald, S. Wells, and M. DeNiro. 1994. Factors and processes governing the ^{14}C content of carbonate in desert soils. *Earth and Planetary Science Letters* 125:385–405.

Arnold, Richard W. 1983. Concepts of soils and pedology. In *Pedogenesis and Soil Taxonomy, Part 1:* Concepts and Interactions, edited by L. P. Wilding, N. E. Smeck, and G. F. Hall, pp. 1–21. New York: Elsevier.

Arnold, Richard W. 1984. Impact of *Soil Taxonomy* on the National Cooperative Soil Survey of the United States (1965–1981). In *Soil Taxonomy: Achievements and Challenges,* edited by R. B. Grossman, pp. 61–68. Soil Science Society of America Special Publication 14. Madison, Wis.: Soil Science Society of America.

Arnold, Richard W. 1994. Soil geography and factor functionality: Interacting concepts. In *Factors of Soil Formation: A Fiftieth Anniversary Retrospective,* edited by R. Amundson, pp. 99–109. Soil Science Society of America Special Publication 33. Madison, Wis.: Soil Science Society of America.

Bachman, G. O., and M. N. Machette. 1977. *Calcic Soils and Calcretes in the Southwestern United States.* Open-File Report 77-794. U.S. Geological Survey, Washington, D.C.

Balster, C. A., and R. B. Parsons. 1968. *Geomorphology and Soils, Willamette Valley, Oregon.* Special Report 265. Agriculture Experiment Station, Oregon State University, Corvallis, Ore.

Bartelli, Lindo J. 1984. *Soil Taxonomy:* Its evolution, status, and future. In *Soil Taxonomy: Achievements and Challenges,* edited by R. B. Grossman, pp. 7–13. Soil Science Society of America Special Publication 14. Madison, Wis.: Soil Science Society of America.

Bettis, E. A., III. 1995. The Holocene stratigraphic record of entrenched stream systems in thick loess regions of the Mississippi River Basin. Ph.D. diss., University of Iowa, Iowa City, Iowa.

Bettis, E. A., III, and W. J. Autin. 1997. Complex response of a North America drainage system to late Wisconsinan sedimentation. *Journal of Sedimentary Research* 67:740–748.

Bilzi, A. F., and E. J. Ciolkosz. 1977. A field morphology rating scale for evaluating pedological development. *Soil Science* 124:45–48.

Birkeland, Peter W. 1974. *Pedology, Weathering, and Geomorphological Research.* New York: Oxford University Press.

Birkeland, Peter W. 1984. *Soils and Geomorphology.* 2nd ed. New York: Oxford University Press.

Birkeland, Peter W. 1999. *Soils and Geomorphology.* 3d ed. New York: Oxford University Press.

Birkeland, Peter W., Michael N. Machette, and Kathleen M. Haller. 1991. *Soils as a Tool for Applied Quaternary Geology.* Miscellaneous Publication 91-3. Utah Geological and Mineral Survey, Washington, D.C.

Blackstock, Dan A. 1979. *Soil Survey of Lubbock County, Texas*. Soil Conservation Service, U.S. Department of Agriculture, Washington, D.C.

Boardman, John, ed. 1985. *Soils and Quaternary Landscape Evolution*. Chichester, U.K.: John Wiley and Sons.

Branham, Charles E. 1989. *Soil Survey of Pottawattamie County, Iowa*. Soil Conservation Service, U.S. Department of Agriculture, Washington D.C.

Bryan, Kirk. 1940. The retreat of slopes. *Annals of the Association of American Geographers* 30:254–268.

Bull, William B. 1991. *Geomorphic Responses to Climatic Change*. New York: Oxford University Press.

Buol, S. W., F. D. Hole, R. J. McCracken, and R. J. Southard. 1997. *Soil Genesis and Classification*. 4th ed. Ames, Iowa: Iowa State University Press.

Calvin, S. 1899. Iowan drift. *Geological Society of America Bulletin* 10:107–120.

Catt, J. A. 1986. *Soils and Quaternary Geology: A Handbook for Field Scientists*. London: Oxford University Press.

Cerling, T. E., J. Quade, W. Yang, and J. R. Bowman. 1989. Carbon isotopes in soils and palaeosols, and ecology and palaeoecology indicators. *Nature* 341:138–139.

Dahms, D., and V. T. Holliday. 1998. Soil taxonomy and paleoenvironmental reconstruction: A critical commentary. *Quaternary International* 51/52:109–114.

Dahms, D., V. T. Holliday, and P. W. Birkeland. 1998. Technical comment: Paleosols and Devonian forests. *Science* 279:151.

Daniels, R. B. 1988. Pedology, a field or laboratory science. *Soil Science Society of America Journal* 52:1518–1519.

Daniels, R. B., and E. E. Gamble. 1978. Relations between stratigraphy, geomorphology, and soils in coastal plain areas of southeastern U.S.A. *Geoderma* 21:41–65.

Daniels, R. B., E. E. Gamble, and S. W. Buol. 1973. Oxygen content in the groundwater of some North Carolina Aquults and Udults. In *Field Soil Water Regime*, edited by R. R. Bruce, pp. 153–166. Soil Science Society of America Special Publication 5. Madison, Wis.: Soil Science Society of America.

Daniels, R. B., E. E. Gamble, and J. G. Cady. 1970. Some relationships among coastal plain soils and geomorphic surfaces in North Carolina. *Soil Science Society of America Proceedings* 34:648–653.

Daniels, R. B., E. E. Gamble, and L. A. Nelson. 1967. Relations between A2 horizon characteristics and drainage in some fine loamy Ultisols. *Soil Science* 104:364–369.

Daniels, R. B., E. E. Gamble, and W. H. Wheeler. 1978. Age of soil landscapes in the Coastal Plain of North Carolina. *Soil Science Society of America Journal* 42:98–105.

Daniels, Raymond B., and Richard D. Hammer. 1992. *Soil Geomorphology*. New York: John Wiley and Sons.

Daniels, R. B., and L. A. Nelson. 1987. Soil variability and productivity: Future developments. In *Future Developments in Soil Science Research*, edited by L. L. Boersma, pp. 279–291. Madison, Wis.: Soil Science Society of America.

Easterbrook, D. J. 1993. *Surface Processes and Landforms*. New York: Macmillan.

Effland, Anne B. W., and William R. Effland. 1992. Soil geomorphology studies in the U.S. Soil Survey program. *Agricultural History* 66:189–212.

Effland, William R., and Anne B. W. Effland. n.d. The soil geomorphology studies of the U.S. Soil Conservation Service: Description and applications. Unpublished manuscript.

Fehrenbacher, J. B., I. J. Jansen, and K. R. Olson. 1986a. *Loess Thickness and Its Effect on Soils in Illinois*. Bulletin 782. College of Agriculture, University of Illinois at Urbana-Champaign, Ill.

Fehrenbacher, J. B., K. R. Olson, and I. J. Jansen. 1986b. Loess thickness in Illinois. *Soil Science* 141:423–431.

Gamble, E. E., R. B. Daniels, and W. D. Nettleton. 1970. Geomorphic surfaces and soils in the Black Creek Valley, Johnston County, North Carolina. *Soil Science Society of America Proceedings* 34:276–281.

Gerig, Allen J. 1985. *Soil Survey of the Clackamas County Area, Oregon*. Soil Conservation Service, U.S. Department of Agriculture, Washington D.C.

Gerrard, J. 1992. *Soils and Geomorphology*. London: Chapman and Hall.

Gile, Leland H. 1975. Causes of soil boundaries in an arid region. *Soil Science Society of America Proceedings* 39:316–330.

Gile, L. H. 1979. Holocene soils in eolian sediments of Bailey County, Texas. *Soil Science Society of America Journal* 43:994–1003.

Gile, Leland H. 1985. *The Sandhills Project Soil Monograph*. Las Cruces, N.M.: New Mexico State University, Rio Grande Historical Collections.

Gile, Leland H. 1987. A pedogenic chronology for Kilbourne Hole, Southern New Mexico. *Soil Science Society of America Journal* 51:746–760.

Gile, Leland H., and Robert B. Grossman. 1979. *The Desert Project Soil Monograph*. Soil Conservation Service, U.S. Department of Agriculture, Washington, D.C.

Gile, Leland H., John W. Hawley, and Robert B. Grossman. 1981. *Soils and Geomorphology in the Basin and Range Area of Southern New Mexico: Guidebook to the Desert Project*. New Mexico Bureau of Mines and Mineral Resources Memoir 39. Socorro, N.M.: New Mexico Bureau of Mines and Mineral Resources.

Gile Leland H., John W. Hawley, R. B. Grossman, H. C. Monger, and G. H. Mack. 1995. *Supplement to the Desert Project Guidebook, with Emphasis on Soil Micromorphology*. New Mexico Bureau of Mines and Mineral Resources Bulletin 142. Socorro, N.M.: New Mexico Bureau of Mines and Mineral Resources.

Gile, Leland H., Fred F. Peterson, and Robert B. Grossman. 1965. The K Horizon—A master soil horizon of carbonate accumulation. *Soil Science* 99:74–82.

Gile, Leland H., Fred F. Peterson, and Robert B. Grossman. 1966. Morphological and genetic sequences of carbonate accumulation in desert soils. *Soil Science* 101:347–360.

Hallberg, George R. 1984. The U.S. system of soil taxonomy: From the outside looking in. In *Soil Taxonomy: Achievements and Challenges,* edited by R. B. Grossman, pp. 45–59. Soil Science Society of America Special Publication 14. Madison, Wis.: Soil Science Society of America.

Hallberg, George R. 1986. Pre-Wisconsin glacial stratigraphy of the central plains region in Iowa, Nebraska, Kansas, and Missouri. In *Quaternary Glaciations in the Northern Hemisphere,* edited by V. Sibrava, D. Q. Bowen, and G. M. Richmond, pp. 11–15. New York: Pergamon Press.

Hallberg, George R., Thomas E. Fenton, G. A. Miller, and A. J. Luteneggar. 1978. The Iowan erosion surface: An old story, an important lesson, and some new wrinkles. In 42nd Annual Tri-State Geological Field Conference Guidebook, edited by R. R. Anderson, pp. 2.2–2.94. Iowa City, Iowa: Iowa Geological Survey.

Harden, Jennifer W. 1982. A quantitative index of soil development from field descriptions: Examples from a chronosequence in central California. *Geoderma* 28:1–28.

Harden, Deborah R., Norma E. Biggar, and Mary L. Gillam. 1985. Quaternary deposits and soils in and around Spanish Valley, Utah. In *Soils and Quaternary Geology of the Southwestern United States,* edited by D. L. Weide, pp. 43–64. Geological Society of America Special Paper 203. Boulder, Colo.: Geological Society of America.

Harden, Jennifer W., and Emily Taylor. 1983. A quantitative comparison of soil development in four climatic regimes. *Quaternary Research* 20:342–359.

Hole, Francis D. 1976. *Soils of Wisconsin.* Madison, Wis.: University of Wisconsin Press.

Holliday, Vance T. 1985. Morphology of late Holocene soils at the Lubbock Lake Site, Texas. *Soil Science Society of America Journal* 49:938–946.

Holliday, Vance T. 1990. Soils and landscape evolution of Eolian plains: The southern high plains of Texas and New Mexico. In *Soils and Landscape Evolution,* edited by P. L. K. Knuepfer and L. D. McFadden. *Geomorphology* (special issue) 3:489–515.

Holliday, Vance T. 1994. The "state factor" approach in geoarchaeology. In *Factors of Soil Formation: A Fiftieth Anniversary Retrospective,* edited by R. Amundson, pp. 65–86. Soil Science Society of America Special Publication 33. Madison, Wis.: Soil Science Society of America.

Holliday, Vance T. 1995. *Stratigraphy and Paleoenvironments of Late Quaternary Valley Fills on the Southern High Plains.* Geological Society of America Memoir 186. Boulder, Colo.: Geological Society of America.

Huckle, H. F., H. D. Dollar, and R. F. Pendleton. 1974. *Soil Survey of Brevard County, Florida.* Soil Conservation Service, U.S. Department of Agriculture, Washington, D.C.

Hunt, Charles B. 1972. *Geology of Soils: Their Evolution, Classification, and Uses.* San Francisco: W. H. Freeman and Company.

Jacob, John S., and Lee C. Nordt. 1991. Soil and landscape evolution: A paradigm for pedology. *Soil Science Society of America Journal* 55:1194.

Jenny, Hans. 1941. *Factors of Soil Formation: A System of Quantitative Pedology.* New York: McGraw-Hill.

Jenny, Hans. 1946. Arrangement of soil series and types according to functions of soil-forming factors. *Soil Science* 61:375–391.

Johnson, Donald L. 1993. Dynamic denudation evolution of tropical, subtropical, and temperate landscapes with three tiered soils: Toward a general theory of landscape evolution. *Quaternary International* 17:67–78.

Kay, G. F., and E. T. Apfel. 1929. The pre-Illinoian Pleistocene geology of Iowa. *Iowa Geological Survey Annual Report* 34:1–304.

King, L. C. 1949. The Pediment landform: Some current problems. *Geological Magazine* 86:245–250.

King, L. C. 1950. The world's plainlands: A new approach in geomorphology. *Geological Society of London, Quarterly Journal* 106:101–131.

King, L. C. 1953. Cannons of landscape evolution. *Bulletin of the Geological Society of America* 64:721–751.

Knox, E. G. 1965. Soil individuals and soil classification. *Soil Science Society of America Proceedings* 29:79–84.

Knuepfer, P. L. K., and L. D. McFadden, eds. 1990. *Soils and Landscape Evolution. Geomorphology* (special issue) 3:197–575.

Ku, T. L., W. B. Bull, S. T. Freeman, and K. G. Knauss. 1979. Th 230–U 234 dating of pedogenic carbonates in gravelly desert soils of Vidal Valley, southeastern California. *Geological Society of America Bulletin* 90:1063–1073.

Lehman, T. M. 1989. Upper Cretaceous (Maastrichtian) paleosols in trans-Pecos Texas. *Geological Society of America Bulletin* 101:188–203.

Leighton, M. M. 1933. The naming of the subdivision of the Wisconsin glacial age. *Science,* new series 77:168.

Leverett, F. 1942. Note by Frank Leverett. *Journal of Geology* 50:1001–1002.

Maat, P. B. 1992. Eolian stratigraphy, soils and geomorphology in the Hudson dune field. M.S. thesis, University of Colorado, Boulder, Colo.

Machette, Michael N. 1978. Dating quaternary faults in the southwestern United States by using buried calcic paleosols. *U.S. Geological Survey Journal of Research* 6:369–381.

Machette, Michael N. 1985. Calcic soils of the southwestern United States. In *Soils and Quaternary Geology of the Southwestern United States*, edited by D. L. Weide, pp. 1–21. Geological Society of America Special Paper 203. Boulder, Colo.: Geological Society of America.

Machette, Michael N. 1988. *Quaternary Movement along the La Jencia Fault, Central New Mexico.* U.S. Geological Survey Professional Paper 1440. Washington, D.C.: Government Printing Office.

Mack, G. H., W. C. James, and H. C. Monger. 1993a. Classification of paleosols. *Geological Society of America Bulletin* 105:129–136.

Mack, G. H., S. L. Salyards, and W. C. James. 1993b. Magnetostratigraphy of the plio-Pleistocene Camp Rice and Palomas formations in the Rio Grande Rift of southern New Mexico. *American Journal of Science* 293:49–77.

Mandel, Rolfe D. 1995. Geomorphic controls of the archaic record in the central plains of the United States. In *Archaeological Geology of the Archaic Period in North America,* edited by E. A. Bettis, III, pp. 37–66. Geological Society of America Special Paper 297. Boulder, Colo.: Geological Society of America.

Marion, G. M., W. H. Schlesinger, and P. J. Fonteyn. 1985. Caldep: A regional model for soil CaCO3 (caliche) deposition in southwestern Deserts. *Soil Science* 139:468–481.

Markewich, H. W., W. C. Lynn, M. J. Pavich, R. G. Johnson, and J. C. Meetz. 1988. *Analyses of Four Inceptisols of Holocene Age, East-central Alabama.* Bulletin 1589-C. U.S. Geological Survey, Washington, D.C.

Markewich, H. W., and M. J. Pavich. 1991. Soil chronosequence studies in temperate to subtropical, low-latitude, low-relief terrain with data from the eastern United States. In *Weathering and Soils,* edited by M. J. Pavich. *Geoderma* (special issue) 51:213–239.

Markewich, H. W., M. J. Pavich, and G. R. Buell. 1990. Contrasting soils and landscapes of the Piedmont and Coastal Plains, eastern United States. In *Soils and Landscape Evolution,* edited by P. L. K. Knuepfer and L. D. McFadden. *Geomorphology* (special issue) 3:417–447.

Markewich, H. W., M. J. Pavich, M. J. Mausbach, R. L. Hall, R. G. Johnson, and P. P. Hearn. 1987. *Age Relations between Soils and Geology in the Coastal Plain of Maryland and Virginia.* Bulletin 1589-A. U.S. Geological Survey, Washington, D.C.

Markewich, H. W., M. J. Pavich, M. J. Mausbach, R. G. Johnson, and V. M. Gonzales. 1989. *A Guide for Using Soil and Weathering Profile Data in Chronosequence Studies of the Coastal Plain of the Eastern United States.* Bulletin 1589-D. U.S. Geological Survey, Washington, D.C.

Markewich, H. W., M. J. Pavich, M. J. Mausbach, B. N. Stuckey, R. G. Johnson, and V. Gonzales. 1986. *Soil Development and Its Relation to the Ages of Morphostratigraphic Units in Horry County, South Carolina.* Bulletin 1589-B. U.S. Geological Survey, Washington, D.C.

Mason, Joseph A. 1992. Loess distribution and soil landscape evolution, southeastern Minnesota. M.S. thesis, University of Minnesota, St. Paul, Minn.

Mason, Joseph A., and Edward A. Nater. 1994. Soil-morphology—Peoria loess grain size relationships, southeastern Minnesota. *Soil Science Society of America Journal* 58:432–439.

Mason, Joseph A., Edward A. Nater, and Howard C. Hobbs. 1994. Transport direction of Wisconsinan loess in southeastern Minnesota. *Quaternary Research* 41:44–51.

Mathers, A. C. 1963. *Some Morphological, Physical, Chemical, and Mineralogical Properties of Seven Southern Great Plains Soils.* Agricultural Research Service 41-85. U.S. Department of Agriculture, Washington, D.C.

Mayer, L., L. D. McFadden, and J. W. Harden. 1988. Distribution of calcium carbonate in desert soils: A model. *Geology* 16:303–306.

McDonald, E., and L. D. McFadden. 1994. Quaternary stratigraphy of the Providence Mountains Piedmont and preliminary age estimates and regional stratigraphic correlations of Quaternary deposits in the eastern Mojave Desert, California. In *Geological Investigations of an Active Margin,* edited by S. F. McGill and T. M. Ross, pp. 205–210. Geological Society of America Cordilleran Section Guidebook. Redlands, Calif.: San Bernardino County Museum.

McDonald, E. V., F. B. Pierson, G. N. Flerchinger, and L. D. McFadden. 1996. Application of a soil-water balance model to evaluate the influence of Holocene climatic change on calcic soils, Mojave Desert, California, U.S.A. *Geoderma* 74:167–192.

McDowell, P. F., and M. C. Roberts. 1987. *Field Guidebook to the Quaternary Stratigraphy, Geomorphology and Soils of the Willamette Valley, Oregon.* Field Trip 3, Association of American Geographers Annual Meeting, Portland, Ore.

McFadden, Leslie D. 1981. Geomorphic processes influencing the Cenozoic evolution of the Canada del Oro Valley, southern Arizona. In *Arizona Tectonic Digest*, edited by Claudia Stone and J. P. Jenney, pp. 13–20. Tuscon: Arizona Geologic Bureau.

McFadden, Leslie D. 1982. The impacts of temporal and spatial climatic changes on alluvial soils genesis in southern California. Ph.D. diss., University of Arizona, Tucson, Ariz.

McFadden, Leslie D. 1988. Climatic influence on rates and processes of soil development in Quaternary deposits of southern California. In *Paleosols and Weathering through Geologic Time: Principles and Applications*, edited by J. Reinhardt and W. R. Sigleo, pp. 153–177. Geological Society of America Special Paper 216. Boulder, Colo.: Geological Society of America.

McFadden, Leslie D. 1993. Integrating soil science and Quaternary geology: Recent advances, problems and philosophical issues. Abstracts with Programs, Geological Society of America Annual Meeting, Boston, 25–28 October, 25:A306-A307.

McFadden, Leslie D., and Peter L. K. Knuepfer. 1990. Soil geomorphology: The linkage of pedology and surficial processes. In *Soils and Landscape Evolution*, edited by P. L. K. Knuepfer and L. D. McFadden. *Geomorphology* (special issue) 3:197–205.

McFadden, Leslie D., Eric V. McDonald, Steven G. Wells, K. Anderson, Jay Quade, and Stephen L. Forman. 1998. The vesicular layer and carbonate collars of desert soils and pavements: Formation, age and relation to climatic change. *Geomorphology* 24: 101–145.

McFadden, Leslie D., and John C. Tinsley. 1985. The rate and depth of accumulation of pedogenic carbonate accumulation in soils: Formulation and testing of a compartment model. In *Soils and Quaternary Geology of the Southwestern United States*, edited by D. L. Weide, pp. 23–42. Geological Society of America Special Paper 203. Boulder, Colo.: Geological Society of America.

McFadden, Leslie D., Steven G. Wells, and John C. Dohrenwend. 1986. Influences of Quaternary climatic changes on processes of soil development on desert loess deposits of the Cima volcanic field, California. *Catena* 13:361–389.

McFadden, Leslie D., Steven G. Wells, and M. J. Jercinovic. 1987. Influences of Eolian and pedogenic processes on the evolution and origin of desert pavements. *Geology* 15:504–508.

McGee, W. J. 1891. The Pleistocene history of northeastern Iowa. *U.S. Geological Survey* 11th Annual Report, pp. 189–577. U.S. Geological Survey, Washington, D.C.

Milne, G. 1935a. Some suggested units for classification and mapping, particularly for east African soils. *Soils Research* (Berlin) 4:183–198.

Milne, G. 1935b. Composite units for the mapping of complex soil associations. *Transactions of the Third International Congress of Soil Science* 1:345–347.

Monger, H. C., and H. P. Adams. 1996. Micromorphology of calcite-silica deposits, Yucca Mountain, Nevada. *Soil Science Society of America Journal* 60:519–530.

Monger, H. C., L. A. Daugherty, W. C. Lindemann, and C. M. Liddell. 1991. Microbial precipitation of pedogenic calcite. *Geology* 19:997–100.

Morrison, Roger B. 1967. Principles of Quaternary soil stratigraphy. In *Quaternary Soils*, edited by R. B. Morrison and H. E. Wright, Jr., pp. 1–69. Reno, Nev.: Desert Research Institute, University of Nevada.

Morrison, R. B. 1978. Quaternary soil stratigraphy-concepts, methods, and problems. In *Quaternary Soils*, edited by W. C. Mahaney, pp. 77–108. Norwich, U.K.: Geo Abstracts.

Nettleton, W. D., B. R. Brasher, and R. J. Ahrens. 1998. A classification system for buried paleosols. *Quaternary International* 51/52:175–183.

North American Commission on Stratigraphic Nomenclature (NACOSN). 1983. North American stratigraphic code. *Association of Petroleum Geologists Bulletin* 67:841–875.

Olson, Carolyn G. 1989. Soil geomorphic research and the importance of paleosol stratigraphy to Quaternary investigations, midwestern USA. In *Paleopedology: Nature and Applications of Paleosols*, edited by A. Bronger and J. Catt, pp. 129–142. *Catena* Supplement 16. Cremlingen, Germany: Catena Verlag.

Olson, Carolyn G. 1997. Systematic soil-geomorphic investigations—Contributions of R. V. Ruhe to pedologic interpretation. In *History of Soil Science*, edited by D. H. Yaalon and S. Berkowicz, pp. 415–438. Advances in Geo-Ecology 29. Reiskirchen, Germany: Catena Verlag.

Parsons, R. B. 1978. Soil-geomorphology relationships in mountains of Oregon, U.S.A. *Geoderma* 21:25–39.

Parsons, R. B., and C. A. Balster. 1967. Dayton—A depositional planosol, Willamette Valley, Oregon. *Soil Science Society of America Proceedings* 31:255–258.

Parsons, R. B., C. A. Balster, and O. A. Ness. 1970. Soil development and geomorphic surfaces, Willamette Valley, Oregon. *Soil Science Society of America Proceedings* 34:485–491.

Parsons, R. B., G. H. Simonson, and C. A. Balster. 1968. Pedogenic and geomorphic relationships of associated aqualfs, albolls, and xerolls in Western Oregon. *Soil Science Society of America Proceedings* 32:556–563.

Paton, T. R., G. S. Humphreys, and P. B. Mitchell. 1995. *Soils: A New Global View*. New Haven, Conn.: Yale University Press.

Patterson, P. E. 1991. Differentiation between the effects of diagenesis and pedogenesis in the origin of color banding in the Wind River formation

(Lower Eocene), Wind River Basin, Wyoming. Ph.D. diss. University of Colorado, Boulder, Colo.

Peterson, F. F. 1981. *Landforms of the Basin and Range Province Defined for Soil Survey.* Technical Bulletin 38. Nevada Agricultural Experiment Station.

Phillips, J. D., J. Gosweiler, M. A. Tollinger, R. Gordon, S. Mayeux, M. Witmeyer, and T. Altieri. 1994. Edge effects and spatial variability in coastal plain Ultisols. *Southeastern Geographer* 34:125–137.

Prior, J. C. 1991. *Landforms of Iowa.* Iowa City, Iowa: University of Iowa Press.

Quade, Jay, T. E. Cerling, and J. R. Bowman. 1989. Systematic variation in the carbon and oxygen isotopic composition of Holocene soil carbonate along elevation transects in the southern Great Basin, USA. *Geological Society of America Bulletin* 101:464–528.

Rector, D. D. 1981. *Soil Survey of Rhode Island.* Soil Conservation Service, U.S. Department of Agriculture, Washington, D.C.

Reheis, Marith C., J. C. Goodmacher, Jennifer W. Harden, Leslie D. McFadden, Thomas K. Rockwell, Ralph R. Shroba, Janet M. Sowers, and Emily M. Taylor. 1995. Quaternary soils and dust deposition in southern Nevada and California. *Geological Society of America Bulletin* 107:1003–1022.

Reheis, Marith, and Rolf Kihl. 1995. Dust deposition in southern Nevada and California, 1984–1989: Relations to climate, source area, and source lithology. *Journal of Geophysical Research* 100,D5:8893–8918.

Retallack, Gregory J. 1986. The fossil record of soils. In *Paleosols: Their Recognition and Interpretation,* edited by V. P. Wright, pp. 1–57. Princeton, N.J.: Princeton University Press.

Retallack, Gregory J. 1988. Field recognition of paleosols. In *Paleosols and Weathering through Geologic Time: Principles and Applications,* edited by J. Reinhardt and W. R. Sigleo, pp. 1–20. Geological Society of America Special Paper 216. Boulder, Colo.: Geological Society of America.

Retallack, Gregory J. 1990. *Soils of The Past.* Boston: Unwin-Hyman.

Retallack, Gregory J. 1993. Classification of paleosols: Discussion. *Geological Society of America Bulletin* 105:1635–1636.

Retallack, Gregory J. 1997. Early forest soils and their role in Devonian global change. *Science* 276:583–585.

Ritter, Dale F., R. Craig Kochel, and Jerry R. Miller. 1995. *Process Geomorphology.* 3d ed. Dubuque, Iowa: W. C. Brown Publishing.

Ruhe, Robert V. 1954a. *Erosion Surfaces of Central African Interior High Plateaus.* Ser. Sci. 59. Publications de L'Institut National Pour L'Etude Agronomique du Congo Belge (I.N.E.A.C.), Bruxelles.

Ruhe, Robert V. 1954b. Pleistocene soils along the Rock Island relocation in southwestern Iowa. *American Railway Engineering Association Bulletin* 514:639–645.

Ruhe, Robert V. 1956a. *Landscape Evolution in the High Ituri, Belgian Congo.* Ser. Sci. 66. Publications de L' I.N.E.A.C., Bruxelles.

Ruhe, Robert V. 1956b Geomorphic surfaces and the nature of soils. *Soil Science* 82:441–455.

Ruhe, Robert V. 1960. Elements of the soil landscape. *7th International Congress of Soil Science Transactions* 4:165–170.

Ruhe, Robert V. 1962. Age of the Rio Grande Valley in southern New Mexico. *Journal of Geology* 70:151–167.

Ruhe, Robert V. 1964a. An estimate of paleoclimate in Oahu, Hawaii. *American Journal of Science* 262:1098–1115.

Ruhe, Robert V. 1964b. Landscape morphology and alluvial deposits in southern New Mexico. *Annals of the Association of American Geographers* 54:147–159.

Ruhe, Robert V. 1965. Relation of fluctuations of sea level to soil genesis in the Quaternary. *Soil Science* 99:23–29.

Ruhe, Robert V. 1967. *Geomorphic Surfaces and Surficial Deposits in Southern New Mexico*. New Mexico Bureau of Mines and Mineral Resources Memoir 18. Socorro, N.M.: New Mexico Bureau of Mines and Mineral Resources.

Ruhe, Robert V. 1969. *Quaternary Landscapes in Iowa*. Ames, Iowa: Iowa State University Press.

Ruhe, Robert V. 1970. Soil-geomorphology studies 1953–1970. Manuscript, Iowa State University Agronomy Library, Ames, Iowa.

Ruhe, Robert V. 1975. *Geomorphology*. Boston: Houghton Mifflin.

Ruhe, Robert V., R. B. Daniels, and J. G. Cady. 1967. *Landscape Evolution and Soil Formation in Southwestern Iowa*. Technical Bulletin 1349. U.S. Department of Agriculture, Washington, D.C.

Ruhe, Robert V., W. P. Dietz, T. E. Fenton, and G. F. Hall. 1968. *Iowan Drift Problem Northeastern Iowa*. Report of Investigations 7. Iowa Geological Survey, Iowa City, Iowa.

Ruhe, Robert V., and Carolyn G. Olson. 1980. Soil welding. *Soil Science* 130:132–139.

Ruhe, Robert V., J. M. Williams, and E. L. Hill. 1965a. Shorelines and submarine shelves, Oahu, Hawaii. *Journal of Geology* 73:485–497.

Ruhe, Robert V., J. M. Williams, R. C. Shuman, and E. L. Hill. 1965b. Nature of soil parent materials in Ewa-Waipahu Area, Oahu. *Soil Science Society of America Proceedings* 29:282–287.

Runge, E. C. A., and R. J. McCracken. 1984. The role of soil classification in research planning. In *Soil Taxonomy: Achievements and Challenges,* edited by R. B. Grossman, pp. 15–28. Soil Science Society of America Special Publication 14. Madison, Wis.: Soil Science Society of America.

Schaetzl, R. J., and D. L. Mokma. 1988. A numerical index of podzol and podzolic soil development. *Physical Geography* 9:232–246

Shipman, John A. 1997. *Soil Survey of the Lincoln County Area, Oregon*. Soil Conservation Service, U.S. Department of Agriculture, Washington D.C.

Simonson, Roy W. 1987. *Historical Aspects of Soil Survey and Soil Classification*. Madison, Wis.: Soil Science Society of America.

Simonson, Roy W. 1997. Evolution of soil series and type concepts in the United States. In *History of Soil Science*, edited by D. H. Yaalon and S. Berkowicz, pp. 79–108. Advances in GeoEcology 29. Reiskirchen, Germany: Catena Verlag.

Soil Survey Division Staff. 1993. *Soil Survey Manual*. Agriculture Handbook 18. U.S. Department of Agriculture, Washington, D.C.

Soil Survey Staff. 1951. *Soil Survey Manual*. Agriculture Handbook 18. U.S. Department of Agriculture, Washington, D.C.

Soil Survey Staff. 1960. *Soil Classification: A Comprehensive System, 7th Approximation*. Soil Conservation Service, U.S. Department of Agriculture, Washington, D.C.

Soil Survey Staff. 1975. *Soil Taxonomy*. Agriculture Handbook 436. U.S. Department of Agriculture, Washington, D.C.

Stoner, H. R., and M. L. Dixon. 1974. *Soil Survey of Martin County, Texas*. Soil Conservation Service, U.S. Department of Agriculture, Washington, D.C.

Swanson, David K. 1990a. Soil landform units for soil survey. *Soil Survey Horizons* 31:17–21.

Swanson, David K. 1990b. Landscape classes: Higher-level map units for soil survey. *Soil Survey Horizons* 31:52–54.

Swanson, David K. 1993. Comments on "The soil survey as paradigm-based science." *Soil Science Society of America Journal* 57:1164.

Tandarich, John P. 1998a. Agricultural geology: Disciplinary history. In *Sciences of the Earth: An Encyclopedia of Events, People, and Phenomena*, edited by Gregory A. Good, pp. 23–29. New York: Garland.

Tandarich, John P. 1998b. Pedology: Disciplinary history. In *Sciences of the Earth: An Encyclopedia of Events, People, and Phenomena*, edited by Gregory A. Good, pp. 666–670. New York: Garland.

Tandarich, John P., Robert G. Darmody, and Leon R. Follmer. 1988. The development of pedological thought: Some people involved. *Physical Geography* 9:162–174.

Tandarich, John P. and Steven W. Sprecher. 1994. The intellectual background for the factors of soil formation. In *Factors of Soil Formation: A Fiftieth Anniversary Retrospective*, edited by R. Amundson. pp. 1–13. Soil Science Society of America Special Publication 33. Madison, Wis.: Soil Science Society of America.

Verrecchia, E. P., P. Freytet, K. E. Verrecchia, and J.-L. Dumont. 1995. Spherulites in calcrete laminar crusts: Biogenic $CaCO_3$ precipitation as a major contributor to crust formation. *Journal of Sedimentary Research* A65:690–700.

Vreeken, W. J. 1975. Quaternary evolution in Tama County, Iowa. *Annals of the Association of American Geographers* 65:283–296.

Walker, P. H. 1966. Postglacial environments in relation to landscape and soils. *Iowa Agricultural Research Station Bulletin* 549:838–875.

Walker, P. H., and R. V. Ruhe. 1968. Hillslope models and soil formation: Closed systems. *Transactions of the 9th International Congress of Soil Science* 4:561–568.

Wang, Y., Eric McDonald, Ronald Amundson, Leslie McFadden, and Oliver Chadwick. 1996. An isotopic study of soils in chronological sequences of alluvial deposits, Providence Mountains, California. *Geological Society of America Bulletin* 108:379–391.

Wells, Stephen G., Leslie D. McFadden, John C. Dohrenwend, B. D. Turrin, and K. D. Mahrer. 1985. Late Cenozoic landscape evolution of lava flow surfaces of the Cima volcanic field, Mojave Desert, California. *Geological Society of America Bulletin* 96:1518–1529.

Wells, Stephen G., Leslie D. McFadden, J. Poths, and C. T. Olinger. 1995. Cosmogenic ^{3}He exposure dating of stone pavements: Implications for landscape evolution in deserts. *Geology* 23:613–616.

Yaalon, D. H., and E. Ganor. 1973. The influence of dust on soils in the Quaternary. *Soil Science* 116:146–155.

10

Soil Survey
Interpretations: Past,
Present, and Looking
to the Future

Patricia J. Durana and Douglas Helms

INTRODUCTION

Every day land managers make decisions, the success of which depends wholly or in part on the nature of soils. Understanding the capability, limitations, and potential uses of our soils is fundamental to effective decision making. Soil survey data are among the most important pieces of information used by government units, businesses, and individuals to make land-use decisions that range from development and taxation to farming and natural resource protection.

Interpretations of soil survey data are based on how soils respond for a specific use. Soil interpretations employ a set of rules or criteria based on basic measured soil properties, inferred properties, or classes of properties. Soil interpretations may be developed at different levels of generalization. Highly integrative generalizations may be constructed for *management groups*—groupings of soils that require similar kinds of practices to achieve acceptable performance for a soil use. Many of these broad national-level interpretive groups (e.g., highly erodible lands, hydric soils, prime farmland) find their way into national legislation. National-level generalizations, by their very scope, typically lack the sensitivity necessary to support highly localized decision making. Thus, interpretations useful at the local level generally require adjustment or development of interpretive criteria that reflect local or regional distinctions as well as user needs. Effectively using soil survey interpretations in decision making depends on understanding the level of generalization as well as other pertinent considerations (see Figure 10.1).

Soil survey interpretations provide valuable information on the suitability, potentials, and limitations of an area for identified uses. To make effective use of interpretations, it is important to understand that:

- An interpretation has a specific purpose and rarely is adaptable to other purposes without modification.

- Application of interpretations for a specific land area has an inherent limitation related to the variability in the composition of delineations within a map unit. This may particularly significant for high value - small area uses such as house sites, where the size of the map unit is sufficiently large that dissimilarities in soils and their limitations within the map unit may differ substantially from those of the major components of the unit.

- Inherent variability of soils in nature defines the restraints in soil interpretations and the precision of soil behavior predictions for specific areas. Interpretations based on soil surveys are rarely suitable for such onsite evaluations as home sites without further evaluations at the specific site. They retain substantial value for screening areas for a planned use and expressing this as a suitability or limitation.

- Specific soil behavior predictions are commonly presented in terms of limitations posed by one or a few soil properties. Limitations must be considered along with other properties to determine which poses the greatest limitation to a proposed use. For example, a soil may exhibit a shrink swell limitation as well as shallow depth to bedrock, however the depth to bedrock may be the greater limitation to the proposed use.

- Considerations that determine economic value of land may not be a part of interpretations but are integral in developing soil potentials for a certain landuse. Location of an area in relation to road systems, markets, or other services may be considered when developing ratings.

- Some interpretations are more sensitive to changes in technology and landuse. For example, new technologies may reduce limitations to certain uses, such as the development of reinforced concrete slab construction, which reduces the limitation posed by shrink swell for small building construction. New uses of land will require new prediction models for soil interpretations.

- Interpretations based on soil properties are only appropriate if land characteristics have remained constant or similar to their condition at the time that soil mapping was accomplished.

FIGURE 10.1 Considerations in the use of soil survey interpretations. (Adapted from U.S. Department of Agriculture, Soil Survey Division 1993)

Today, soil information that meets current standards for users has been prepared on about 1.7 billion acres (75 percent) of the nation, and about 10 percent of completed soil surveys are available in digital form. The Natural Resources Conservation Service (NRCS) is accelerating digitization of completed soil surveys and other federal, state, and local entities are providing funds or services to assist in this effort.

Soil interpretations, however, were not always used in land-management decision making. Interpretations of soil survey data that were construed to be "negative" were often soundly criticized, or even repressed, during the early days of soils interpretation. It took time to finally persuade people that all land was not the same and that it could differ, often significantly, even over very short distances. Even more difficult to convey was the concept that physical soil characteristics could be more important to success in planning development than geographic location.

The history of understanding the range of potential uses of our land resources began over 100 years ago with the authorization of a national effort to investigate soils. On 3 May 1899, Congress authorized and appropriated $16,000 for the "investigation of the relation of soils to climate and organic life . . . [and] . . . investigation of the texture and composition of soils in the field and in the laboratory . . ." (Statutes at Large 1899, 334–335). With this humble beginning, operations commenced to survey the nation's soil resource. Through the tireless efforts of the early soil survey pioneers, a National Cooperative Soil Survey (NCSS) was ultimately realized. Originally, Congress and the U.S. Department of Agriculture (USDA) supported the soil survey for its agricultural interpretations. Today, the soil survey is valued for a broad range of interpretations beyond agriculture.

Over the course of the twentieth century, the soil survey and soil science developed together. A number of factors limited the interpretive value of early surveys. Survey scales were too small to permit the highly localized interpretations that would have benefited farmers and other land users. Soil series were too unrefined or broad to provide a basis for interpretive decisions. At the time, surveys focused on soil texture with little or no emphasis on geological, mineralogical, or chemical soil attributes. And, perhaps most importantly, early surveys did not have the benefit of the body of knowledge on plant-soil relationships that developed in later decades (Jenny 1961).

Charles Kellogg, who became chief of the Soil Survey Division in 1935, questioned how much the earliest soil surveyors could recommend because of the limited knowledge of plant-soil relationships and elementary ability to analyze soils. Advances in soil science, agronomy, animal industry, entomology, forestry and other associated disciplines, along with those in soils mapping and classification, were necessary to improving and expanding interpretations. As scientists discovered significant soil characteristics, the survey was better able to classify and map those characteristics for interpretation and use.

DEVELOPMENT OF SOIL INTERPRETATIONS

Soil interpretations have been emphasized to varying degrees throughout the history of the soil survey. Milton Whitney, the first chief of the

Bureau of Soils, started the systematic soil survey in four areas of the country in 1899. These early surveys concentrated on soil characteristics that could be measured and analyzed and correlated with the properties Whitney thought important: soil texture, soil moisture, soil temperature, and concentration of soluble salts in soils.

The ability to identify soluble salts in the soil and water of western states was perhaps the most accurate and financially valuable interpretation the early soil surveyors could make. This interpretation identified areas where irrigated agriculture might result in salt accumulation on the soil surface and in the water table, conditions that adversely affected plant growth. While the problem could often be corrected, it meant an added expense that needed to be considered in the opening of land to agriculture. Predictably, making such information publicly available was not always popular with landowners, developers, or others. Eugene Hilgard, of the University of California, articulated the problem of local pressures to suppress results in a letter to Secretary of Agriculture James Wilson:

> The reasons why we have not published a good many of our results are partly financial, partly because of the silly opposition of landowners, as well as others, to having the subject even mentioned, much less thoroughly discussed with reference to localities. (Hilgard to Wilson 1900)

Curtis Marbut assumed leadership of soil survey operations in 1911, but he was under Whitney's general supervision, as Whitney remained chief of the Bureau of Soils until 1927. Marbut worked assiduously at establishing the Russian concepts of soil genesis, developing a soil classification system, and establishing the scientific reputation of the published soil survey. However, he generally relinquished interpretations to the states. In a statement that reflected his attitude about the survey in the post-World War I period, Marbut wrote in 1924:

> There is a constant tendency to limit more and more the amount of agricultural advice given. This is being done partly because the soil survey report is being regarded more and more as a scientific publication and should not attempt to give practical advice. The soil surveyor is being regarded more and more as primarily a scientific man concerned with the scientific investigation of the soil, and one who is not primarily con-

cerned with the use or the treatment of the soil, except in so far as his wide experience may suggest. (Gardner 1998, 110)

Marbut proposed that once a soil was described, the researcher could devise agricultural management recommendations for the mapped soils and then transfer the management techniques to similar soils elsewhere. Marbut's attitude certainly influenced the priorities of the federally employed soil surveyors.

The cooperating soil surveyors and other agricultural scientists employed by the state agricultural experiment stations or other state institutions, however, were much more attuned to the need to develop interpretations for agriculture. Cooperators in the states knew better than anyone that financial support from the state authorities depended upon the usefulness of the surveys (Gardner 1998).

With the federal leadership de-emphasizing interpretations in published reports, states took the lead in preparing and publishing interpretations of completed surveys. A common theme was to identify soils particularly suited to certain crops, for example, apples or tobacco. In acquiring soils information for their particular uses, other federal agencies such as the Bureau of Reclamation and the Bureau of Public Roads established cooperative studies with the Bureau of Soils.

Despite Marbut's point of view, there was some federal effort to develop soil erosion interpretations to support development of soil conservation practices. In the late 1920s, H. E. Middleton of the Bureau of Chemistry and Soils made substantial progress toward understanding the complex chemical and physical properties and processes related to erodibility (Middleton 1930; Middleton et al. 1932). Simultaneously, Hugh Hammond Bennett was moving his campaign for soil conservation beyond the confines of the Soil Survey Division. Bennett had identified areas where the blend of geology and agricultural practices combined to produce serious soil erosion, and he took this discovery to the public and politicians.

Largely as a result of Bennett's campaign for research on erosion and conservation measures, Congress authorized a series of state soil erosion experiment stations. Interdisciplinary teams of researchers established plots to measure erosion conditions under different types of crops, soils, rotations, and various agricultural management practices and structures (Borst et al. 1945; Browning et al. 1948; Copley et al. 1944; Daniel et al. 1943; Hays et al. 1949; Hill et al. 1944; Horner et al. 1944; Musgrave et al. 1937; Pope et al. 1946; Smith et al. 1945). Federal support of these state research efforts ultimately led to the accumulation of national-level data on soil erodibility. The origins of the erodibility data that supports current conservation planning tools, such as the Universal Soil Loss Equation (and the revised equation), reach back to these pioneering studies (Lyles 1985; Meyer and Moldenhauer 1985).

The period of scant federal interest in soil interpretation ended when Charles E. Kellogg became chief of the Soil Survey Division in 1935. Kellogg brought new vigor to developing soil interpretations. His appointment also coincided with H. H. Bennett's success in influencing Congress to create the Soil Conservation Service (SCS)—an event that would have later significance on the continuing development of the soil survey.

Kellogg reversed Marbut's policy and proclaimed that "pedologists themselves must devote more attention to summarizing the results of their investigations and making them readily available for public use" (Gardner 1998). Kellogg's new view on interpretations was understandable, even predictable, as he had participated in some of the early events shaping soil survey interpretations. As a student at Michigan State University, Kellogg had worked on the 1922 Michigan Land Economic Survey to determine the suitability of lands of northern Michigan for agriculture, recreation, or other uses (Kellogg 1943; Schoenmann 1923). Geographers recognize the survey as a landmark in multidisciplinary research and inventorying for land classification, while soil scientists point to the centrality of interpreting soils as one component in land classification. The survey greatly influenced the land classification scheme and survey of the Tennessee Valley Authority area (James and Jones 1954). During this time, Kellogg also studied the relationship of soil type to potential sites for highway construction as a result of difficulties the Michigan highway department had with road buckling (Kellogg 1928).

As chief, Kellogg established a policy that all new soil surveys would include productivity ratings. The division's earlier attempts to develop natural or inherent productivity ratings had been abandoned in the face of criticism by detractors who pointed out an obvious inconsistency— soil productivity could be profoundly influenced by management. Thus, interpreters were faced with the question of how to rate a soil that is very productive under irrigation but is nonarable in its absence, or a soil that is not inherently fertile but is very responsive to soil amendments. Productivity as an interpretation was henceforth to be relative to the technology and management applied (Gardner, 1998). The invigorated emphasis on interpretations that Kellogg brought to the soil survey occurred in an opportune era, when the combination of new agricultural technologies matched to refined differentiation of soil types held potential for substantial productivity increases.

The outset of the 1940s found the USDA in the unique position of having two soil survey efforts, each working cooperatively with state agricultural experiment stations and other agencies. The federal component of the NCSS was housed in the Division of Soil Survey under the Bureau of Plant Industry, Soils, and Agricultural Engineering (BPISAE). The SCS maintained its federal soil survey effort for the purposes of making recommendations for the control of soil erosion. While the general soil classification framework was similar between these two survey efforts, scale, legends, and interpretations differed considerably. The USDA worried about controversy and potential for duplication of effort and tried unsuccessfully to reconcile the two systems. Finally, in 1945, the department excluded "SCS farm planning surveys" from the requirements it had instituted in 1942 for BPISAE legend preparation, field reviews, and soil correlation. Each soil map produced for farm planning would become part of the farm plan but would not be integrated into the national portrait of soil resources (Simonson 1987).

In 1952, the two efforts were finally united. The BPISAE soil survey division was integrated into the SCS, thus bringing the SCS's financial resources and public support to the soil survey. With this reorganization, mapping was accelerated and interpretations benefited from the expertise of a large cadre of conservationists with training in range science, agronomy, forestry, biology, engineering, and other disciplines. This new group supplemented the contributions of the state experiment stations and the state extension services. The merger linked the soil survey to a major user group—the landowners with whom the SCS conservationists worked directly.

During this period, a new soil classification system was being developed. In an atmosphere where previous precepts were to be abandoned, or at least questioned, one decision was out of the hands of the development team. Guy D. Smith, leader of the effort to develop the new classification system, recalled that:

The classification of soils into series was entrenched by 1951, not only with those who made soil survey, but with those who used them. (Smith 1983)

The Highway Research Board asked Smith to allay any fears that the soil classification development effort would abandon the soil series (Smith 1983). The soil survey leaders understood that dropping the series names that had been known for decades would damage the ultimate objective of the surveys—their acceptance and use. The new system—*Soil Taxonomy*—retained the concept of soil series, made classification more quantitative, and provided a more accurate, clearly defined basis for making interpretations (Soil Survey Staff 1975).

Interpretations, especially those for nonagricultural uses, flourished after World War II. Rapid urbanization brought forth example after example of building on sites poorly suited for homes. Soil scientists made the point that they could help predict, and thus avoid, some problems by mapping soil characteristics in urbanizing areas. Among other things, the surveys could identify hillside slip, shrink-swell potential, high water tables, and flooding potential.

Soil scientists found ready allies among planners. Planners and soil scientists joined forces to make local government officials and the public aware that soil surveys could be used to predict problem areas. By the late 1950s, some states and counties had begun to integrate soil survey data and interpretations into their land-use planning functions. Fairfax County, Virginia, a rapidly urbanizing suburb of Washington, D.C., became perhaps the first county in the United States to hire a full-time soil scientist (Pettry and Coleman 1974).

Soil Surveys and Land Use Planning (Bartelli et al. 1966), a publication based on a special symposium at the 1965 annual meeting of the Soil Science Society of America, highlighted the rapid evolution of soil interpretations in the postwar period, especially of those related to land-use planning. The publication included authors from widely ranging disciplines—soil scientists, highway engineers, architects, and city and regional planners, among others. The symposium revealed the cost effectiveness of integrating soil survey interpretations into planning efforts. As the cost savings to communities became increasingly apparent, local communities began providing funds to expedite the survey (Bartelli et. al. 1966).

To clarify the legal basis for conducting soil surveys, Congress passed a bill defining the authorities of the soil survey activities in 1966 (P.L. 89-560, 80 Stat.706, 42 U.S.C. 3271–3274, 7 September 1966). The language stated that the survey could be conducted on nonfarm areas:

> ... to provide soils surveys to assist States, their political subdivisions, soil and water conservation districts, towns, cities, planning boards and commissions, community development districts, and other public agencies in community planning and resource development ...

The decades of Kellogg's leadership after the merger of the soil survey into the SCS were quite active for nonfarm interpretations but were also a time of great advances in interpretations to benefit farm and ranch management. The land capability classification system designed to help plan conservation for the farm identified soil classes beginning with those having no limitations to cultivation, moving to classes with some limitations and requiring conservation practices, and ending with classes that should not be used for cropping but might provide valuable pasture or woodland (see Figure 10.2).

The limitations the classification system addressed were not necessarily significant for tree growth, but neither did the absence of limitations correlate to superior forestland. It was this revelation that prompted the first soil site correlations in the nation (Helms 1997, 1999). Ted Plair and other SCS foresters completed pioneering work in developing forest site indexes. The work on rangelands and forestlands became firmly integrated into the soil survey in the postwar decades. Biologists also developed criteria and ratings for soils related to food and plant cover for wildlife. Clearly, by the mid-1900s the soil survey had moved well along the transition from an agricultural perspective to one with a much broader view of natural resource management with a major focus on engineering properties.

SOIL INTERPRETATIONS—MIXED RECEPTIONS

From the beginning, soil interpretations were highly valued (as evidenced by increasing requests for surveys and the steady expansion of the program). Yet vociferous opposition was encountered along the way. In most cases, this opposition resulted from real or perceived

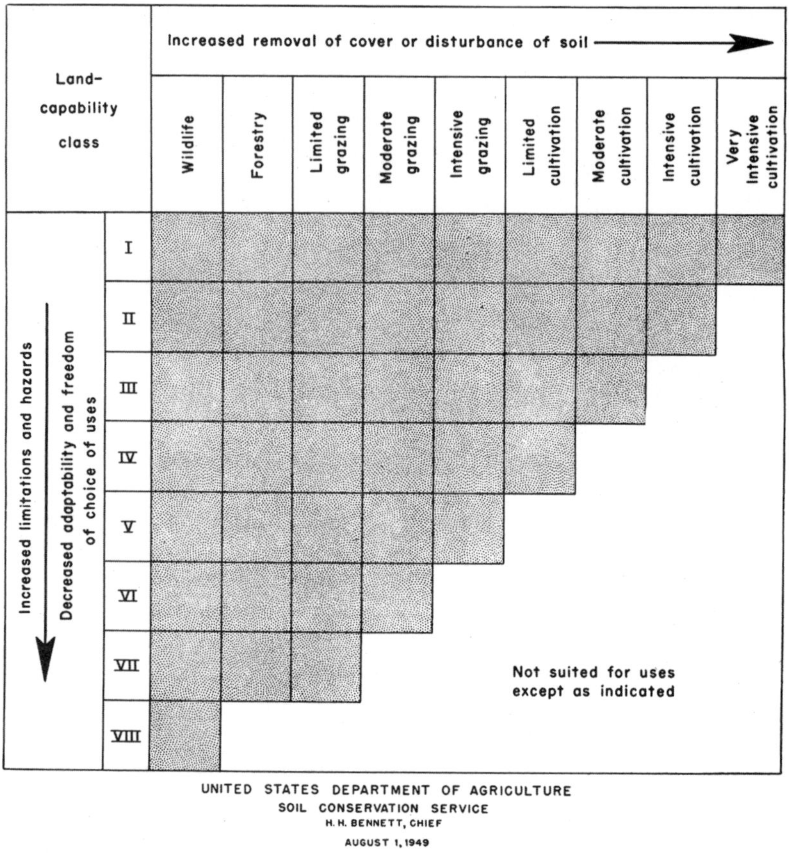

Increased removal of cover or disturbance of soil ⟶

Land-capability class

Wildlife | Forestry | Limited grazing | Moderate grazing | Intensive grazing | Limited cultivation | Moderate cultivation | Intensive cultivation | Very Intensive cultivation

Increased limitations and hazards
Decreased adaptability and freedom of choice of uses

I
II
III
IV
V
VI
VII
VIII

Not suited for uses except as indicated

UNITED STATES DEPARTMENT OF AGRICULTURE
SOIL CONSERVATION SERVICE
H. H. BENNETT, CHIEF
AUGUST 1, 1949

FIGURE 10.2 Land capability classification. (Courtesy of the Natural Resources Conservation Service)

impacts on land values that might result from soil survey publications.

In 1901 and 1903, interpretations from work in southern California's Imperial Valley accurately identified large areas of heavy textured impervious soils of high salt content that would pose substantial limitations for drainage and reclamation. This report coincided with a large-scale irrigation project that was promoted by an influential stock company. The reaction?

These reports were construed as condemning the larger part of the Valley as unfit for cultivation, and put the financial stability of the project

Land capability classification. (Courtesy of the National Archives at College Park, Maryland; Print 114-H-Cal.6571; Record Group 114, Records of the Natural Resources Conservation Service)

in question. This brought forth a storm of protest, and threat of political activity adverse to the Bureau and the future of the Soil Survey. (Day 1955)

Not much later, a survey for the Modesto-Turlock area of the San Joaquin Valley in California (1909) generated similar dissatisfaction. Although local citizenry had requested the survey, when the results indicated "alkali" presence, the reports were "kept safely under lock and key and withheld from the public for which they were intended" (Lapham 1945).

Similar outcries resulted from other surveys that identified limitations to agricultural use of land. A survey in Minnesota elicited the following protest from a land agent who found his land had been identified as having "low" agricultural value:

> There will be no more land classification or soil surveys in Minnesota if that map is published! (Gardner 1998)

A 1908 report for Richland County, North Dakota, described in detail some of the poorer soils in the county as unsuitable for cultivation. Local real-estate interests raised such a furor that the reports were destroyed. Ironically, years later local citizens were requesting federal assistance to address the very problems identified in the report.

An interpretation describing the poorly drained peat soils of the Florida Everglades warned that the land, untried for agriculture and much under water, was being sold for $20 to $65 per acre. Land agents were irate and the report was contained. Subsequently, however, some land agents were prosecuted and convicted of fraud in Florida. This example illustrates, however, that interpretations are only relevant for the technology and other conditions present at the time of the interpretation. Years later the same peat soils discredited by the Florida survey were producing high yields as a result of new technology.

INTERPRETATIONS TODAY

From the very beginning, soil interpretations have been based on how soils respond for a specific use. Over its 100-year history, soil surveys made continuing advances that overcame the factors limiting the interpretive value of early surveys. Adoption of larger scales increased the value of surveys to farmers and other users in need of highly localized interpretations. Advances in soil science supported the refinement of soil classification and for understanding the relationship of soil chemistry or ecology to plant nutrition (e.g., the importance of microbes, ion exchange and hydrolysis, nutrients absorbed on clays, and pH, and their relationship to plant health) (Jenny 1961).

Today, soil survey interpretations are becoming increasingly dynamic, with the application of information technology to support decision making. The demand for digitized soil surveys is increasing exponentially as the use of Geographic Information Systems (GIS) has increased. Advances in information technology, using digital orthophotography maps and digital data for computer manipulation and retrieval, enhance the efficiency of delivery of soil survey information. Detailed resource information on specific land areas to help landowners, communities, and others in their land-use decision making can now be done quickly and interactively. The orthophotographic maps provide the geographic reference points needed to make soils, farm field boundary, rivers, roads, and other information contained in computerized databases useful to land users, policy makers, and others.

Digital data are provided at three levels of generalization: 1) The Soil Survey Geographic Data Base (SSURGO) provides the greatest detail and is used primarily by landowners, local government units, and watershed hydrologic units for planning and resource management; 2) The State Soil Geographic Data Base (STATSGO) is used primarily for multicounty, state, and river basin planning and resource management and monitoring; and 3) The National Soil Geographic Data Base (NATSGO) is used primarily for multistate, regional, and national planning and resource appraisal and monitoring. The use of these digitized forms of soil surveys in GIS also increases the demand for new kinds of interpretations.

For those familiar with the tradition of innovation in the soil survey, it comes as no surprise that the NCSS is building the capacity to shift from static soil survey reports to a dynamic resource of soils information for a wide range of needs. The National Soil Information System (NASIS) is designed to manage and maintain soils data from collection to dissemination. The NASIS will support the collection of new information in compliance with standards, application of expert knowledge to make information usable for a variety of purposes, and dissemination of information to a wide variety of users.

Irrespective of the method used to develop and distribute soil survey interpretations, their purpose has remained constant—to provide the public with local information on the uses and capabilities of their soil resource based on scientific analyses and classification of the soils. While standard interpretive categories for soils data exist (see Figure 10.3), the range of potential applications seems limitless. The following examines a few important or unique ways in which soil survey interpretations have been and are used by agencies, institutions, organizations, and individuals.

FEDERAL AND STATE PROGRAMS

Historically, soil survey interpretations focused on predictions for a specific land use, such as agriculture. More and more, however, soils information answers a much wider range of questions and is put to a much broader range of uses. Today, interpretations support all aspects of the NRCS Conservation Operations activities, including reducing excessive soil erosion, reducing agricultural nonpoint-source contamination of water bodies, improving irrigation efficiency, increasing water

Soil survey data may be interpreted for a variety of purposes. The Soil Survey Handbook (1993) identifies 12 standard interpretive applications:

1. **National Inventory Groupings**–Various taxonomic and nontaxonomic map unit criteria, along with interpretive soil properties have been employed to construct definitions for national inventory purposes. Groupings may pertain Prime and unique farmland, hydric soils, and highly erodible land., such as criteria for the application of national legislation on environment and commodity programs (e.g., highly erodible lands, wetlands)

2. **Land Use Planning**–Interpretations for land use planning may be done through groupings or ratings of soils according to their limitations, suitabilities, and potentials for specified uses. Soil surveys can provide a source of information for the evaluation of the environmental and economic effects of proposed land uses.

3. **Farmland**–Interpretations for farming involve placement of soils in management groups and identification of the important soil properties that relate to crop production, conservation needs, among other aspects of agriculture (e.g., yield potential, erodibility, depth to hardpan, etc.). Soils information may be interpreted for: productivity ratings; soil fertility capability classification; resiliency ratings, among many others.

4. **Rangeland**–Rangeland interpretations are given as ecological sites, which are a distinctive kind of rangeland that differs by its ability to produce a characteristic natural plant and animal community. Soil-ecological site correlation identifies the suitability of the soil to produce different kinds, proportions, and amounts of plants.

5. **Forestland**–Soil-ecological site correlation within a soil survey gives the suitability of the soil to produce wood products. Interpretations may consider erosion hazard, equipment limitations, seedling mortality, windthrow hazard, plant competition, and trees to plant.

6. **Windbreaks**–Interpretations provide a list of species for each kind of soil where windbreaks will serve a useful purpose. Correlation of soil properties and adaptable tree and shrub species is essential to ensure successful establishment of windbreaks.

7. **Recreation**–Interpretations are made for a variety of recreational development opportunities (e.g., golf fairways, picnic sites, playgrounds, paths and trails, campsites, ski slopes and snowmobile trails). Ratings are generally made on the basis of restrictive soil interpretive properties such as slope, occurrence of internal free water, and texture of surface horizons.

8. **Wildlife Habitat**–Interpretations for wildlife show where management for wildlife can be applied most effectively and which practices are most appropriate. Descriptions of the soil for wildlife habitat is provided as: 1) habitat elements (the suitability class for different vegetation groups), and 2) ratings for several kinds of wildlife based on the type of system (e.g., woodland, rangeland, openland, etc.).

9. **Construction Materials**–Interpretations rate soils as probable sources of desired materials (e.g., gravel and sand, organic soil material, high strength and low shrink-swell, etc.) but site quality generally cannot be specified.

10. **Building Sites**–Interpretations provide the screening necessary for preliminary site selection, comparing alternative sites, and land use planning for construction of small buildings; installation of roads, streets, and utilities; and establishment of lawns and landscaping. These types of development also generally require onsite evaluation. Since construction often alters soil properties (i.e., upper horizons may be removed by land leveling), some interpretive soil properties must be applied cautiously.

11. **Waste Disposal**–Interpretations consider whether the waste will be placed in a relatively small area or distributed at low rates over a large area and identify the suitabilities and limitations based on the type of system.

12. **Water Management**–Interpretations for water management focus on the construction of small to medium impoundments, control of waterways, installation of drainage and irrigation systems, and control of surface runoff. Onsite evaluations are required to design engineered projects, however, the soil survey can be helpful in the evaluation of alternative sites.

FIGURE 10.3 Standard soils data interpretive categories.

use efficiency, reducing upstream flood damages, improving range condition, and restoring, maintaining, and improving wetlands. Soil surveys are the basis for evaluating conservation alternatives by predicting how soil will respond to use, management, and treatment.

Technical soil groupings have been developed as criteria for the application of environment and commodity programs at the national level. For example, interpretations identifying highly erodible land and wetlands were necessary to support the compliance provisions of the Food Security Act of 1985. Nearly three decades earlier, the Federal Housing Administration used soil maps in some localities to help determine eligibility of residential subdivisions to receive guaranteed loans (Hunter 1966; Mausbach 1994, 2001).

State and local governments may also use interpretations to support programs. Farmland Protection Programs, which serve to keep prime and unique farmland in an agricultural land use, may use national inventory soil groupings to identify eligible farmland. Since the 1920s, local governments have used interpretations to support local tax assessment (Kellogg 1935; Gardner 1998).

STATE AND LOCAL PLANNING

One of the earliest uses of the soil survey information for planning was in Fairfax, Virginia, in the 1950s. So useful were the survey and interpretations that the county retained a soil scientist to provide consultation to planners. Within four years, the county had a soil scientist on permanent staff to assess septic tank disposal systems, rezoning cases, floodplain extent and areas, best trees and shrubs for planting, type of and depth to bedrock, slope stabilization and soil slippage problems, and new school sites (Obenshain 1966; Pettry and Coleman 1974). Interpretations formed the foundation for land-use planning ordinances across the nation (e.g., floodplain setbacks; slope protection; agricultural zoning) (Bartelli 1966).

Local planners, engineers, zoning commissions, tax commissions, homeowners, and developers, among others, use soil surveys to guide decisions on proposed land uses, including:

1) *Planning urban or suburban growth*—interpretations can identify potentials and limitations of areas for buildings, highway construction, and waste disposal, among other features important to land-use planning (Doyle 1966; Klingebeil 1966).

San Antonio city officials study soil survey information for planning utility extensions. (Courtesy of the National Archives at College Park, Maryland; Print 114-H-TEX-51,085; Record Group 114, Records of the Natural Resources Conservation Service)

2) *Identifying water recharge areas*—interpretations can identify soil associations where groundwater recharge occurs (Doyle 1966).

3) *Supporting local zoning ordinances*—interpretations can identify inventory groupings of interest for development of local zoning laws (Doyle 1966). For example, inventory groupings may be constructed to identify prime land that a community wishes to protect, or areas of high quality that may support a variety of uses. This information can assist zoning boards to plan toward the most desirable land use. Many examples exist of development failures due to a lack of attention to the capability of the soil resource to support the activity installed (Klingebiel 1966).

4) *Identifying environmentally sensitive areas*—some soils are more prone to degradation than others, and soil maps can identify areas where any development is likely to create adverse environmental impacts. These areas can then be conserved in native vegetation, as

wildlife habitat or as buffers, among other possible uses of undeveloped land (Doyle 1966).

NATURAL RESOURCE MANAGEMENT

AGRICULTURAL USES

Early interpretations focused on the capabilities of land for agricultural production. Indeed, the productivity ratings produced by the Bureau of Soils and the land capability classification effort of the SCS were devoted to agricultural interpretations of soils data. Over time, interpretations have become much broader, covering a wide array of environmental effects.

Agricultural interpretations traditionally have involved placement of soils in management groups and identification of the important soil properties that relate to crop production and conservation needs, among other aspects of agriculture (e.g., yield potential, erodibility, depth to hardpan). Interpretations could be made to determine areas for specialty crops, matching crops to capable soils, size fields for soil homogeneity, and identify needs for management practices to mitigate soil conditions that might constrain production. One of the earliest agronomic interpretations was identifying "tobacco soils" in the southeast to guide expansion of tobacco production to other suitable areas.

With the emergence of precision farming, the NCSS is challenged to meet a new range of data needs. Precision agriculture systems employ Global Positioning System (GPS) and GIS technology to map and compare variability patterns in fields for the purpose of evaluating best management opportunities. Through the use of GPS, operators "georeference" highly specific data on soils and crop yield from sampling the field. These data are used to create layers in GIS to inform variable rate application plans that are tailored to specific field conditions (Blackmore, 1994). The resolution of data used in precision agriculture generally exceeds that provided by soil surveys, limiting the applicability of general soil survey data to this technology. However, the National Academy of Sciences (NAS) identifies a significant role for the NCSS. The NCSS provides information infrastructure through the development of data quality standards for detailed work, methods for data collection, testing, and interpretation and for procedures for private soil consultants to access and archive data. In addition, the NCSS shift toward characterizing soil-landscape relations in soil surveys, along with the development of the NASIS, is expected to benefit some kinds of precision agriculture (NAS 1997).

FOREST AND RANGELAND USES

Unlike agricultural land, where fertilizers and other inputs are used to improve productivity, foresters must work with natural soil potentials (Howell 1998). Thus, soil-vegetation relationships identified through the soil survey interpretations are significant tools for managing natural forest vegetation. The National Park Service also uses soils information to support vegetation management in the park system (Popenoe 1998). Similarly, these soil-ecological site interpretations may be used to identify forest ecosystems that have a high potential for containing specialty nontimber forest products such as truffles or ginseng.

Interpretations for forests and rangeland have continued to evolve over time. Because of the agricultural focus of early interpretations, soil suitability, limitations, or potential for nonagricultural uses often was overlooked. For example, sites not suited to intense cultivation under the land capability classification system were interpreted as class IV or greater, with such suggested uses as forests, range, and recreation. But a limitation for cropland did not necessarily mean a limitation for forests, and some of these sites were high-quality forestland. Once this shortcoming was identified, the SCS began efforts in 1945 to correlate tree growth to soils. The interpretive material that developed as a result of this work became a standard part of the soil survey, and by 1970, forest site-soil vegetation correlations were integrated into the soil survey (Helms 1999).

Today, rangeland and forestland correlations are being enhanced to provide a broader array of information and interpretations. Under the planned revisions, interpretations will be soil-ecological site correlations that distinguish sites by their ability to produce a characteristic natural plant and animal community. This ecological site data for forestlands and rangelands will be available through the Ecological Site Information System (ESIS) and linked to the NASIS to allow users to obtain data describing the ecological site, plant and animal composition, history, and steady state details of the community over time for that system (Thompson 1999).

WILDLIFE CONSERVATION USES

Another expanding application of soil survey interpretations is the conservation of threatened and endangered wild populations. Many plants listed as endangered or threatened under the Endangered Species Act of 1973 are rare or narrowly endemic. These endemic species are often restricted to unusual soils that are produced from substrates like gypsum, marine clays, volcanic cinders, or to unusual soils that result from the interactions of substrates and dynamic geomorphic processes to produce deep sands, peat bogs, or talus slopes.

Soils information is an important tool in the recovery of endemic endangered plants. It is useful for locating new populations as well as additional habitat suitable for reintroductions to promote species recovery. When accurate soils information is lacking, the development of such information is usually a relatively high-priority task in species recovery plans.

Soil maps help botanists identify new localities to search for endemic species. There are many instances of such searches leading to the discovery of additional populations and thus leading to greater flexibility in management of the species. Since plant distribution is closely tied to particular soil types or associations, soil survey maps can help identify potential habitat for rare plants. For any particular species, all the known locations are plotted on a soils map, the soil association /type is determined, and any correlation between plant species and soil association is identified. This method has proved successful for locating new populations of green pitcher plant, Mohr's Barbara's button, and geocarpon on Lafe soils.

Chemical soil analyses also can provide clues to why a plant is restricted to one soil type and does not occur on nearby soils that appear similar. Some endangered plants have been found to be associated with particular elements in the soil such as high levels of selenium or copper.

HISTORIC SITE IDENTIFICATION

Interpretations have also proven useful for identifying sites of potential historic or archeological interest. Many early settlements developed because of access to rivers or their tributaries, productive soils, or other landscape features. Soil maps and interpretations provide a resource to identify many of these features, as well as physiographic changes over time. Comparison of older surveys with current aerial photos can be used to identify shifts that may help target potential sites. Some uses have included verifying property boundaries, which frequently were associated with stream channels. Many early survey maps also included farmstead boundaries, rural cemeteries, and other settlement features (e.g., churches, towns, city buildings). Because of these potential historical values, soil survey resources have found their way into genealogical research (Family Puzzlers 1994).

Soil-vegetation interpretations also have proven useful for identifying potential historic sites. Species that were considered to have some value in predicting high-potential sites included exotic or unusual plants that would have been planted around buildings or vegetation indicative of disturbance (which tend to indicate over-tilled or otherwise poor soils). For example, at Ft. Bragg, North Carolina, historic sites appear to be correlated with black walnut trees (Kaczor 1998).

Knowledge of soils and the soil classification system, particularly identification of areas with base-rich soils favorable for production of grasses and staple grains like wheat, has been combined with knowledge of history to improve understanding of agricultural settlement patterns and the development of agricultural systems. While soils may not have determined historical developments, they clearly set limitations on and defined opportunities for development throughout history. Soil surveys and *Soil Taxonomy* have even been used to explain the location of major colonial transportation routes such as the Great Wagon Road (Helms 2000).

REAL ESTATE AND INSURANCE

Knowing as much as possible about the suitability, limitations, and potentials of land is essential for wise real-estate transactions, as well as for making good judgments on insurance needs. Soil interpretations historically have had significant influence on the sale or transfer of property. Indeed, the most vociferous opposition of soil surveys arose in response to interpretations that identified limitations on land potentials, primarily of an agricultural nature. Soils maps and interpretations also have been used as supporting evidence in court cases. In one case, a community effectively argued lower land valuation as a result of the unsuitability of the soils to support building foundations in their natural condition (Morris 1966).

From a liability perspective, soil data and interpretations can be used to evaluate insurance needs. Insurance underwriters commonly use information and interpretations from soil surveys. Soil interpretations can identify high-flood potentials, shrink-swell soils unsuitable for construction, and other important characteristics upon which development and insurance decisions depend. For example, the suitability of an area for a fire pond, which is required for insurance purposes in many rural areas not served by public water supplies, can be derived from soil data.

Soil interpretations can also be used to screen home sites for potential radon gas exposure. The need for radon testing for homes can be identified based on soil composition of the building site. This application was used to implement radon testing on the St. Regis Mohawk Reservation to identify new homes that should be tested for radon levels. Twenty-five buildings on soil with a high correlation to radon emissions were targeted for testing, and of these, four were identified as exceeding the standard (Iroquois Environmental Newsletter, Spring 1997).

ENGINEERING AND CONSTRUCTION

Soil maps and interpretations are widely applied in siting development such as highways, buildings, and other construction. These data allow planners to evaluate decisions based on their associated economic and environmental costs. One example of the use of these data was presented in 1965 at the annual meeting of the Soil Science Society of America. County officials in Illinois used soil data to choose between two proposed sites for highway development. Soil interpretations provided decision makers with the information needed to select the site with the fewest limitations. In this instance, the county saved an estimated $2,250 per acre in reduced excavation and construction costs (Klingebiel 1966). Selecting sites with soils of sufficient load-bearing quality, not subject to flooding, and suitable for on-site sewage disposal results in substantial costs savings and reduces the potential for unforeseen failures.

Soil survey interpretations also are used to estimate costs of laying buried pipelines, including identifying needs for types of pipes (e.g., shrink/swell soils making use of rigid pipes unsuitable), mitigation required (e.g., extremely wet, low resistancy, or differentially draining soils leading to more rapid corrosion and the need for protective coatings or cathodic protection), and costs of excavation (e.g., short depth to bedrock or float rocks potentially increasing excavation costs).

MILITARY

Military uses of the soil survey have been extensive and varied, ranging from responding to strategic needs and site assessments to interpretations for more pragmatic engineering purposes. Some of the larger scale survey efforts, in fact, were intended to fulfill military needs. For example, the world soils map project, conceived and supported by the Department of Defense, was intended to provide detailed soil information on strategic locations worldwide.

Soil survey data have been interpreted to evaluate the capability and limitations of terrain for military training. Interpretations have been made for uses from artillery training to evasive maneuver training. For example, soils data interpreted for a military training facility identified limitations for artillery practice due to high ricochet potential (Laux 1997). Soil interpretations also support military engineering efforts. Great differences in soil properties can occur over short distances and affect how well they can be used in military maneuvers. Understanding the load-bearing capacity of an area is essential to planning ground movements of heavy equipment and troops. The U.S. Army Corps of Engineers once used soil data in 1948 in a military engineering effort that investigated the potential for swamp stabilization. The idea for this effort arose from an encounter between Russian and German troops during World War II, in which powder from Russian howitzer fire stabilized a swamp separating the two armies. As a result, the Russian army was able to engage the German troops at their most vulnerable point because the Germans had felt defended by a natural and impenetrable boundary—the swamp (Laux 1997).

HAZARDOUS WASTE, BROWNFIELDS, AND REMEDIATION

Soils data increasingly have a role in identifying subsurface features of interest through the use of high technology ground-penetrating radar systems. Ground-penetrating radar gives rapid and accurate readings of any subsurface density changes without disturbing the soil, which, when combined with the information provided by soil survey data, allows assessors to locate buried foreign substances. Ground-penetrating radar is being used successfully to locate underground storage tanks, which are classified as hazardous waste and create liabilities under the

Comprehensive Environmental Response, Compensation, and Liability Act of 1980 (CERCLA) (Laux 1997). Such information is particularly important in property transfers since the CERCLA liability transfers with the property.

Evaluation for toxic substances, such as heavy metals, may also become integrated into soil surveys. Soil properties chosen for urban surveys depend in part on the land use and intensity of use that is planned by customers in urban areas. The prototype survey for LaTourette Park on Staten Island, New York, for example, evaluates heavy metals in soils in addition to the more traditional interpretations for playgrounds and picnic areas. These kinds of interpretations for very specific nonfarm customers are new to the soil survey, but they indicate the survey is beginning to integrate urban interpretations into its work (Soil Survey of LaTourette Park 1998).

Military base closures during the last decades bring another new use of the soil survey—supporting redevelopment efforts. Seven years before the 1995 closing of Plattsburgh (New York) Air Force Base (PAFB), the base's soil survey had been completed to help the military plan for future construction or development. The completed soil survey has helped guide redevelopment efforts by providing the information needed for environmental impact statements and planning remediation for reclamation of contaminated sites (e.g., dumps, fuel spills) (Trevail 1998).

RESEARCH AND INVESTIGATIONS

Not surprisingly, soil survey data and interpretations constitute a substantial resource for a variety of research efforts. Crop research, land evaluation and appraisal, statistical studies, and even sociological investigations build on soil survey data. Soil surveys help historians and historical geographers interpret historical patterns of settlement, agriculture, transportation, and other human activities (Helms 2000).

In the agronomic arena, soil survey interpretations have been fundamental to developing improved management practices to optimize productivity. Interpretations also have been used to identify sites to collect soils having certain characteristics required for experiments (e.g., soils of varying acidity for testing materials durability). Similar applications exist for identifying crop fields with desired characteristics for specific research activities. This use of interpretations was first seen in 1911 when the state of Alabama requested a detailed survey and interpretation of its agricultural experiment stations in order to evaluate site suitability and limitations for agricultural research (Gardner 1998).

During World War II, soil survey interpretations were called upon to help identify areas of high potential for the production of rubber, oils, fiber, and other products in short supply nationally. Interpretations identified areas suitable for the production of guayule as a substitute for rubber; of oil-producing alternatives such as castor bean; and of fiber crops such as American Hemp. Wild American Hemp can still be found in some agricultural areas as a result of wartime production, and guayule continues to be produced on a small scale in the Southwest.

The array of research-related uses of soils data continues to expand. Today, soil survey data are being used to evaluate potential of lands for carbon sequestration, as well as other soil conditions related to global climate change. Research on the role that soils play in mediating the effects of agriculture and forestry on the global atmospheric composition of greenhouse gases has been a major component of the USDA's Global Change Program since 1990. Research undertaken by the NRCS helps us understand terrestrial soil carbon and its interactions with the biogeochemical fluxes in the atmosphere (Kimble 1995).

LOOKING AHEAD

The more than 100-year history of the soil survey is replete with examples of the codevelopment of soil science and soil survey interpretation to meet society's needs. Interpretations for the future will continue to be molded by the nature and range of the economic, environmental, social, and political conditions that influence our lives. David R. Gardner captured the dynamic role of soils data and the soil survey when he wrote in his study of the NCSS in 1955:

> As a nation, we are still learning how to make, and use, soil surveys. There is a great deal yet to be learned and there always will be. (Gardner 1998)

REFERENCES

Bartelli, L. J., A. A. Klingebiel, J. V. Baird, and M. R. Heddleson, eds. 1966. *Soil Surveys and Land Use Planning.* Madison, Wis.: Soil Science Society of America and American Society of Agronomy.

Blackmore, S. 1994. Precision farming: An introduction. *Outlook on Agriculture* 23(4):275–280.

Borst, Harold, A. G. McCall, and F. G. Bell. 1945. *Investigations in Erosion Control and the Reclamation of Eroded Land and the Northwest Appalachian Conservation Experiment Station, Zanesville, Ohio.* USDA Technical Bulletin Number 888. U.S. Department of Agriculture, Washington, D.C.

Brady, Nyle C. 1974. *The Nature and Properties of Soils.* 8th ed. New York: Macmillan

Browning, G. M., R. A. Norton, and A. G. McCall. 1948. *Investigation in Erosion Control and the Reclamation of Eroded Land at the Missouri Valley Loess Conservation Experiment Station, Clarinda, Iowa, 1931.* USDA Technical Bulletin Number 959. U.S. Department of Agriculture, Washington, D.C.

Buckman, H. O., and N. C Brady. 1969. *The Nature and Properties of Soils.* 7th ed. New York: Macmillan Company.

Copley, T. L., Luke A. Forrest, A. G. McCall, and F. G. Bell. 1944. *Investigation in Erosion Control and Reclamation of Eroded Land at the Central Piedmont Conservation Experiment Control Station, Statesville, North Carolina, 1930–1940.* USDA Technical Bulletin Number 873. U.S. Department of Agriculture, Washington, D.C.

Daniel, Harley, Harry Elwell, and Maurice Cox. 1943. *Investigations in Erosion Control and Reclamation of Eroded Land at the Red Plains Conservation Experiment Station Guthrie, Oklahoma, 1930–40.* USDA Technical Bulletin Number 837. U.S. Department of Agriculture, Washington, D.C.

Day, T. 1955. Soils surveys in the Soil Conservation Service. Unpublished term paper, Harvard University, Cambridge, Mass.

Doyle, R. H. 1966. Soil surveys and the regional land use plan, In *Soil Surveys and Land Use Planning,* edited by L. J. Bartelli, A. A. Klingebiel, J. V. Baird, and M. R. Heddleson, pp. 8–14. Madison, Wis.: Soil Science Society of America and American Society of Agronomy.

Family Puzzlers. 1994. *Heritage Papers,* Athens, Georgia. Number 1347, p.3.

Gardner, David Rice. 1998. *The National Cooperative Soil Survey of the United States.* Historical Notes Number 7. Natural Resources Conservation Service, U.S. Department of Agriculture. (Reprint, unpublished dissertation, 1955, Harvard University, Cambridge, Mass.)

Hackensmith, R.D., and J. G. Steele. 1949. Recent trends in the use of the land capability classification. *Proceedings of the Soil Science Society of America* 14:383–388.

Hays, O. E., A. G. McCall, and F. G. Bell. 1949. *Investigation in Erosion Control and Reclamation of Eroded Land at the Upper Mississippi Valley Conservation Experiment Station Near La Crosse, Wisconsin.* USDA Technical Bulletin Number 973. U.S. Department of Agriculture, Washington, D.C.

Helms, Douglas. 1997. Land capability classification: The U.S. experience. In *History of Soil Science: International Perspectives. Advances in GeoEcology* 29, edited by Dan H. Yaalon and S. Berkowicz, pp. 159–176. Reiskirchen, Germany: Catena Verlag.

Helms, Douglas. 1999. Contributions of the Soil Conservation Service to forest science. In *Forest and Wildlife Science in America: A History*, edited by Harold K. Steen, pp. 66–85. Durham, N.C.: Forest History Society.

Helms, Douglas. 2000. Soil and southern history. *Agricultural History* 74(4, Fall):723–758.

Hilgard, E. W. to Secretary of Agriculture James Wilson. 16 January 1900. Letters Received; Division of Soils; Record Group 54, Records of the Bureau of Plant Industry, Soils, and Agricultural Engineering. National Archives at College Park, Maryland.

Hill, H. O., W. J. Peevy, A. G. McCall, and F. G. Bell. 1944. *Investigations in Erosion Control and the Reclamation of Eroded Land at the Palouse Conservation Experiment Station, Pullman, Washington, 1931–42.* USDA Technical Bulletin Number 859. U.S. Department of Agriculture, Washington, D.C.

Horner, Glenn, A. G. McCall, and F. G. Bell. 1944. *Investigations in Erosion Control and Reclamation of Eroded Land at the Central Piedmont Conservation Experiment Station Statesville, North Carolina, 1930–40.* USDA Technical Bulletin Number 860. U.S. Department of Agriculture, Washington, D.C.

Howell, David. 1997. Area Resource Soil Scientist, California, Natural Resources Conservation Service, U.S. Department of Agriculture. Personal communication with P. Durana.

Hunter, W. R., C. W. Tipps, and J. R. Coover. 1966. Use of soil maps by city officials for operational planning. In *Soil Surveys and Land Use Planning*, edited by L. J. Bartelli, A. A. Klingebiel, J. V. Baird, and M. R. Heddleson, pp. 31–36. Madison, Wis.: Soil Science Society of America and American Society of Agronomy.

Iroquois Environmental Newsletter. 1997. St. Regis Mohawk Tribe, Environmental Division, Hogansburg, N.Y. Spring.

James, Preston E., and Clarence F. Jones, eds. 1954. *American Geography: Inventory & Prospect.* Syracuse, N.Y.: Syracuse University Press for the Association of American Geographers.

Jenny, Hans. 1961. *E. W. Hilgard and the Birth of Modern Soil Science.* Pisa, Italy: Collana Della Rivista "Agrochimica."

Kaczor, Michael. 1997. National Cultural Resources Specialist, Natural Resources Conservation Service, U.S. Department of Agriculture, Washington, D.C. Personal communication with P. Durana.

Kellogg, Charles E. 1928. Soil type as a factor in highway construction in Michigan. *Michigan Academy of Science, Arts, and Letters* 10:169–177.

Kellogg, Charles E. 1935. Report of the Land-Use Committee. *American Association of Soil Survey Workers Bulletin.* 16:147.

Kellogg, Charles E. 1943. *The Soils That Support Us.* New York: Macmillan Company.

Kellogg, Charles E. 1966. Soil surveys for community planning. In *Soil Surveys and Land Use Planning*, edited by L. J. Bartelli, A. A. Klingebiel, J. V. Baird, and M. R. Heddleson, pp. 1–7. Madison, Wis.: Soil Science Society of America and American Society of Agronomy.

Kellogg, Charles E., and J. Kenneth Ableiter. 1935. *A Method of Rural Land Classification.* USDA Technical Bulletin Number 469. U.S. Department of Agriculture, Washington, D.C.

Kimble, J. M., ed. 1995. *Global Change Activities of the Natural Resources Conservation Service, Soil Survey Division—Progress Report.* National Soil Survey Center, U.S. Department of Agriculture, Natural Resources Conservation Service, Lincoln, Neb.

Klingebiel, A. A. 1966. Costs and returns of soil surveys. *Soil Conservation* 32:3–6.

Lapham, M. J. 1945. The soil survey from the horse and buggy days to the modern age of the flying machine. *Soil Science Society of America Proceedings* 10:344–350.

Laux, Burton R. 1997. BRL Environmental Assessment Associates, Garrison, N.Y. Personal communication with P. Durana.

Lyles, Leon. 1985. Predicting and controlling wind erosion. *Agricultural History* 59(2):205–214.

Mausbach, M. J. 1994. Classification of wetland soils for wetland identification. *Soil Survey Horizons.* 35(1):17–25.

Mausbach, Maurice J., and W. Blake Parker. 2001. Background and history of the concept of hydric soils. In *Wetland Soils: Genesis, Hydrology, Landscapes, and Classification,* edited by J. L. Richardson and M. J. Vepraskas, pp. 19–33. Boca Raton, Fla.: Lewis Publishers.

Meyer, L. Donald, and William C. Moldenhauer. 1985. Soil erosion by water: The research experience. *Agricultural History* 59(2):192–204.

Middleton, H. E. 1930. *Properties of Soils which Influence Soil Erosion.* USDA Technical Bulletin Number 178. U.S. Department of Agriculture, Washington, D.C.

Middleton, H. E., C. S. Slater, and H. G. Byers. 1932. *Physical and Chemical Characteristics of the Soils from the Erosion Experiment Stations.* USDA Technical Bulletin Number 316. U.S. Department of Agriculture, Washington, D.C.

Morris, John G. 1966. The use of soils information in urban planning and implementation. In *Soil Surveys and Land Use Planning,* edited by L. J. Bartelli, A. A. Klingebiel, J. V. Baird, and M. R. Heddleson, pp. 37–41. Madison, Wis.: Soil Science Society of America and American Society of Agronomy.

Musgrave, G. W., and R. A. Norton. 1937. *Soil and Water Conservation Investigations at the Soil Conservation Experiment Station Missouri Valley Loess Region, Clarinda, Iowa.* USDA Technical Bulletin Number 558. U.S. Department of Agriculture, Washington, D.C.

National Academy of Sciences, Board on Agriculture. 1997. *Precision Agriculture in the 21st Century: Geospatial Information Technologies in Crop Management.* Washington, D.C.: National Academy Press.

Obenshain, S. S. 1966. Changes in the need and use of soils information. In *Soil Surveys and Land Use Planning,* edited by L. J. Bartelli, A. A. Klingebiel, J. V. Baird, and M. R. Heddleson, pp. 175–179. Madison, Wis.: Soil Science Society of America and American Society of Agronomy.

Pettry, D. E., and C. S. Coleman. 1974. Two decades of urban soil interpretations in Fairfax County, Virginia. In *Non-Agricultural Applications of Soil Surveys*, edited by R. W. Simonson. New York: Elsevier.

Pope, J. B., James Archer, P. R. Johnson, and A. G. McCall. 1946. *Investigation in Erosion Control and Reclamation of Eroded Sandy Clay Lands of Texas, Arkansas and Louisiana at the Conservation Experiment Station*. USDA Technical Bulletin Number 916. U.S. Department of Agriculture, Washington, D.C.

Popenoe, James. 1997. Soil Scientist, National Park Service, U.S. Department of the Interior. Personal communication with P. Durana.

Schoenmann, L. R. 1923. Description of field methods followed by the Michigan land and economic survey. *American Association of Soil Survey Workers Bulletin* 4:44–52.

Simonson, R. W., ed. 1974. *Non-Agricultural Applications of Soil Surveys*. New York: Elsevier.

Simonson, R. W. 1987. Historical aspects of soil survey and soil classification. Special reprint issue of *Soil Survey Horizons*. Madison, Wis.: Soil Science Society of America.

Smith, D. D., D. M. Whitt, Austin Zingg, and A. G. McCall. 1945. *Investigations in Erosion Control and Reclamation of Eroded Shelby and Related Soils at the Conservation Experiment Station Bethany, MO, 1930–42*. USDA Technical Bulletin Number 883. U.S. Department of Agriculture, Washington, D.C.

Smith, Guy D. 1983. Historical development of soil taxonomy—Background. In *Pedogenesis And Soil Taxonomy*, edited by L. P. Wilding, N. E. Smeck, and G. F. Hall. New York: Elsevier, pp. 23–49.

Soil Survey Staff. 1975. *Soil Taxonomy: A Basic System of Soil Classification for Making and Interpreting Soil Surveys*. U.S. Department of Agriculture Handbook Number 436. U.S. Department of Agriculture, Washington, D.C.

Statutes at Large. 1899. Vol. 30, pp. 334–335. Agriculture Appropriation Act of 1899.

Thompson, Dennis. 1998. National Grazing Land Ecologist, Natural Resources Conservation Service, U.S. Department of Agriculture, Washington, D.C. Personal communication with P. Durana.

Trevail, Theodore. 1997. Soil Resource Specialist, Natural Resources Conservation Service, U.S. Department of Agriculture, Plattsburg, N.Y. Personal communication with Douglas Helms.

U.S. Department of Agriculture, Cornell University Agricultural Experiment Station, and New York City Soil & Water Conservation District. 1998. Soil Survey of South LaTourette Park, Staten Island, New York City, NY. U.S. Department of Agriculture, Washington, D.C.

U.S. Department of Agriculture, Soil Survey Division. 1993. *Soil Survey Manual*. USDA Handbook Number 18 (revised). U.S. Department of Agriculture, Washington, D.C.

Yolo County Planning and Public Works Department. 1997. *Willow Slough Integrated Management Plan*. Woodland, Calif.

11

THE AMERICAN SOIL SURVEY IN THE TWENTY-FIRST CENTURY

Horace Smith and Berman D. Hudson

From its very inception, the U.S. Soil Survey has been oriented towards specific applications. A major objective of the earliest surveys was to identify lands that were suitable for the introduction of specific crops, such as tobacco (Gardner 1998). Later, after the droughts of the 1930s caused great ecological damage and severe human suffering, soil surveys were increasingly used to plan conservation and erosion control systems (Dumanski 1993). More recently, soil surveys have been used to provide information to use in protecting wetlands, reducing pollution of waterways and groundwater, controlling stormwater runoff, and many other environmental objectives.

Changing uses of soil surveys reflect—and often require—changes in our knowledge, concepts, and models of soil. Dumanski (1993) recently presented five models or conceptions of soil, which he credited Meurisse for originally suggesting to him. Each of these models represents a different framework or viewpoint to use in organizing knowledge about a soil system. The five models or viewpoints include soil as 1) a natural body, 2) a medium for plant growth, 3) a structural material, 4) a water-transmitting mantle, and 5) an ecosystem component.

All of these models have had and will continue to have a place in the U.S. Soil Survey. Three of them—soil as a natural body, soil as a medium for plant growth, and soil as a structural mantle—have been used for many years. The other two—soil as a water-transmitting mantle and soil as an ecosystem component—were recognized more recently and still are evolving rapidly as concepts. A strategy to guide the U.S. Soil Survey in the twenty-first century must reflect each of these models. Accordingly, the ensuing discussion will address the potential role of each.

SOIL AS A NATURAL BODY

The concept of soil as a natural body has its origin in the soil factor equation outlined by Dokuchaev (Glinka 1927) and Hilgard (Jenny 1961). This well-known equation characterizes soil as a function of parent material, climate, organisms, relief, and time. Although simple and qualitative in nature, the soil factor equation has served as a powerful and highly useful model. It led early pedologists to reason correctly that by looking for changes in one or more of these factors as the landscape was traversed, one could accurately plot the boundaries separating different kinds of soil. This simple idea attracted a large number of adherents; the idea that this qualitative model could be used to delineate bodies of soil anywhere in the world was extremely compelling.

The promise inherent in the original soil factor equation was given a more precise focus by Milne's (1936) introduction of the catena concept—the idea that soils are predictably related to landscape position. As a result of many years of soil survey experience, the soil factor equation and the catena concept evolved into a more applied and refined conceptualization—the soil-landscape model (Hudson 1992). This model now serves as the primary medium for soil-based technology transfer. It also serves as the foundation for the other models to be discussed.

MAPPING SURFACE SHAPE

The soil-landscape model is the scientific underpinning of the soil survey program, so improving the way we apply it can have immense benefits to soil science. There are several promising areas for improvement in the next century. Hall and Olson, in a very insightful article, presented this disturbing but, we believe, accurate criticism of current soil maps:

> Much effort has been expended on taxonomic classification of soils during the last few years but the importance of proper representation of landscape relations within and between soil mapping units has been virtually ignored. The same mapping unit is often delineated on convex, concave and linear slopes. This mapping results in the inclusion of areas of moisture accumulation, moisture depletion and uniform moisture flow within a given mapping unit. (Hall and Olson 1991, 21)

Hall and Olson make an important point: convexity and concavity of slopes are major determinants of water movement across the landscape and of the relative availability of water to affect soil formation and plant growth (Aandahl 1948; Pennock and Acton 1989; Sinai et al. 1981; Stone et al. 1985). The relative convexity or concavity of the land surface often affects plant growth more than degree of slope. However, most existing soil maps do a poor job of reflecting the shape of the land surface, a result of relying almost exclusively on aerial photograph interpretation to delineate landforms and slopes. With practice, one can do a reasonable job of identifying the degree of slope, but even with a stereoscope, it is virtually impossible to determine land surface shape from an aerial photograph.

Fortunately, very recent technology will enable us to delineate areas differing in convexity and concavity very accurately and with great precision. For example, recently available digital elevation models (DEM) will yield a number of useful derivative maps, including slope percent, rate of slope change, aspect, and diurnal variation of sun energy on a slope. Wider use of DEMs and the recognition and delineation of soil map units based on surface shape—in addition to degree of slope—will result in more useful soil maps in the twenty-first century.

EXPERT KNOWLEDGE SYSTEMS AND "FUZZY LOGIC"

Hudson (1992) argued that application of the soil-landscape model suffers from excessive reliance on tacit knowledge. In the course of their work, soil scientists acquire much detailed knowledge about the soils and their distribution on the landscape. However, most of this knowledge is used only to prepare soil maps; very little of it is written down. As Hanson (1969) asserted, if information is not written down, it will not affect the general body of scientific knowledge. A large body of scientific information about soil now exists only on soil maps and in the minds of a large number of soil classifiers, which means that each generation of soil scientists must relearn much of what previous generations have already discovered. Furthermore, this knowledge cannot be transferred outside the discipline. A way must be found to avoid this. One possible solution is the application of expert knowledge systems using "Fuzzy Logic."

Zadeh (1965) introduced the concept of Fuzzy Logic in the 1960s, and it has already been applied successfully in such diverse areas as home appliances, manufacturing plants, and subway systems (McNeil and Freiberger 1993). The concept expands the exact *if/then* rationale of Boolean logic to conditions of continuous variation in which classes may overlap (Burrough 1993). Recently, Zhu and others (Zhu and Band 1994; Zhu et al. 1996; and Zhu et al. 1997) have provided convincing arguments that this concept is applicable to soil survey. The proposed application method consists of a soil-land inference model combining the knowledge of local soil scientists (a knowledge base) with layers of information in a Geographic Information System (GIS) (a physical database). A computer program is written to link the local knowledge base with the physical database in order to predict soil properties at any selected point.

The application of expert knowledge systems to the mapping of soils holds great promise. Their use can improve the accuracy and precision of maps, and building local knowledge bases will force soil scientists to write down all of the information and concepts that are needed to make decisions while mapping soils. This will reduce the reliance on tacit knowledge, and as a result, enhance the efficiency of knowledge transfer among individuals and between generations. Research and development into the use of expert knowledge systems to apply the soil-landscape model more effectively must be a priority in the twenty-first century.

SOIL AS A MEDIUM FOR PLANT GROWTH

This model has been used to predict the potential and limitations of land areas for growth of plants; principally those used for food or fiber. Properties that can be used to predict suitability of soil as a medium for root growth have been emphasized, for example, texture, available nutrients, available water, and soil density. This traditional model has been applied with such overall success that overproduction of food and lack of adequate markets are primary concerns in many developed countries.

But the bounty has not been without its costs. Soil erosion on farmland has long been recognized as a problem. Recently, agricultural runoff has been recognized as a source of phosphorus, nitrogen, pesticides, microbial organisms, and other constituents that degrade the quality of aquatic systems. In the future, the focus of this model increasingly will shift from maximizing crop production to emphasizing an environmentally benign agriculture, which will require that fertilizer and other inputs be optimized to meet, but not greatly exceed, crop needs. Assuming current trends continue, developing soil (and landscape) threshold levels for nitrogen, phosphorus, and other agricultural inputs will be a major task for soil survey in the next century.

PRECISION AGRICULTURE

Precision agriculture is growing rapidly in the United States (NRC 1997). This development was made possible by the convergence of several technologies, including yield monitors on harvesters, precise Global Positing Systems (GPS), and computers capable of storing large amounts of digital data. Adapting to the needs of precision agriculture will be a major challenge for the U.S. Soil Survey. For example, we must develop procedures and quality control criteria for preparing very detailed soil maps; we have already developed the first draft of standards for Order 1 soil surveys (Soil Survey Staff 1996), which will be reviewed and revised as needed in the future. And precision agriculture includes temporal as well as spatial precision, making it necessary to gain a better understanding of temporal and use-dependent soil properties in the twenty-first century. The U.S. Soil Survey will be required to develop and maintain a comprehensive database of such properties.

Some investigators engaged in site-specific management have expressed concerns that the current U.S. Soil Survey database does not support site-specific management. There was never an intent for this database to support such a detailed level of management; it was developed from data gathered from medium-intensity soil surveys—scales of 1:24,000, 1:20,000, 1:15,840, and 1:12,000. However, with the use of GIS and Fuzzy Logic technologies, it is very probable that this database can be adapted to support most versions of site-specific management. These adaptations will be a major challenge to the U.S. Soil Survey in the twenty-first century.

MONITORING OF SOIL QUALITY

Worldwide, about 25 percent of cultivated land is being degraded. Processes of degradation include erosion, compaction, acidification, accumulation of toxic elements, and salinization. In recent years, there has been increasing emphasis both on understanding what constitutes soil quality and on devising national and local programs to monitor it (Karlen et al. 1996). In this sense, soil quality is recognized and monitored in the same way as air and water quality. The soil survey of the twenty-first century will increasingly emphasize research to identify both indicators and criteria to assess soil quality.

Emphasis on soil quality is part of a larger global movement to develop an international framework to evaluate sustainable land management (Dumanski 1993). The Natural Resources Conservation Service (NRCS) recently established a Soil Quality Institute to help meet these needs. In addition to focusing on soil quality, it is very probable that emphasis in the U.S. Soil Survey will shift away from traditional, systematic data collection and towards outcome-oriented project research and environmental monitoring.

SOIL AS A STRUCTURAL MANTLE

This model relates to the use of soils for the infrastructure necessary to modern societies. It has direct application to such diverse activities as urban and suburban development; construction of dams, highways, and airports; and onsite waste disposal (Smith 1976). The model relies heavily on estimations of soil strength and plasticity, as well as the soil's ability to transmit heat, water, and energy. The soil survey of the twenty-first century will require enhancements to the way this model is applied. Burrough (1993) stated that, compared with other resource monitoring sciences, soil survey is notable because most of the information collected and presented to users remains qualitative in nature. That criticism is especially applicable to information used to support the model of soil as a structural mantle. Therefore, it is important that we identify the kinds of quantitative data needed to support this model and begin collecting them.

In addition to collecting more quantitative data, it is important that the soil and the underlying material be characterized to a greater depth. The needs of the twenty-first century will no longer allow us to restrict our zone of investigation to the top 2 meters of the earth's surface; such artificial boundaries, whether in our minds or on the landscape, can no longer be tolerated. The soil survey of the twenty-first century will require a concerted effort to study and characterize the soil and underlying material to whatever depth is needed to meet our scientific needs. All appropriate technology, including ground-penetrating radar and geomagnetic studies, will be employed.

SOIL AS A WATER-TRANSMITTING MANTLE

The soil is recognized as a major component of the hydrologic cycle. It is a complex, highly organized, porous medium permeable to both atmospheric gases and water. Soil plays a vital role in the partitioning of water on the landscape. This model has major implications for public health, watershed management, and environmental quality. The soil survey historically has provided measures or estimates of soil physical properties to be used in various models dealing with water movement. Pedotransfer functions, which use easily measured soil properties to predict other soil properties that are more difficult to measure, are commonly used to derive input values for models. There is at least one major problem with this process, however. The soils information provided to modelers typically is limited to pedon data—measurements or estimates made at a single point on the landscape. This will not suffice in the future. A focused research program to understand water-soil interactions at the landscape level and at whatever depths are needed must be an integral part of soil survey in the twenty-first century.

SOIL AS AN ECOSYSTEM COMPONENT

The ecosystem component is an emerging model of soil and a very promising one for the future of the U.S. Soil Survey. It views soil as a dynamic, living mantle or membrane in constant interaction with the atmosphere, biosphere, and geosphere. This model postulates that the pedosphere is an essential part of all land-based life-support systems on Earth. The global carbon cycle is a definitive example supporting this as true. In 1957 Revelle and Suess put forth the then novel idea that humans are conducting a planetary scale "experiment" with carbon. For millions of years plant life removed carbon from the atmosphere and stored it in sedimentary rocks. Recently, mostly through industrial activity, humans have reversed this trend. We are witnessing a dramatic increase in carbon dioxide in the atmosphere—an increase of more than 25 percent in the last 100 years—there is now widespread alarm concerning the potential effect of atmospheric carbon dioxide enrichment on climate and on human life. Soil has come to be recognized as a major source or sink for the primary greenhouse gases.

Because of soil's importance to the global ecosystem, it is imperative that we gain a better understanding of how it functions as one of the major world ecospheres. Unfortunately, as Burrough (1993) has pointed out, soil survey organizations currently do not relate well to this major soil function. Soil interacts with the biosphere, atmosphere, lithosphere, and hydrosphere mainly through biologically driven processes. Historically, the U.S. Soil Survey has placed little emphasis on soil biology. As a result, our knowledge is very limited. Gaining a better understanding of soil biological processes at the landscape, regional, continental, and global levels must be a major goal of the U.S. Soil Survey in the twenty-first century.

SOIL SURVEY: OPPORTUNITIES AND CONSTRAINTS

When viewed through the five operative models, or views, of soil, the challenges and opportunities for soil survey in the next century are virtually unlimited. One cannot help being optimistic about the future. Advances in GIS, remote sensing technology, and other fields, for example, have the potential to revolutionize many aspects of the soil survey. We also are experiencing a revolution in communications; the Internet will drastically change the way we store and access soil maps, data, and interpretive information. Learning to take advantage of all the technology available to us will be a major challenge. However, it also is important to recognize possible constraints and prepare to cope with them. Potential constraints to the success of soil survey will vary in kind, but they can be grouped into three major classes—technological, human, and financial.

TECHNOLOGICAL CONSTRAINTS

Soil survey will be faced with a number of temporary technological constraints in the next century, although the development of new concepts, the adaptation of new technology, or new combinations of concepts and technologies inevitably will solve most or all of them. One possible constraint deserves special mention, perhaps because it is as much institutional as technological and will be largely self-imposed. The wide availability of databases and GIS, coupled with the fact that soil surveys have been completed for many areas, poses a paradoxical danger to soil survey organizations (Burrough 1993).

Recent technological innovations and a large archive of completed soil maps will enable the U.S. Soil Survey to supply users with a wide range of information in many formats. However, over time the cost and expense of maintaining a massive digital database could limit innovation and reduce the incentive to respond to the changing needs of users. If care is not taken, the soil survey of the twenty-first century could become merely a broker for its own existing data. As a result, users would seek new kinds of data elsewhere. This must not happen. The U.S. Soil Survey must continue to be perceived as the preferred organization to collect, analyze, and supply the data to meet the nation's soil information needs.

HUMAN AND FINANCIAL CONSTRAINTS

Soil survey requires the application of high levels of knowledge and technology in many areas. For some years the practicing soil scientist has needed, in addition to an understanding of soil science, knowledge of such diverse fields as plant ecology, geology, agronomy, engineering, and hydrology. In recent years the required knowledge base has expanded. The soil scientist of today must also understand the nature of digital data, how to utilize global positioning technology, and how to use large database management systems. Because of the high level of education and training required, soil survey staffs already are very expensive. Staff costs can only increase in the future. Furthermore, the tasks of the soil survey are becoming more technology intensive; the soil survey of the twenty-first century will require an expanding array of expensive technology, including GPS, GIS, digital photography, ground-penetrating radar, field data recorders, and satellite imagery.

A knowledge- and technology-centered soil survey offers a number of challenges. One, of course, will be to acquire the funding to pay for the needed staff and equipment. A second challenge will be to achieve the optimum mix of human talent and technology. This will require a high level of insight and organizational skill. A third challenge will be to lead and manage successfully a diverse, highly skilled group of scientists, who employ a growing array of complex technology.

In recent years, some pessimistic views have been expressed about the future of soil survey. For example, Burrough wrote, "In many countries, not only in western Europe, there are indications that conventional soil survey has finished its task and is no longer needed as a major day-to-day activity" (Burrough 1993, 15). But we do not believe this statement applies to the United States. On the contrary, the American soil survey of the twenty-first century will be professional, productive, and respected, with an increasingly strong science base. Though potential constraints exist, the opportunities for achievement are unlimited.

REFERENCES

Aandahl, A. R. 1948. The characterization of slope positions and their influence on the total nitrogen content of a few virgin soils in western Iowa. *Soil Science Society of America Proceedings* 13:449–454.

Burrough, P. A. 1993. The technological paradox in soil survey: New methods and techniques of data capture and handling. In *Soil Survey: Perspectives and Strategies for the 21st Century*, edited by J. A. Zinck, pp. 15–22. ITC Publication 21. International Institute for Aerospace Survey and Earth Sciences, Enschede, The Netherlands.

Dumanski, J. 1993. Strategies and opportunities for soil survey information and research. In *Soil Survey: Perspectives and Strategies for the 21st Century*, edited by J. A. Zinck, pp. 36–41. ITC Publication 21. International Institute for Aerospace Survey and Earth Sciences, Enschede, The Netherlands.

Gardner, D. R. 1998. *The National Cooperative Soil Survey of the United States*. USDA-NRCS Historical Notes 7. Natural Resources Conservation Service, U.S. Department of Agriculture, Washington, D.C.

Glinka, K. D. 1927. *The Great Soil Groups of the World and Their Development*. Translated by C. F. Marbut. Ann Arbor, Mich.: Edwards Bros.

Hall, G. F., and C. G. Olson. 1991. Predicting variability of soils from landscape models. In *Spatial Variabilities of Soils and Landforms*, edited by M. J. Mausbach and L. P. Wilding, pp. 9–24. Special Publication Number 28. Proceedings of the Symposium, SSSA and ISSS, Las Vegas, Nevada, 17 October 1989. Madison, Wis.: Soil Science Society of America.

Hanson, N. R. 1969. *Perception and Discovery—An Introduction to Scientific Inquiry.* San Francisco: Freeman, Cooper, and Co.

Hudson, B. D. 1992. The soil survey as paradigm-based science. *Soil Science Society of America Journal* 56(3):836–841.

Jenny, H. 1961. *E. W. Hilgard and the Birth of Modern Soil Science.* Berkeley, Calif.: Farallo.

Karlen, D. L., M. J. Mausbach, J. W. Doran, R. G. Cline, R. F. Harris, and G. E. Schuman. 1996. Soil quality: a concept, definition, and framework for evaluation. *Soil Science Society of America Journal* 61:4–10.

McNeil, D., and P. Freiberger. 1993. *Fuzzy Logic: The Discovery of a Revolutionary Computer Technology—and How it Is Changing Our World.* New York: Simon and Schuster.

Milne, G. 1936. *A Provisional Soil Map of East Africa.* Amani Memoirs Number 28. Eastern African Agricultural Research Station, Tanganyika Territory.

National Research Council (NRC). 1997. *Precision Agriculture in the 21st Century: Geospatial and Information Technologies in Crop Production.* Washingon, D.C.: National Academy Press.

Pennock, D. J., and D. F. Acton. 1989. Hydrological and sedimentological influences on Boroll Catenas, Central Saskatchewan. *Soil Science Society of America Journal* 53:904–910.

Revelle, R., and H. Suess. 1957. Carbon dioxide exchange between atmosphere and ocean and the question of an increase of atmospheric CO_2 during the past decade. *Tellus* 9:18–27.

Sinai, G., D. Zaslavsky, and P. Golany. 1981. The effect of soil surface curvature on moisture and yield—Beer Sheba observation. *Soil Science* 132:367–375.

Smith, H. 1976. *Soil Survey of District of Columbia. USDA-SCS in Cooperation with USDI-NPS.* Washington, D.C.: Government Printing Office.

Soil Survey Staff. 1996. *National Soil Survey Handbook.* USDA-NRCS 430-VI-NSH. Washington, D.C.: Government Printing Office.

Stone, J. R., J. W. Gilliam, D. K. Cassel, R. B. Daniels, L. A. Nelson, and H. J. Kleiss. 1985. Effect of erosion and landscape position on productivity of Piedmont soils. *Soil Science Society of America Journal* 49:987–991.

Zadeh, L. 1965. Fuzzy sets. *Information and Control* 8(3):338–353.

Zhu, A. X., and L. E. Band. 1994. A knowledge-based approach to data integration for soil mapping. *Canadian Journal of Remote Sensing* 20:408–418.

Zhu, A. X., L. E. Band, B. Dutton, and T. Nimlos. 1996. Automated soil inference under Fuzzy Logic. *Ecological Monitoring* 90:123–145.

Zhu, A. X., L. E. Band, R. Vertessy, and B. Dutton. 1997. Derivation of soil properties using a Soil Land Inference Model (SoLIM). *Soil Science Society of America Journal* 61:523–533.

Appendixes

Appendix A: Chronology of the U.S. Soil Survey
Patricia J. Durana

1890

Farmers make up 43 percent of the population. They operate 4.5 million farms averaging 136 acres in size.

1894

Secretary of Agriculture J. Sterling Morton establishes the Division of Agricultural Soils in the U.S. Department of Agriculture's Weather Bureau. Milton Whitney is named chief of the new division.

1896

The Agricultural Appropriations Act of 1896 designates the Division of Agricultural Soils as distinct from the Weather Bureau and appropriates $15,000 for its budget that year.

1897

The Agricultural Appropriations Act of 1897 renames the Division of Agricultural Soils simply the Division of Soils.

1898

A trial soil survey is undertaken on 250 square miles near Hagerstown, Maryland.

1899

The National Cooperative Soil Survey (NCSS) is established as a cooperative undertaking of the U.S. Department of Agriculture and the State Agricultural Experiment Stations, with local cooperation from state geological surveys, boards of agriculture, and other local institutions.

Soil mapping is initiated, but without benefit of a systematic framework of soil classification.

Texture is recognized as the key soil characteristic for classifying soil type, resulting in the evolution of a naming convention based on location and texture (Jordan sandy loam, Podunk fine sandy loam, etc.).

Soil type remains the lowest category of soil classification, with variations within soil type identified as "phases."

1901

The Division of Soils is reorganized as the Bureau of Soils and granted an appropriation of $109,140 for 1902.

One hundred soil types have been identified.

1903

The concept of the soil series is introduced, but it is found to have greater utility in the natural classification of soils than in the interpretive use of soils information, for which soil type remains the dominant category of classification.

1904

Four hundred soil types have been identified.

Additional soil characteristics, color and organic matter content, are used to differentiate soil series.

1906

Surveys become more detailed and are mapped at a larger scale, providing greater potential for interpretation.

Soil structure, lime content, alkali presence, drainage, erodibility, physiographic position, nature of subsoil, lithology, and origin and age of parent material are identified as criteria for soil characterization.

Sixty-one soil series are recognized.

The concept of soil province is identified. Soil provinces are large physiographic regions without clearly defined boundaries, which because of assumed geological and geomorphological homogeneity are thought to delimit the extent of the soil series occurring within them. Thirteen soil provinces are recognized.

The Alabama Agricultural Experiment Station requests a soil survey of its experimental farm at a large scale (1 square inch per 1 acre) to provide a basis for further study of the soils of Alabama.

1907

The influence of natural drainage conditions on soils is recognized.

In the Colusa Area, California, the Glenn County board of supervisors uses the soil survey for land appraisal and tax assessment.

1908

Soil phases begin to be delineated on maps, and are considered subdivisions of certain soil types. Soil slope and soil erosion are the two most important criteria for recognizing phases.

1909

Seven hundred and fifteen soil types in 16 series have been identified.

The principle of interpreting soil survey data as a means of classifying land by land-use capability is developed and begins to be applied.

1910

The Bureau of Soils establishes an organizational unit charged with interpreting the "Use of Soils."

Extensive and systematic mapping of current land use is first included in a soil survey when it is used for the Reconnaissance Soil Survey of Puget Sound Basin, Washington.

Farmers make up 38 percent of the total population. They operate 6.4 million farms averaging 138 acres in size.

1911

Soil provinces are considered a category of soil classification and mapping.

The Uses of Soils unit publishes its first circular—*Soils of the Eastern United States and Their Use.*

The Texas Agricultural Experiment Station requests the same kind of small-scale soil survey of its experimental farm that was provided to Alabama in 1906.

1912

Use of the soil survey for land appraisal and tax assessment receives official recognition and support from the National Tax Association.

One thousand six hundred and fifty soil types in 534 soil series have been identified.

George Nelson Coffey proposes a new system of classifying soils based on five categories and subclasses. This approach recognized soil as a "natural body having a definite genesis and distinct nature of its own."

Soil survey costs are about $0.25 per acre.

1912–1920

Soil survey undergoes a period of de-emphasizing interpretations. Under Chief Curtis F. Marbut, the survey adopts as its role ". . . to get the facts about soils, to classify them, and to map them in ways that would furnish a sound basis for interpretation by other people."

1920

Increases in the quality and precision of base maps and the adoption of aerial photography for use in soil mapping increase the precision of plotting boundaries.

Michigan uses soil survey data to plan road and highway development.

North Dakota uses soil surveys for tax assessment purposes.

1930

Soil survey interpretations for land appraisal in rural tax assessment become increasingly refined as a result of improved mapping techniques and soil productivity ratings.

The Bureau of Reclamation begins to use soil surveys in planning large-scale irrigation and reclamation projects.

Soil surveys provide the foundation for the majority of the state and federal land-use activities.

Soil surveys provide the science-based information used as a guide in conservation planning to prevent erosion.

1933

Productivity ratings for specific soil types and other mapping units begin to be developed.

1935

The Soil Erosion Service is established within the Department of Agriculture.

Charles E. Kellogg is named director of the Soil Survey Division.

The Soil Conservation Act of 1935 (PL 74–46, 49 Stat. 163, 16 U.S.C. 590a-f) establishes a "Soil Conservation Service" within the Department of Agriculture. The new agency is responsible for the development and implementation of a continuing program of soil and water conservation and is created from extant facilities. The Soil Erosion Service becomes the Soil Conservation Service and absorbs the erosion control experiment stations of the Bureau of Chemistry and Soils and the Bureau of Agricultural Engineering and the erosion control nurseries of the Bureau of Plant Industry.

The interpretation of soils for soil and water conservation uses is firmly established and accepted.

1939

The Agricultural Appropriations Act of 1940 formally provides for a program of Soil Information Assistance for Community Planning and Resource Development. The act provides for the secretary to conduct a soil survey program that makes available soil survey information to states and other public agencies, including community development districts, for guidance in community planning and resource development, and for other purposes.

1941

All field maps are being made on a base of aerial photographs.

Soil surveys are being undertaken by the Division of Soil Survey in the Bureau of Plant Industry, Soils, and Agricultural Engineering and by the newly formed Soil Conservation Surveys Division of the Soil Conservation Service.

1950

Application of soil surveys for supporting land-use planning and development efforts increases.

Soil survey interpretations are made for correlating residual radioactivity from strontium 90 with soil type. The work explains the observed variability of fallout effects and offered a means to predict future intensities and effects.

1952

Secretary's Memorandum 1318 consolidates all soil survey work of the Department—including mapping, classification, correlation, interpretation, laboratory services, map compilation, and publication—in the Soil Conservation Service.

1966

PL 89-560, 80 Stat.706, 42 U.S.C. 3271-3274, reiterates the language of the 1940 Appropriations Act that the secretary will conduct a soil survey program that will make available soil survey information to public and private entities for land-use planning and resource development.

1985

Soil interpretations are used to support land designations for the 1985 Farm Act conservation compliance provisions.

Soil survey costs are about $1.50 per acre.

1990

Farmers make up 2.6 percent of the total population. They operate 1.8 million farms averaging 461 acres in size.

1999

The Soil Survey celebrated its centennial. At this time, 941 soil surveys had been digitized and made available in the Soil Survey Geographic Database (SSURGO).

APPENDIXES

APPENDIX B: History of the U.S. Soil Survey:
A Bibliography

John P. Tandarich

This bibliography is limited to works about the history of the U.S. Soil Survey, oral or written items about personalities connected with the soil survey, and bibliographies of soil science subjects in which the soil survey is a prominent feature and which can be used as source material for writing history. The author gratefully acknowledges the assistance of Michael G. Ulmer and Douglas Helms in the preparation of this bibliography.

Amundson, Ronald, and Dan H. Yaalon. 1995. E. W. Hilgard and John Wesley Powell: Efforts for a joint agricultural and geological survey. *Soil Science Society of America Journal* 59(1):4–13.

Arnold, Richard W. 1983. Concepts of soils and pedology. In *Pedogenesis and Soil Taxonomy, Part I: Concepts and Interactions*, edited by Lawrence P. Wilding, Neil E. Smeck, and George F. Hall, pp. 1–21. New York: Elsevier.

Bailey, George D. 1979. A brief history of the Miami soil series. *Soil Survey Horizons* 19(3):9–13.

Baldwin, Mark, Charles E. Kellogg, and James Thorp. 1938. Soil classification. In *Soils and Men*, pp. 979–1001. Yearbook of Agriculture 1938. Washington, D.C.: Government Printing Office.

Brasfield, James. 1982. More soil scientists identified with the great past era of Soil Survey. *Soil Survey Horizons* 23(3):27–30.

Brevik, Eric C. 1999. George Nelson Coffey, early American pedologist. *Soil Science Society of America Journal* 63(6):1485–1493.

Buol, Stanley W. 1997. Beginnings of *Soil Survey Horizons*. *Soil Survey Horizons* 38(2):39–40.

Coffey, George N. 1912. The development of soil survey work in the United States with a brief reference to foreign countries. *Proceedings of the American Society of Agronomy* 1911 3:115–129.

Coffey, George N. 1913. *A Study of the Soils of the United States*. U.S. Department of Agriculture Bureau of Soils Bulletin 85. U.S. Department of Agriculture, Washington, D.C.

Cline, Marlin G. 1949. Basic principles of soil classification. *Soil Science* 67:81–91.

Cline, Marlin G. 1961. The changing model of soil. *Soil Science Society of America Proceedings* 25:442–446.

Cline, Marlin G. 1977. Historical highlights in soil genesis, morphology, and classification. *Soil Science Society of America Journal* 41:250–254.

Cline, Marlin G. 1979. *Soil Classification in the United States*. Agronomy Mimeo Number 79–12. Cornell University, Department of Agronomy, Ithaca, N.Y.

Dokuchaev, Vasilli V. 1893. *The Russian Steppes: Study of the Soil in Russia, Its Past and Present*. St. Petersburg, Russia: Department of Agriculture Ministry of Crown Domains for the World's Columbian Exposition at Chicago.

Dokuchaev, Vasilli V., and Nikolai M. Sibirtsev. 1893. *Short Scientific Review of Professor Dokuchaev's and His Pupil's Collection of Soils Exposed in Chicago in the Year 1893*. St. Petersburg, Russia: Department of Agriculture Ministry of Crown Domains for the World's Columbian Exposition at Chicago.

Effland, Anne B. W., and William R. Effland. 1992. Soil geomorphology studies in the U.S. Soil Survey Program. *Agricultural History* 66(2):189–212.

Fippin, Elmer O. 1912. The practical classification of soils. *Proceedings of the American Society of Agronomy 1911* 3:76–89.

Gardner, David R. 1998. *The National Cooperative Soil Survey of the United States*. Historical Notes Number 7. Soil Survey Division, Resource Economics and Social Science Division, Natural Resources Conservation Service, U.S. Department of Agriculture, Washington, D.C. Original publication, doctoral thesis, 1957, Graduate School of Public Administration, Harvard University, Cambridge, Mass.

Helms, Douglas. 1997. Land capability classification. In *History of Soil Science: International Perspectives*, edited by Dan H. Yaalon and Simon Berkowicz, pp. 159–175. Advances in Geoecology 29. Reiskirchen, Germany: Catena Verlag.

Hilgard, Eugene W. 1891. Soil studies and soil maps. *Overland Monthly* 18:607–616.

Hopkins, Cyril G. 1904. The present status of soil investigation. U.S. Department of Agriculture Office of Experiment Stations Bulletin 142:95–104.

Hudson, Berman D. 1999. Some interesting historical facts about the American soil survey. *Soil Survey Horizons* 40(1):21–26.

Jenny, Hans. 1961. *E. W. Hilgard and the Birth of Modern Soil Science*. Pisa, Italy: Collana Della Rivista "Agrochimica."

Kellogg, Charles E. 1937. *Soil Survey Manual*. U.S. Department of Agriculture Miscellaneous Publication Number 274. Washington, D.C.: Government Printing Office.

Kellogg, Charles E. 1938. Soil and society. In *Soils and Men*, pp. 863–886. Yearbook of Agriculture 1938. Washington, D.C.: Government Printing Office.

Kellogg, Charles E. 1943. *The Soils That Support Us: An Introduction to the Study of Soils and Their Use by Men.* New York: Macmillan Company.

Kellogg, Charles E. 1957. We seek; We learn. In *Soil*, pp. 1–11. The Yearbook of Agriculture 1957. Washington, D.C.: U.S. Department of Agriculture.

Kellogg, Charles E. 1974. Soil genesis, classification, and cartography: 1924–1974. *Geoderma* 12(4):347–362.

Krusekopf, Henry H., ed. 1942. *Life and Work of C. F. Marbut.* Madison, Wis.: Soil Science Society of America.

Kuhl, Arthur D. 1989. The 1957 conservation needs inventory: A historical aspect of soil survey. *Soil Survey Horizons* 30(4):84–87.

Lankford, Nancy, Lynn Gentzler, Darrell Garwood, Tom Norris, Randy Roberts, and Cindy Stewart. 1985. Unpublished materials of Dr. Curtis Fletcher Marbut. *Soil Survey Horizons* 26(1):36–40.

Lapham, Macy H. 1949. *Crisscross Trails: Narrative of a Soil Surveyor.* Berkeley, Calif.: Willis E. Berg.

Marbut, Curtis F. 1921. The contribution of soil surveys to soil science. *Proceedings of the Society for the Promotion of Agricultural Science* 41:116–142.

Marbut, Curtis F. 1924. The United States Soil Survey. In *Etat de l'etude et de la cartographie des sols dans divers pays*, edited by Gheorghe Murgoci, pp. 215–224. Bucharest, Romania: Comite Internationale de Pedologie.

Marbut, Curtis F., Hugh H. Bennett, James E. Lapham, and Macy H. Lapham. 1913. *Soils of the United States.* U.S. Department of Agriculture Bureau of Soils Bulletin 96. U.S. Department of Agriculture, Washington, D.C.

McCormack, Donald E., and Andrew H. Paschall. 1982. The 1934 national reconnaissance erosion survey. *Soil Survey Horizons* 23(4):13–15.

McCracken, Ralph J. 1987. Soils, soil scientists, and civilization. *Soil Science Society of America Journal* 51(6):1395–1400.

Newman, Allen L. 1979. Diamond anniversary of the Austin series. *Soil Survey Horizons* 20(4):3–10.

Nikiforoff, Constantin C. 1931. History of A, B and C. *American Soil Survey Association Bulletin* 12:67–70.

Olson, Carolyn G. 1997. Systematic soil-geomorphic investigations—Contributions of R. V. Ruhe to pedologic interpretation. In *History of Soil Science: International Perspectives*, edited by Dan H. Yaalon and Simon Berkowicz, pp. 415–438. Advances in Geoecology 29. Reiskirchen, Germany: Catena Verlag.

Olson, Kenneth R., and Joseph B. Fehrenbacher. 1998. Illinois Agricultural Experiment Station contributions to Soil Survey (1902–1995). *Soil Survey Horizons* 39(2):31–33.

Ruhe, Robert V. 1970. *Soil-Geomorphology Studies: 1953–1970.* Ames, Iowa: n.p. (privately printed).

Seaton, Charles H. 1921. Uses of the Soil Survey. In *United States Department of Agriculture Yearbook*, 1920, pp. 413–419. Washington, D.C.: Government Printing Office.

Simonson, Roy W. 1952a. Lessons from the first half-century of Soil Survey: I. Classifications of soils. *Soil Science* 74:249–257.

Simonson, Roy W. 1952b. Lessons from the first half-century of Soil Survey: II. Mapping of soils. *Soil Science* 74:323–330.

Simonson, Roy W. 1962. Soil classification in the United States. *Science* 137:1027–1034.

Simonson, Roy W. 1963. Soil correlation and the new soil classification system. *Soil Science* 96:23–30.

Simonson, Roy W. 1964. Soil series as used in the USA. In *Proceedings of the 8th International Congress of Soil Science,* Volume 2, pp. 17–24. Bucharest, Romania: International Society of Soil Science.

Simonson, Roy W. 1979. Soil survey: The study of our earthen looms. *Soil Survey Horizons* 20(2):9–16.

Simonson, Roy W. 1980. Soil survey and soil classification in the United States. In *Proceedings 8th National Congress, Soil Science Society of South Africa, 1978,* pp. 10–21. Technical Communication Number 165. Department of Agriculture Technical Service, South Africa.

Simonson, Roy W. 1985. Soil classifications in the past—Roots and philosophies. In *Annual Report 1984, International Soil Reference and Information Centre,* pp. 6–18. Wageningen, The Netherlands: ISRIC. Reprint, 1989, *Soil Morphology, Genesis, and Classification,* edited by Delvin S. Fanning and Mary C. Fanning, pp. 139–152. New York: John Wiley and Sons.

Simonson, Roy W. 1987. *Historical Aspects of Soil Survey and Soil Classification.* Madison, Wis.: Soil Science Society of America.

Simonson, Roy W. 1989. *Historical Highlights of Soil Survey and Soil Classification with Emphasis on the United States, 1899–1970.* International Soil Reference and Information Centre Technical Paper Number 18. Wageningen, The Netherlands: ISRIC.

Simonson, Roy W. 1991. The U.S. Soil Survey—Contributions to soil science and its application. *Geoderma* 48:1–16.

Simonson, Roy W. 1992a. Historical footnote—Factors of soil formation. *Soil Survey Horizons* 33(3):70.

Simonson, Roy W. 1992b. Beginnings of the soil type and soil series in the USA. *Soil Survey Horizons* 33(4):75–80.

Simonson. Roy W. 1993. Soil color standards and terms for field use—History of their development. In *Soil Color,* edited by Jerry M. Bigham and Edward J. Ciolkosz, pp. 1–20. Soil Science Society of America Special Publication Number 31. Madison, Wis.: Soil Science Society of America.

Simonson, Roy W. 1996. My initial encounter with aerial photographs. *Soil Survey Horizons* 37(2):55–58.

Simonson, Roy W. 1997. Rendzina—Origin, history, and descendants. *Soil Survey Horizons* 38(3):76–78.

Simonson, Roy W., and Michael G. Ulmer. 1989. Initial soil survey and resurvey of McKenzie County, North Dakota. *Soil Survey Horizons* 30(1):1–11.

Smith, George D. 1983. Historical developments of soil taxonomy—Background. In *Pedogenesis and Soil Taxonomy: I. Concepts and Interactions,* edited by Lawrence P. Wilding, Neil E. Smeck, and George F. Hall, pp. 23–49. New York: Elsevier.

Smith, George D. 1986. *The Guy Smith Interviews: Rationale for Concepts in Soil Taxonomy*. U.S. Department of Agriculture, Soil Conservation Service, Soil Management Support Services Technical Monograph Number 11. Cornell University, Department of Agronomy. Ithaca, N.Y.

Smith, George D., and Michael Leamy. 1980. Conversations in taxonomy. *Soil Survey Horizons* 21(1):10–12.

Smith, Raymond S. 1942. Dr. C. F. Marbut's contribution to Soil Survey. In *Life and Work of C. F. Marbut*, edited by Henry H. Krusekopf, pp. 50–53. Madison, Wis.: Soil Science Society of America.

Soil Conservation Service Oral History Program. 1982. *William M. Johnson*. Soil Conservation Service, U.S. Department of Agriculture, Washington, D.C.

Tandarich, John P., Robert G. Darmody, and Leon R. Follmer. 1988. The development of pedologic thought: Some people involved. *Physical Geography* 9:162–174.

Tandarich, John P., Chris J. Johannsen, and William E. Wildman. 1985. James Thorp talks about Soil Survey, C. F. Marbut, and China. *Soil Survey Horizons* 26(2):5–12.

Tandarich, John P., Randall J. Schaetzl, and Robert G. Darmody. 1988. Conversations with Francis D. Hole. *Soil Survey Horizons* 29(1):9–21.

Tanner, Champ B., and Roy W. Simonson. 1993. Franklin Hiram King—Pioneer scientist. *Soil Science Society of America Journal* 57(1):286–292.

Thompson, Kenneth W. 1992. *A History of Soil Survey in North Dakota*. Dickinson, N.D.: King Speed Printing.

Thorp, James. 1985. Impressions of Dr. Curtis Fletcher Marbut, 1921–1935. *Soil Survey Horizons* 26(1):26–30.

Weber, George A. 1928. *The Bureau of Chemistry and Soils: Its History, Activities, and Organization*. Service Monographs of the United States Government Number 52, Brookings Institution Institute for Government Research. Baltimore, Md.: Johns Hopkins University Press. Reprint, 1974, New York: AMS Press.

Whitney, Milton. 1900. Soil investigations in the United States. In *Yearbook of the United States Department of Agriculture, 1899*, pp. 335–346. Washington, D.C.: Government Printing Office.

Whitney, Milton. 1909. *Soils of the United States*. U.S. Department of Agriculture, Bureau of Soils Bulletin 55. U.S. Department of Agriculture, Washington, D.C.

Whitney, Milton. 1924. The future of the Soil Survey in our national agricultural policy. *Journal of the American Society of Agronomy* 16(7):409–412.

Wilding, Lawrence P., Neil E. Smeck, and George F. Hall. 1983. *Pedogenesis and Soil Taxonomy. I. Concepts and Interactions*. New York: Elsevier.

Wildman, William E. 1985. Dr. James Thorp, a close Marbut associate—A memorial tribute. *Soil Survey Horizons* 26(1):25–26.

INDEX

Italic page numbers refer to figures.